OIL ECONOMICS AND POLICY

OIL ECONOMICS AND POLICY

by

Alberto Clô
Bologna University, Italy

EUROPEAN SECRETARIAT FOR SCIENTIFIC PUBLICATIONS

The translation of this book has been funded by the
European Secretariat for Scientific Publications (SEPS)
Via Val D'Aposa 7, 40123 Bologna, Italy
Tel: +39 051/271992, Fax: +39 051/265983
E-mail: seps@alma.unibo.it, http://www.seps.it

Kluwer Academic Publishers
Boston / Dordrecht / London

Distributors for North, Central and South America:
Kluwer Academic Publishers
101 Philip Drive
Assinippi Park
Norwell, Massachusetts 02061 USA
Telephone (781) 871-6600
Fax (781) 871-6528
E-Mail <kluwer@wkap.com>

Distributors for all other countries:
Kluwer Academic Publishers Group
Distribution Centre
Post Office Box 322
3300 AH Dordrecht, THE NETHERLANDS
Telephone 31 78 6392 392
Fax 31 78 6546 474
E-Mail <orderdept@wkap.nl>

Electronic Services <http://www.wkap.nl>

Library of Congress Cataloging-in-Publication Data
Clô, Alberto
[Economia e politica del petrolio. English]
Oil economics and policy / by Alberto Clô.
p. cm.
Includes bibliographical references.
ISBN 0-7923-7906-3
1. Petroleum industry and trade-History. 2. Petroleum industry and trade-Economic aspects. 3. Petroleum products--Prices. 4. Energy policy. I. Title.

HD9560.5 .C568 13 2000
333.8'232--dc21 00-055994

Original title: *Economia e politica del petrolio,* Editrice Compositori, Bologna, Italy
English translation of the Italian edition sponsored by "The European Secretary for Scientific Publications" (SEPS).

Printed on acid-free paper.

Printed in the United States of America

TABLE OF CONTENTS

TABLES OF STATISTICS

Introduction

THE PURPOSE OF THIS STUDY

There are various reasons why the international oil industry and market constitute one of the study *cases* of particular interest for industrial organisations and for economic politics in general. In particular:

- with about 40% of the total, oil is still the most widely consumed energy source in the world and is therefore a production factor of essential input for the development of the economies;
- entire industrial sectors, such as the chemical industry, and a large part of the transport system depend fundamentally on this resource, and with the current state of technology, are inextricably bound to it;
- the economies of many countries, who have huge surpluses or great shortages of oil, are considerably influenced by its price dynamics and by its physical availability;
- the trading values of other sources of energy, particularly those of natural gas (representing approximately another 25% of total consumption) are directly correlated to the price of oil;
- oil has always represented and still today represents a political factor of primary importance in international relations and within individual countries. Economic and political reasons have conditioned the reactions of the various factors which are interwoven in such a way that they are often not easy to unravel.

Due to more than a century of experience, to the large dimensions that are structurally inherent to it, and to the large daily sales (2-3 billion dollars at current prices for refined products), the oil industry has always been a precursor to and amplifier of dynamics and phenomena which eventually become evident in other industrial sectors. This explains why the pioneer studies of Alfred Chandler on the dynamics of industrial capitalism, or those of Joe Bain and Frederic Scherer on economies of scale or on the entry barriers, or even those of Edith Penrose on the large multinationals have inevitably made reference to the oil industry. There is practically no topic dealt with by an industrial organisation that is not given a complete and original evolvement by oil.

This book aims to present much of the subject matter treated in the industrial organisations by following the historical development of the oil industry, and inevitably also covering many aspects of energy resource economy. In doing this, we will be paying particular attention to one aspect, namely, the fixing of oil prices. This is mainly in order to attempt to understand *whether* and *by how much* the structural transformations that the oil industry has undergone during the various phases of its existence – and the various market structures deriving from them – have influenced the dynamics of oil prices. This has always been a crucial aspect in the study of energy economics, and increasingly important for the work of economists in general.

DOES THEORY HELP US TO UNDERSTAND?

The central importance of the oil industry in the world economy and the traumatic events of the late Seventies have often caused the debate on oil prices to assume strongly emotive tones. There are two main theoretic approaches. An initial line of reasoning, known as *consumer's logic*, considers every level higher than the marginal costs (in the long-term) to be an unjustified profit for the producers. According to this logic, first the large multinationals and then OPEC acted as a cartel to keep prices well above the marginal costs of the huge Middle Eastern oil fields, thus reaping immense differential rents. The other line of reasoning, defined as *producer's logic*, rejects the idea that the price of crude oil must be determined with reference to its cost, and claims that, since it is an exhaustible resource, it must implicitly contain a shortage value since its full value derives from the opportunity cost of its alternative sources, while every lower level would *de facto* represent an exploitation of the producer and an avoidable waste of a finite resource.

With commendable commitment, many economists have approached the themes of energy and oil, attracted by the great interest aroused in the Seventies, *each time* with the intention of finding the successful theoretic model which, more than any others, would be able to take into account the complexities involved, obtaining better results in forecasts with respect to those achieved so far. It would be ungenerous, though perhaps not useless, to justify this disappointing and frenzied race. To be brief, it pursued the following approaches:

(a) *models of optimisation*, where the collusive or competitive behaviour of the producer countries fixes prices and quantities so as to maximise the current value of the flows of profit which may be generated by the control of an exhaustible resource, according to Hotelling's theory (1931). The works of Pindyck (1978) and Salant (1982), among others, belong to this line of thought;

(b) *simulative models*, in which OPEC is considered as behaving as a not perfectly disciplined cartel, in an uncertain world (as regards the demand-supply parameters) and aiming to reach some objective of utilisation of production capacity. This approach, which refers to H. Simon's theory of limited rationality, has been applied in particular in the works carried out at Stanford (EMF, 1982);

(c) *competitive models*, doggedly supported particularly by Morris A. Adelman, where the low production costs of the richest oil fields exercise an inevitable pressure on any monopolist practice, forcing prices in the long term to level out on the marginal costs.

The reasons behind the "failures" of these approaches are many and their analysis is not within the scope of this work. Of some, however, it is worth mentioning: (i) the difficulty of taking into account and quantifying *geopolitical factors* that have played such a part in conditioning the choices and behaviours of the players in the market, and not just of the governments; (ii) the methodological, and even more, the empirical difficulties in separating *factors of the short and middle terms* and thus in identifying and isolating the corresponding mechanisms of behaviour; (iii) difficulties in operating the *simplifications* of the oil system made necessary by the mathematical elaborations. In particular, the choice of extracting from all the variables involved some to be considered as exogenous; (iv) finally, the difficulties in reasoning on the dynamics of the oil market in isolation, or only partially connected to those of the other energy markets, given the close interdependence between them. All these reasons could basically be summarised in one way only: the impossibility of simplifying things in a highly *complex* system such as the oil industry, an industry in constant *development* due to the economic, political, technical and environmental values affecting it, according to cycles of priority which can rarely be predicted.

UNCERTAINTY AND INSTABILITY

Beyond the unquestionable importance of these theoretic approaches, the fact remains that levels and actual dynamics of oil prices do not seem to have been very consistent with them, except for limited periods. Their interpretative and predictive capacity has proved to be insufficient given the complexity of the functional relations involved, the multiplicity and divergence of the objectives pursued by the various economic agents, and the impossibility of writing off their behaviour as the mere result of economic rationality.

It is certain however that of all the determining elements that could have influenced the price trend, none could be considered in itself the univocal key to interpretation. Wars, for example, have sometimes led to a rise in price, but in other cases they have had totally insignificant effects. Even variations in demand, which normally condition variations in the price of raw materials, have produced effects which have been considerably different at times from the *expected* effects. Nor are we aided by another order of determining features: the *level of collusion* of the sellers, since it is very doubtful that this reached its peak, as many claim, during the expansive phase of prices during the Seventies, when each seller seemed to proceed on his own account, or instead in the recessive phase, when a classic manoeuvre of control of supply was imposed to prevent further problems.

The fact is that no theoretic paradigm seems to have stood up to the test of facts, although theoretic economists do not seem to be much aware or worried about this. It is perhaps difficult to find another sector of the economy in which the *real dynamics* have more capriciously followed routes different from the expected ones, as has instead been seen in the oil sector. On the contrary, the more reliable the forecasts appeared, due to the theoretic or econometric apparatus which had been implemented, or the wider the consensus they gathered, the further off target they were. This has given rise to the situation of lack of preparation and confusion that has accompanied every rise or collapse in prices over the last three decades. Of the six predictive "vintages" which have occurred in this period, three got the *sign* of the variations wrong and three their *timing*. The costs of incorrect forecasts and incorrect expectations, in a sector of high capital intensity like the oil sector, are truly bewildering. It has been estimated (Arthur Andersen, Cera, 1984) that in the two-year period 1980-81 alone, 500 billion dollars were invested throughout the world on the fundamental expectation that oil prices, on which their convenience was based, would continue their upward course from the 30-32 doll./bbl of that time to 50-100 doll./bbl by the end of the decade, compared with the 20 doll./bbl and lower which was actually observed.

Uncertainty is in fact the core element of the whole history and economy of oil. Company strategies have always been aimed at reducing this uncertainty, facing it and overcoming it, whether it was a question of delineating their internal organisational layout, defining their relations on the market, or setting new contractual mechanisms. This uncertainty was determined, on one hand, by the basic conditions of oil economics and on the other, by the oligopolist market structure which inevitably derived from these. The interdependence of company decisions, in a highly uncertain context, led to conditions of unbalance and instability in the markets, all the more evident the more non-cooperative and competitive behaviour prevailed.

This is dramatically evident as regards prices. Even a superficial analysis of their long-term trend shows that *instability* is a physiological element. Figures 1 and 2 clearly show this. Figure 1 shows the historical series of oil prices – in current and real terms – from 1859, the year of the first exploitation of oil on an industrial scale, to 1995. Figure 2 shows the annual variations in percentage of current prices. Only 41 years out of 135 in fact show relative stability in price, with variations within a minimum band of 5%; 49 years report increases of more than 5%, more than half greater than 20%, while the other 45 years show decreases exceeding 5%, on 21 occasions greater than 20%. The only continuous period of stability in nominal prices was registered for the

Figure 1. *Oil Prices: 1861-1995 (doll./bbl)*

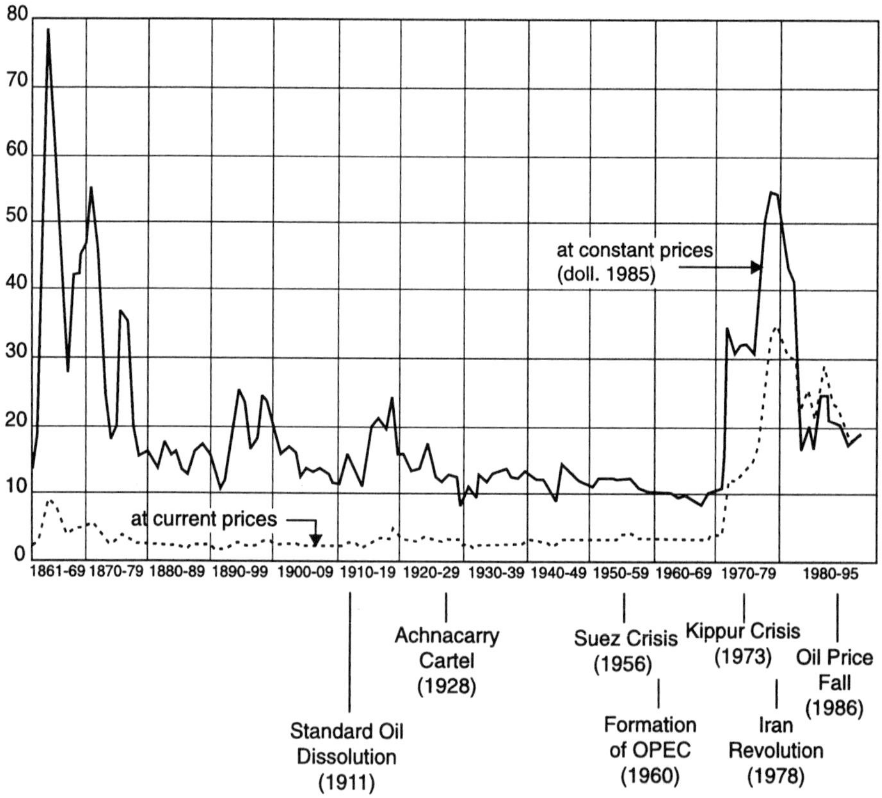

Source: BP, "Statistical Review of World Energy", 1996.

Figure 2. *Annual Oil Price Change: 1860-1996 (%)*

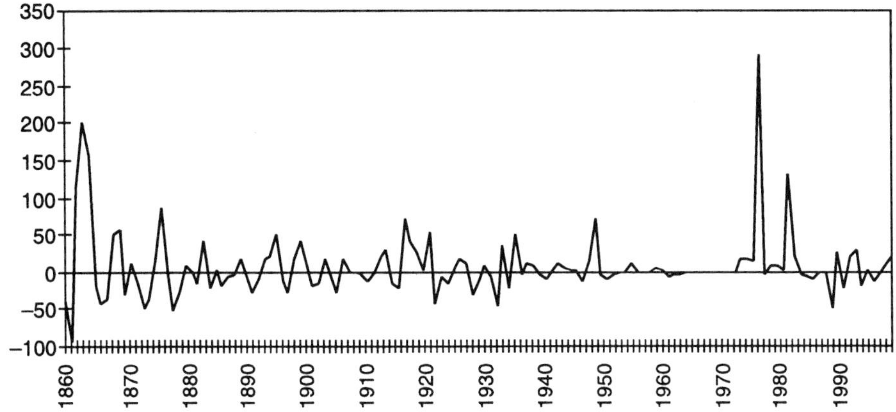

years from the end of the Second World War to the end of the Sixties. Part of the aim of this book is to attempt to understand the *reason* for the physiological instability of oil prices, the correlations that may be identified between industrial market structure and criteria of price formation, and the influence of problems of a political nature, especially in the relations between states and companies and between producer and consumer states.

OUTLINE OF WORK

The analysis follows the historical development, going over the various stages taking place since the beginnings of the oil industry up to today. The reason is that each of these manifests structural characteristics, relations of strength, economic and political dynamics which cannot be immediately attributed to the traditional life cycle of an industry and which indeed can no longer be repeated. There are two aspects which run horizontally through these historical dynamics. The first is the way in which the basic conditions of the oil economy (considered in Chapter 1) have influenced the behaviour of the various players (Chapter 2) and in particular of the companies, according to the specific external context in which they were operating. The second aspect is the way the companies have attempted to respond to the question that has run through the entire history of the oil industry: how to reconcile (short-term) competition with (long-term) stability of the markets. The initial phases (Chapters 3-4) in which first conditions of ruinous competition prevailed followed by conditions of close oligopolist co-ordination, were followed by others (Chapters 5-9) where the disappearance of unequal relations of strength between producer states and companies led to the predominance in international markets of conditions of broad, though not perfect, competition.

This book has been written principally with university students in mind, since they have shown the greatest curiosity in the fascinating world of the oil and energy industry, but it may also be of interest to anyone involved in or intrigued by the subject. For me, as for many others, this interest was aroused during my university years and led to a decision to continue to devote much time and energy to it. Many people have contributed to my knowledge of the subject and thus helped me to write this book (with the exception of any mistakes it may contain). Among all these people I particularly wish to thank Paul Frankel, Hans H. Landsberg, Jean Masseron, Lello Pezzoli, Marcello Colitti and Davide Pastorino. I also owe much to those who have long shared this commitment with me: Pino Cassi, Patrizia Bassani, Alessandro Bianchi, Giovanni Goldoni, Davide Tabarelli, Giovanni Ceccaroni and Morena Grilli (who has been my main support in this work).

1 BASIC CONDITIONS FOR PETROLEUM ECONOMICS

1.1. INTRODUCTION

In order to understand the market dynamics, company strategies, policies and price trends of petroleum economics, a number of important "basic conditions" need to be considered. Each of these has a *different* influence according to the aspect that we intend to investigate, and each *changes* with time according to technological developments and market situations. Only a thorough analysis will help us to isolate the effect of each individual condition as well as, more importantly for analytical purposes, their combined effect. Table 1.1. shows the basic conditions that are the most relevant in petroleum economics as regards supply and demand.

Before we analyse these conditions individually, let us first remind ourselves of the sequence of events which collectively form the oil industry: (a) *mining*, including all activities from the first geophysical prospecting through to production, the abstraction and storage of oil; (b) *marketing* of the oil and/or its refined products; (c) *transport*, from production areas to the oil refineries and then on to areas of consumption of its products; (d) *refining* of oil; (e) *distribution* of its products. As we shall see in the chapters to follow, the majority of firms have been forced to operate in all phases of the oil cycle for reasons of economic convenience and strategic opportunity, though many of them have consolidated their businesses solely in upstream or downstream activities, and in a marginal position with respect to the leading companies, although often with no less profit.

Finally, for a full understanding of the basic conditions of oil economics, we must refer to some fundamental concepts and definitions. The aim of this is to dispel certain popular misconceptions and simplifications that have persistently clouded the general perception of the world of petroleum.

Table 1.1. *Basic economic conditions of the oil industry*

Supply	Demand
High capital intensity and high risk	Low price elasticity
High scale and scope economies	High income elasticity
Increasing plant specificity	Differentiated elasticity
Low price elasticity	Cross elasticity

1.2. OIL: SOME BASIC CONCEPTS AND DEFINITIONS

1.2.1. Origin, composition and nature

Oil is a dense, viscous mixture of hydrocarbons, varying considerably in its chemical composition. Only within certain limits may oil be called a *homogenous* commodity, and this is even more so as the quality demands for both the intermediate and final product has tended to increase, mainly for environmental reasons.

Over the years different theories have emerged time to explain the genesis of oil. The most accredited theory today is the "organic" theory, whereby oil is the product of processes of the anaerobic carbonisation of mainly aquatic animal organisms. This process may be summarised in four phases: (a) algae and marine fauna are deposited on underwater beds (seas, lagoons, marshes, etc.), mixing with sands and clay sediments; during the process of diagenesis a *source rock* is gradually formed which favours the process of the fermentation of organic materials within it. Sedimentation and fermentation are therefore simultaneous; (b) given time, the anaerobic bacteria present in the sludge containing the organic remains (sapropel) take oxygen away from the organic substances, which are enriched with carbon, hydrogen and fats by reduction. In this way various hydrocarbons (oil and gas) are formed through a series of transformations and through the combined action of temperature and pressure; (c) with the deep transformations taking place to the earth's crust, the source rocks are covered, fragmented, compressed and often dispersed by the accumulation of new sediments. In the meantime the oil and methane which has been formed filters through the fissures and permeable rocks until they find a natural impermeable barrier known as *petroliferous trap*; (d) at this point, the hydrocarbons accumulate in the porous rocks inside traps known as *reservoir rocks*. An oil field may not therefore be represented, as it commonly is, as a subterranean cavern containing gas under pressure, below which there is a liquid mass of oil. Instead, it is a layer of rock, impregnated (in its porous cavities) with hydrocarbons. Although all oil is generated through similar phases, these do not always manifest themselves in the same way and can therefore be radically different. This is the reason why oil prospecting is such a high-risk business.

Petroleum is formed through processes which may take millions of years. The rapidity of its depletion rate in relation to the time of its "reproduction" has made this a limited and exhaustible resource. This finiteness, given an increase of 60-70 times in annual consumption since the exploitation of oil on an industrial scale began in 1859, has periodically fuelled fears and worries about its "imminent exhaustion". Indeed, even in the mid-nineteenth century the same fears were being voiced about the fossil source which dominated the period: coal (Clô, 1993). There are several reasons for these inaccurate forecasts and prophecies. In particular: (a) the lack of trust in, or knowledge of, the technological improvements which have allowed for the exploitation of mineral resources where once unimaginable; the development of new sources of energy to replace the previous ones; replacing energy with other sources or increasing the efficiency of its use; (b) the uncertainty that surrounds every element and all information regarding the availability of oil. Indeed, the actual size of any oil field is only known *after* the oil has been extracted, and is limited by the technology used.

1.2.2. The availability of oil

As soon as the very costly tip of the drill first bores through the reservoir rock to find oil and/or natural gas, geologists, engineers and managers begin to debate the extent of the oil field and its potential for technological and economical exploitation. All exploration and development activities preceding the extraction of oil are designed to increase the level and reliability of the information on which subsequent decisions are to be based. Though geochemical and geophysical techniques have greatly improved, the data and information collected remain simple scientific conjectures featuring strong elements of subjectivity and bias. In advancing their (brief) estimates on the size of the oil field that is *presumed* to have been found, geologists and engineers in some way project into the future their experiences of the past. Their estimates will be inaccurate since the past will never reoccur identically in the future. "Their profession," writes Paul Essley, "is more an art than an exact science [because] the majority of the data, phenomena, symptoms are subject to more than one interpretative logic. It is like having more unknown factors than equations and reaching multiple solutions" (Essley, 1965, p. 21).

Three concepts define the quantitative limits of the theoretic availability of oil: resources, reserves and production capacity. The sequence expresses a gradual certainty of these limits, as the result of investment made in the upstream phase (prospecting, development and extraction).

The term *ultimate resources* refers to the stock of oil that it is presumed possible to extract in an undefined future. The estimates are not drawn up (or at least only partly) on the basis of the results of the drilling activities where oilfields are discovered, but taken from geological information, speculation and from deductions based on previous findings. This is basically the last and quantitatively wider limit, but it is also more subjective, hypothetical, and uncertain. This helps to explain why the estimates of the ultimate resources have varied enormously over the last fifty years, although the variation in the estimates has gradually diminished in a continuing upward trend. In 1942 the geologists Pratt, Weeks and Stebinger estimated ultimate reserves as 600 billion barrels (bill. bbl). This figure was re-evaluated to more than double, at 1,500 bill. bbl, by Levorsen. In 1968, just seven years later, the figure once again more than doubled to 3,550 bill. bbl according to Weeks. This was one of the highest estimates in history but since then values have oscillated, mainly between 1,700 and 2,000 billion bbl. There are several reasons to explain this convergence: more and better knowledge is available due to the gradual accumulation of experience; methods of evaluation have improved; geophysical exploration has made great progress, allowing oil prospecting to be extended to almost the entire earth's crust; the "mapping" of all potentially productive areas, including those where little or no investigation has been made.

According to the classification of the most used mineral resources – known by the name of "McKelvey's box" – total ultimate resources include all materials that have already been investigated and discovered through prospecting activities, and others – undiscovered – which could be found in the future. Within the first category the subset, of those which have already been identified with relative certainty, in dimension, location and quality, and for which extraction is actually considered possible, given current prices and technology, is known as *reserves* (proven or presumed according to the extent of the information available). Although the level of certainty in this estimate is greater than that for the resources, the estimate is similarly changeable, since it might diminish as the evidence of current production continues or it might increase due to improvements in extraction techniques (reaching greater depths, for example) or with the identification of new oil fields. On the other hand, the price changes act in both directions. When prices increase, marginal resources are transformed into reserves, and when they fall, oil fields which are exploitable today become marginal deposits (within the limits of using the same technology).

From all this we may see that resources and reserves are part of a *dynamic* system which at no time permits "stocktaking", or counting as if they were boxes on a shelf. This is not only because of the uncertainty surrounding any conjecture of a geological nature but also because changes in technology and economic conditions continuously influence their entity in a trend which has so far always been an increasing one. It is unthinkable however that this process will continue indefinitely, and thus nullify the finite nature of fossil energy sources and of oil in particular. The geological and economic depletion of the oilfields is anything but imaginary and it may be extremely short-sighted to deceive ourselves that this depletion will never happen, particularly for future generations. As Darwin wrote in 1953 in his *The Next Million Years*, "it is much harder to live off income rather than the capital accumulated in geological eras". This was true then and is even more so now, with a fivefold increase in the consumption of fossil resources.

Resources and reserves are concepts identifying the production limits – expressed in terms of stock – whose levels and variations influence market dynamics and long-term cost-price relationships. Short to medium term dynamics is instead conditioned by a third, more restricting limit: the *production capacity* available at any time and sustainable for a given period of time without compromising the mineral potential. Expressed in terms of flow (barrels per day), this is a function both of the mineral dimension of the discovered oil fields and of the volume of investment made in exploration/development (number of wells) as well as the infrastructure necessary to take a barrel of extracted oil to the consumer market. In market strategy and for the consumer the only supply that counts at any time is that which can actually be made available, not the potential supply as expressed by the reserves and even less by the resources.

The inability to distinguish correctly between oil resources, reserves and production capacity – and the improper use made of these data to support any desired objective – has always confused any interpretation of what was happening on the markets or what would have happened if something had not been done. The resource data were conditioned by the dominant opinion at any time, and extremes were chosen from the wide range of estimated values or the conjectures on which they were based, according to whether the intention was to spread a message of abundance or scarcity. In the same way, the proven reserves (part of the already identified resources) went to form the most immediate empirical indicator of energy "scarcity", with respect to other more economically appropriate indicators, such as the costs of discovering new units, mining revenue or prices themselves.

In particular, the *reserves-production* ratio is often improperly used as an indicator of the residual life of the oil fields. It should instead be regarded as a result of the companies policies of exploitation and not as a factor beyond their control. The high and constantly increasing costs of exploration have pushed oil companies to invest in this only as much as it is necessary to cover an adequate production horizon: trying to optimise the (bell-shaped) curve of terrestrial exploitation of the oil fields according to the mineral conditions (especially the level of natural pressure), the forecasts of price and demand, and the policies of their competitors. This also explains why the reserves-production ratio over the last fifty years has always been between 30 and 40 years: the average lifetime of oil fields in production. To conclude, proven reserves are an empirical indicator of the extraction potential of oil and not a certain measure of the total supply for the future. It is wrong and misleading to interpret them as the remaining life of the oil fields, in relation to current or expected production, at least until the residual resources coincide with the proven reserves.

1.3. HIGH CAPITAL INTENSITY AND RISK FACTOR

The oil industry is by nature an industry with high capital intensity. This feature may be found in each of its phases and is particularly significant in mining activity, because of the high *level of risk* that characterises it structurally. Particularly in its pioneering period, mining research is comparable to a "gamble", where the cost of the "chips" may amount to several hundred million dollars[1] with "jackpots" which can run to several billions.

But not everyone can play. Many luckless gamblers will be out of the game once their available funds have been used up unsuccessfully; the few lucky ones will manage instead to accumulate colossal fortunes, thus acquiring a dominant position not only in production but also in the subsequent phases. "The very fact that crude production is always a leap in the dark...", wrote Paul Frankel in 1946 in his book *Essentials of Petroleum*, which remains unsurpassed for some of its intuitions, "affects the stability of all subsequent phases. No part of the oil trade, however far removed, can be altogether untouched by its capricious origins" (Frankel, 1946, p. 3). Investment in the research phase may limit the margin of risk, increasing the likelihood of success and in any case providing data and information useful for subsequent research, but the fact remains that "in spite of all the knowledge and experience gained over almost a century, the occurrence of oil in profitable quantities at any given point cannot be deduced theoretically but can ultimately be proved only by the act of boring a well" (idem, p. 10). Hence the need to make a large number of bores, once a potentially interesting area has been identified, since the cost of many fruitless attempts has to be offset against the few which will be successful. And even once the actual presence of oil has been ascertained, there is no certainty that its extraction and marketing will be profitable, since this depends on the costs that will be sustained globally and on the current prices. To spread the risk, oil firms have developed a strategy of horizontal integration, where possible, operating on a vast scale, in many countries with different levels of mining and political risk, on different research themes extending over entire geographical areas. All this led inevitably to co-operation between various firms. In this way, the areas of the research compensate each other in probability and the few positive results are sufficient to repay the many negative ones.

A number of studies have been done to determine the "probability of success" of upstream activities in the petroleum industry. In the Sixties, according to Jean Masseron of the Institut Français du Pétrole, the probability of discovering new oil fields in Western Europe and the United States was 13%-14% (the equivalent of about 1 successful attempt in every 8); 25%-35% in Western Canada, Libya and the Netherlands; 60% in Nigeria and as much as 75% in Saudi Arabia (Masseron, 1982, p. 18). These percentages are based not only on the distribution of oil in the natural state but also on the attitude to risk of the operators and the total amount invested in exploration. They cannot therefore be taken as probability of discovery in the strict sense. These percentages, established from the ratio between the number of fruitful oil fields (or wells) and the number of oil fields (or wells) explored (or drilled), are in fact variable in time (and also in relation to the level of depletion of resources) and give values generally lower in the Seventies and Eighties than in the past.[2]

These considerations refer to two levels of risk of a different nature, in addition to the normal commercial risk. We may define them as *geological risk* (given by the imponderability of the unknown) and *technical risk* (errors in interpreting the data). In the case of petroleum, they will also include *political risk*: the effects of unfavourable and unexpected changes by the host governments of the operative conditions on which the firms had based their investment decisions (changes in property rights, in fiscal policy, etc.). The evaluation of these risks will influence the level of capital available to the firm for a certain investment, and thus its cost and the cost of the remuneration of the capital.

The hazardous nature of mining activities, being linked to the availability of investment capital, helps to place the problem of the optimal decisions of investment, as well as decisions related to the prices, in terms of *survival* in time of the firms, rather than the maximisation of their profits in the short term. For oil companies, who operate with fixed plants and not only in trading activities, the main requirement is to carry out their activities in *relative certainty* as regards the dynamics of the markets, so that they can rely on specific flows of income. The empirical theory of probability may be applied in the upstream sector: here, in a series of tests repeated many times under the same conditions, every possible event appears with a relative frequency r/n – where in n tests the event occurred r times – which is approximately equal to its probability (favourable in relation to possible chances). The company will therefore have to dedicate enough finance towards research in order to ensure that operations continue until a find compensates a series of previous failures. This means they must be able to survive huge losses, though these might well be expected to be transitory. The above conditions have an important influence on the structure and behaviour of the petroleum industry.

Need for high self-financing. Unable to conform to rigid and pre-fixed reimbursement plans and obligations of time returns, oil companies have had to find the largest part (70%-80%) of the capital they require from funds available internally. However, this is also due to the reluctance of the financial institutions to expose themselves in a high-risk sector, as well as the enormous size of oil investments in which the banks could only take a small part.

Large size of the Companies. Mining risk is conditioned by the total capital available, as well as, obviously, the ability to manage it, and – why not? – also by luck. We can thus state that, within certain limits, risk is a decreasing function of the available capital and thus of the size of the company. The risk factor thus introduces a formidable barrier to entering the industry, establishing – in relation to the entity it has in the various production areas and to the trend of the marginal costs of investment – the minimum size that a company must be to take part in the industry. Smaller firms have also entered the industry, though generally this is due to protective legislation mainly with state-controlled or national companies.

Integrated organisation. Only those firms that manage to produce *large and sufficiently stable* flows of income are able to sustain investments in regular rhythms so that exploited reserves may be suitably replaced and their increase match the rise in demand. This is in order to preserve a long-term horizon in the exploitation of the oil fields and the acquired market positions.[3] In order to achieve these objectives, oil companies have tended to operate at all phases of the oil cycle, so that: (i) hazard upstream activity constitutes an important but not determining element of the whole business; (ii) achieving positive results in research and extraction is not affected by difficulties in selling the final product on the market.

Such high capital intensity and high mining risks now influence the size of the companies, their number, their organisational, political and financial structure, and above all, as we shall see, their supply and pricing policies.

Mining activity, both at the R&D phase and during exploitation of discovered oil fields, has indeed always been strongly conditioned by the public policies of the oil-producing countries. These policies were designed to reach different objectives, according to the specific situation of each country and the change in time, according to the state (actual and expected) of the markets. The policies involved facilitating research, offsetting the effects of the "risk factor" or defending the companies from competitors; fostering a rational world-wide policy for the exploitation of mineral resources; controlling the monopolistic power of the companies, including the destination of the profits; maximising the oil revenue of the producer; supporting the price level, limiting conflicts with competitors. These are all aspects which will be dealt with in this book.

1.4. LEVEL AND STRUCTURE OF OIL PRODUCTION COSTS

For the oil producing company, production costs may be separated into *technical costs* and *fiscal costs*. The former refers to all the expenses borne for the production of oil until it is placed on the primary market. The latter concern the amount that firms have to pay the countries as owners of the resources and/or as receivers of taxes on profits (real or presumed).

Cost analysis is enormously important throughout petroleum economics and is particularly influential in conditioning the behaviour of the producers, much more so than in manufacturing or in other mineral production industries. This is due to:

- the enormous *diversity in levels of cost* from area to area of production, or even within the same area;
- the *high ratio between fixed costs/variable costs* because of the enormous investments that must be made, as we have seen, before production can begin;
- the *low ratio between production costs/oil price* due to the high "degree of monopoly" that has always distinguished this market.

Individually and together these phenomena have an enormous influence on market power and the competitivity of producers (to the advantage of those with lower costs); on their supply policies in the short term (inelastic to the prices, as we will see), and on the policies of development. Let us analyse these phenomena more carefully.

Diversity in cost levels. The technical costs depend on a huge number of variables: difficulties encountered during the research stage (average number of drillings necessary to give a productive well), depth of the wells, nature of the rocks to drill, rate of interest to be paid on the investments, distance from the sorting or loading points of the crude oil, etc. Depending on the situation, the costs of exploration may represent, from 10% to 20% of the total costs; development costs from 40% to 60%; and the operating costs of extraction from 20% to 50%. As shown in Table 1.2., the *unit production costs* are between a *minimum* value of less than 1.0 doll./bbl for on-shore oil fields of the Middle East (with levels of just 0.25 for those discovered and in activity since the Twenties and Thirties) and a *maximum* value of up to 25 dollars for the offshore stripper wells of the North Sea. Generally speaking the Middle East, Africa and Latin America are in the lower band, also taking into account the quantities produced, while North America and Northern Europe are in the top band of a scale of values of costs which gives a *ratio from to 1 to 30 for normal fields and up to 1 to 80 for marginal fields.*

Table 1.2. *Technical unit costs of oil production (doll./bbl)*

Geographical area		Costs (doll./bbl)	Productivity (bbl/d/well)	Wells (no.)
MIDDLE EAST			2,330	5,400
On-shore:	old fields	0.4-0.8	4,700-7,000	
	new fields	0.6-3.0	800-1,400	
Off-shore:	large fields	2.0-4.0		
	small fields	3.0-6.0		
AFRICA				
On-shore		1.0-3.0	1,000	4,900
Off-shore		3.0-6.0		
UNITED STATES				
On-shore		2.5-10.0	13	636,000
Off-shore		3.0-5.0		
NORTH SEA			3,700	912
North Area		8.0-20.0		
South Area		4.0-10.0		
Marginal fields		15.0-25.0		

Source: costs: Masseron (1992); productivity and wells: *International Petroleum Encyclopedia*.

The variability of costs is influenced by various factors, but one in particular is especially significant: the *productivity of the wells*, expressed on a world scale between maximum values in the Middle East of up to 7,000 bbl/d per well and average values in the United States of just 13 barrels. The approximately 640,000 wells in the United States in the mid-Eighties, exploited to their full potential, yielded an overall production of 425 million tons, while the 5,400 wells of the Middle East gave a production 30% greater, though exploited to only half their potential. Distributing fairly similar investment over flows of output up to 400 times greater (per unit of time), shows the differential advantage that Middle East producers have always enjoyed. The relative disadvantage of the other areas also increases enormously if we consider that the probabilities of success in research drilling are also much lower here. This represents an investment cost per barrel up to 2.0-2.5 times greater than can be seen in the "easier" areas of the Persian Gulf.

Nor is it much compensation that in economically more developed areas, the absolute costs for each metre drilled may be lower, since, for this differential to be felt, a much greater number of metres would be needed per ton extracted. In fact this does not happen, with even this indicator being unfavourable in the areas outside the Middle East. The difference in costs has particular influence on the competitivity of the producing companies and the prices. With a range of production costs not to be found in any other mining sector, it is clear in fact that prices in the petroleum industry are very unlikely to be fixed based on production costs to reach an international market price. If it were so – in other words if this result were achieved by prevailing conditions of perfect competition – producers with greater disadvantages would be kept out of the by the more efficient producers. As we shall see, this has never happened, thus proving that reasons outside economic rationality and pure competition have influenced the allocation of resources, investment policies, and oil supply policies.

High fixed costs/variable costs ratio. This basic condition derives directly from the high level of investment required in order to start up production and from the relatively low level of operating expenses needed to sustain the mining stage. The more hazardous the extracting conditions (deep under the sea, for example), or the more remote the oil field with respect to the port of shipment, the higher the costs. On average, the fixed to variable ratio is 4 to 1, though in the most unfavourable cases this ratio may be 1 to 1. Thus, in the majority of cases, the marginal cost is much lower than the average cost, and the cost production curve decreases steadily over a long stretch as output increases. The consequence of this, combined with the huge expense sustained by the producer *before* being able to produce, is extremely significant: the producer has every advantage in maximising output (compatibly with the limits deriving from an optimal exploitation of the wells, in other words so as not to compromise their subsequent operations). That he should mine one ton more or less is fairly irrelevant as regards the costs, given the expense that he has so far borne, while the differential advantage derived from the possibility of gaining part of his money back is considerable. If, on the other hand, difficulties in outlet emerge, instead of reducing or halting production, the operator will be interested in selling even at a lower price than the average cost as long as this is higher than the marginal cost. The difference between the two values may be quite considerable. Paul Frankel attributes much importance to this aspect in explaining the supply policy of the producers and the natural instability of the petroleum market. "The main factors the crude producer had to consider , in the circumstances, were:

1. The necessity of rapid production to make his heavy investment pay as soon as possible.
2. Commitments to the property owner who was promised royalties and who usually granted the lease on the condition that it should be exploited within a specified.
3. The danger of 'his' oil being drained away by his neighbours" (Frankel, 1946, p. 19).

All three factors help to select production as the choice to be made instead of any other consideration (at least until the investments have been amortised or new investments are needed).

Figure 1.1. *Crude oil investments and prices*

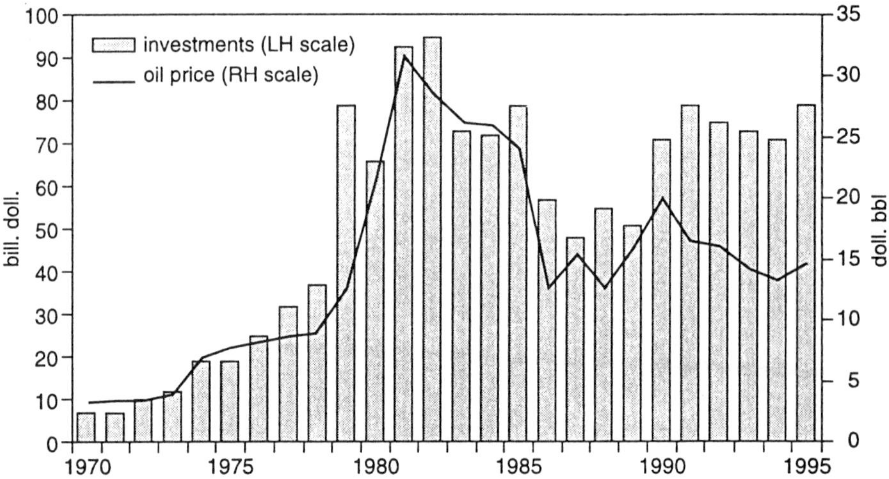

Low supply-price elasticity. To conclude, the (relative) rigidity of petroleum production in the short term is strongly supported by cost analysis and economic calculation. "The decision to exploit an oil field is practically irreversible. An economic calculation that adopts the interest of the producer as criteria means that the production of an active oil well will practically never slacken" (Mainguy, 1967, p. 169). It follows that the *supply is substantially anelastic to price variations caused by the economic trend*. Instead their effects influence the stocks by increasing or reducing them or slowing down or speeding up delivery. When the price variations become relatively stable, the phases *preceding* extraction are the ones influenced. *First of all* there is the development of oil fields already discovered (but not already equipped with development investment); *then* exploration and mining research takes place. Though they have no effect on the current supply, consistent increases or reductions in the prices, expected to be stable, alter the future supply through variations in the volume of investment. *Supply/price elasticity in the long term will thus be much higher than it was in the short term*. These statements are clearly represented in Figure 1.1. where prices and investments are shown for 1970-1995 in mining exploration and development in the world with market economy. The correlation between the two curves is immediately evident.

Variations in the prices will thus influence the future supply in a positively correlated way. The final result is determined by a number of variables, a particularly important one being the *time factor*. The longer it takes to install new production capacities, the easier it will be for the sellers of a commodity (petroleum) to raise prices and keep them high before being affected by the competition of newcomers or of alternative energy sources (coal, gas, nuclear, etc.). The time factor is generally given little importance and when this happens it is difficult to appreciate the real energy options, since the times of the choices may vary in one or two orders of magnitude.

1.5. COST FUNCTIONS (MINING)

Mining activity is very different from traditional manufacturing for various reasons: the uncertainties which never allow truly repetitive operations; the finite quality of hydrocarbon resources; the constant evolution of the external environment in which it develops. The cost functions are also different, since we cannot use the traditional U or L shapes that theory has proposed in reference (mainly) to manufacturing activities. In this section we do not intend to look into the whole theme of cost functions in any great detail (see instead the fundamental texts Campbell, 1959; Lovejoy et al., 1963), but will indicate the most important aspects of mining cost functions for the producers.

An initial clarification concerns the very concept of cost used in the oil mining industry. Indeed, besides the traditional concept of economic cost, referring to the existing production activity, and indicating the sum of costs sustained in order to extract one production unit (barrel), we should also consider the replacement cost – no less important – which indicates the cost of a future production unit and which may be expressed as follows:

$$C_s = C_x + C_d + C_e$$

Where:

C_s = replacement cost (per barrel)
C_x = cost of current production (per barrel)
C_d = development cost (per barrel added to capacity)
C_e = cost of exploration (per barrel discovered).

Those producers who do not intend to cease activity when the wells in production are exhausted will have to bear the replacement costs of the barrels extracted, and judge whether they should invest or not in relation to these and to the expected prices. Current production costs (reflecting the concept of economic cost) and replacement costs thus belong to different long-term horizons: the former reflect *past* investments, the relative mining success rate, the costs of necessary equipment, as well as the current operating costs; the latter consider the *future* situation, with all the margins of risk inherent in this. There might not be much difference between the two values if they did not refer to an activity which (normally) manifests *declining yields*, in other words which operates at growing costs, in a certain state of technique.

If these two concepts of costs are not kept separate, it is easy to confuse the functions connecting the costs with output. Much of the diatribe between economists is caused by lack of precision in specifying the data of reference. To analyse the cost-output functions further, it would be useful to consider the equation (Alchian, 1959), as suggested by P. Bradley (1987, p. 281):

$$C = f(x, V, T)$$

where the total cost C depends on three variables: x, which corresponds to the *rate of production*, in other words the rate of exploitation of the oil field which could be less than, equal to or greater than the *maximum efficient rate*, without detriment to its future activity and making the best use of its physical conformation; V which corresponds to the *total consistency of extractable petroleum*, given the investments made; T which corresponds to the *time spent from the first extraction to delivery*.

Another aspect which should be borne in mind is the production unit to which the cost function refers. In the case of petroleum this is the oil-field or *reservoir*, which could correspond to the plant in manufacturing industries. However any parallels stop here. Whereas in each industrial sector an increase in investment corresponds to an increase (more or less proportional) of production capacity, in oil mining it does not necessarily correspond to any increase in the final quantities of extractable petroleum (V), which may remain unchanged, while their rate of exploitation (x) may increase, with subsequent reduction of the long-term horizon of exploitation. On the other hand, though the nature of the reservoir is fixed, it does not follow that V is rigidly predetermined, since this depends on the same level of C. In general, in fact, the higher the level of expense reached, the greater the total extractable volume (V). Beyond a certain total threshold of production, the incremental cost of V increases, with x constant, since the rate of growth depends on the specific nature of the reservoir.

Bradley identifies 5 ways to increase production:

- *larger recovery*, using more sophisticated techniques, from the existing oil fields;
- *developing new deposits* already identified, but not yet in production;
- *discovery and development of new deposits*, in geological formations already identified;
- *discovery and development of new deposits*, in *new* geological formations;
- development of *new technologies* which affect the above options.

Between one option and another, the supply price elasticity increases, together with the time lag needed for each operation to achieve results. These delays can be indexed by the variable T, linked with a growing uncertainty.

From what we have said so far, we can see that it is impossible to apply to the mining sector a restrictive short-term meaning of the production time to which to refer the costs-output ratio, a period in which the plant and equipment remains constant while the variable factors vary. In petroleum production there is indeed a *continuum* of development expenses for equipping new wells or infrastructures in order to prolong in time the flow of production extractable from an identified reservoir. This increase in expenses might also correspond to an increase in the overall production expectations, as well as in their rate of exploitation per time unit.

Let us now analyse the trend of the functions of unit cost. These will decrease when an invariance of the total development expenses is hypothesised, and when fairly improbably, the producer takes up position on the levels of output (total and per time unit) permitted by the initial investments. This could be permitted only in those geological situations where the oil fields are large enough (as a whole and per well) not to require efforts to increase them. Here we have the typical meaning of the *short term*: with unit costs constantly decreasing due to the possibility of spreading fixed costs over an increasing production and due to the basic invariance of the direct costs of extraction.

It is much more complex to outline the cost functions for the *long term* (when all the factors are variable). This is due to the diversity of situations that may appear, to the difficulty of matching output units to relative expense units (due to the increasing time lags between these and the former) and to the fragmentary nature of the information etc. In general we may say that the *incremental costs of production of petroleum increase*, within each reservoir, when we go from initial production units to those requiring more sophisticated and more costly extraction techniques (especially when natural pressure has exhausted its spontaneous propulsive thrust) and, from reservoir to reservoir, when we go on to those with lower productivity (Fig. 1.2.).

The diversity in production costs of petroleum between various areas or various wells allows production at lower cost differential economic rents – David Ricardo referred to this as tilling soils with different fertility – when the demand, or other factors, requires the production of petroleum at higher cost. A consequence of the diversity of costs is

Figure 1.2. *Oil production marginal costs*[1]

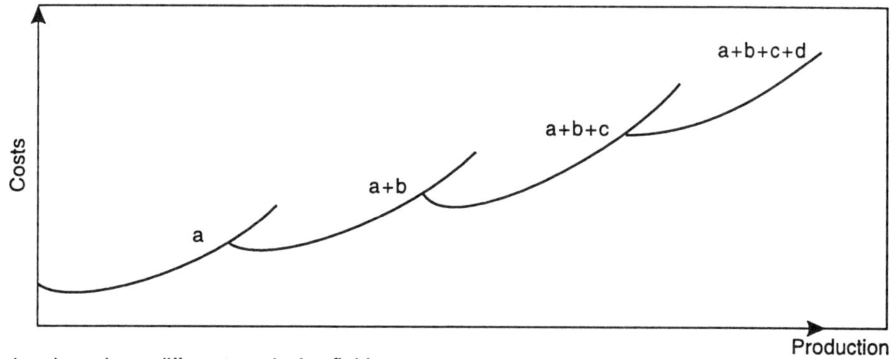

1. a, b, c, d, are different producing fields.

Figure 1.3. *Oil demand and supply*

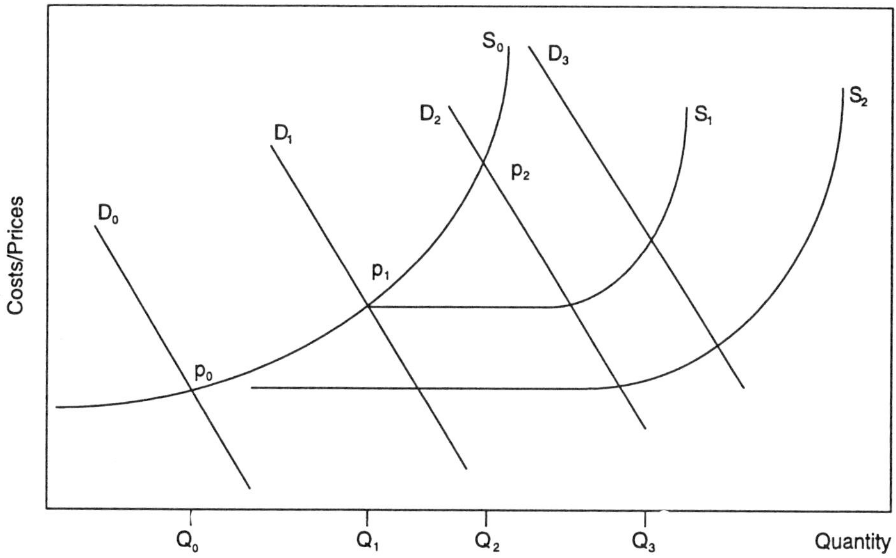

the presence of *discontinuity* in the oil supply curve. This is reflected in a shifting to the right of the supply curve when prices increase to levels sufficient to gain profit from reservoirs which were previously not profitable.

In Figure 1.3. S_0 and S_1 show the trend of petroleum supply from areas or reservoirs with different cost conditions, respectively more and less favourable. If the demand is given by D_0, the production from the more costly reservoirs is zero and the relevant supply curve is S_0. Increases in demand, with the shifting of the relative curve from D_0 to D_1, will be reflected in higher prices and higher production.

If however the demand shifts to D_2, the response is fundamentally different: the production increases with the entry into operation of more costly reservoirs, but the prices are maintained constant for a certain section of the supply curve (from Q_1 to Q_2). Further shifts of the demand, to D_3, will instead accompany increases both in quantity and in price.

Periods of growth of demand will thus see a relative stability in prices. This result may be seen when the process of adjustment is perfect and instantaneous, in other words when the market rapidly reaches a new equilibrium, such as those represented in the figure by the combinations (p_0, Q_0), (p_1, Q_1).

If however the process of adjustment shows sluggishness in timing and in the reactions of the producers to higher cost, the increases in demand may be accompanied by increases in prices at levels (p_2) which are in any case unsustainable as long as productions at higher cost enter the market (returning the prices of equilibrium to p_1) and until the supply curve indicates a growing trend ($> Q_2$). Delays in the response of the producers to higher costs may, therefore, be reflected in momentary peaks of prices, followed by periods of falling or nominally constant prices. The discontinuity in the supply functions may therefore generate periods of constant prices despite the growth of the demand.

The processes of adjustment to increases in demand become much more complicated when they require, as regards the supply, new discoveries and new production capacity (and not just an increase in the production rate of the existing capacity, as in the previous analysis). The time lags between the decision to increase research expenses and the increase in production capacity mean that increases in demand will force the prices rise until new capacity is finally available. In other words, before the capacity represented by S_1 is achieved, shiftings of the demand towards the right will continue to impact along the supply curve S_0. This adjustment will certainly take longer than the one examined above due to variations in the depletion rate of the existing capacity.

Petroleum reserves are not uniformly distributed throughout the world but are concentrated in large oil fields within a limited number of areas and countries. When the prices increase (in a stable way) or when they are expected to increase, the exploration and development of new oil fields increases. This initially takes place inside already identified fields and reserves and only later in areas not identified and at higher risk and expense. The discovery of a reservoir in a field so far unexplored is inevitably accompanied by new exploration and development in the same area in which it has been identified. A new discovery considerably reduces the risks hitherto attributed to that area of exploration: it generates new information on hitherto unknown mineral areas and boosts new investments and new discoveries that bear little relation to the price level which was initially deemed necessary to initiate investment in that area. From this process, additional reserves are generated in a much shorter time than was necessary at the pioneer stage. Outcomes of this type have been observed in all production areas, most recently in the North Sea, Mexico, Brazil and on the West Coast of the United States. The growing productivity of investments in new areas, the cumulative development of the infrastructures for the reception and sorting of crude oil, the greater information available and the lower risks permitted are all factors which help to reduce the costs and therefore the prices necessary to induce increases in supply in the new regions. In short, if the price necessary to boost new supply was initially equal to p_1, and the relative curve was S_1, the limit price for new entries could subsequently be less than p_1 and the supply curve could shift to the right and towards the bottom to S_2.

1.6. ECONOMIES OF SCALE, INTERDEPENDENCIES, SPECIFICITY

The basic conditions we have so far analysed all concern the production phase of the petroleum industry. It would however be wrong to maintain that only these have affected or can affect the behaviour of its sellers (and therefore its prices). In a naturally integrated industry such as that of petroleum, the structure and behaviour of each of the sectors can influence the behaviour and results of the other phases. We therefore have to refer to the whole of the basic economics, attempting each time to assess which conditions, and in which sectors, were the most important.

Throughout the downstream phases two conditions in particular are to be highlighted: (a) the high economies of scale; (b) the close operative interdependence between them. Both conditions lead the companies to reason and decide in terms of

Figure 1.4. *Cost functions*

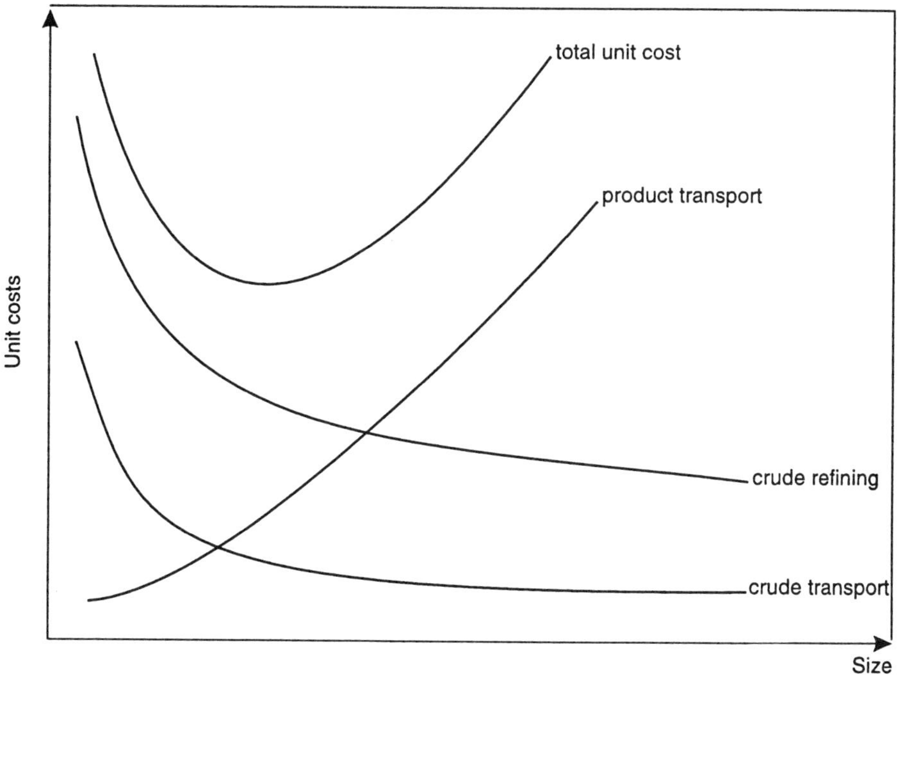

global optimisation of their decisions in both the short and long term: in the sense that the advantages of considering the whole oil business as a unit far outweigh the sum of the single partial optimisations. This does not mean that at certain moments, independent companies – present at only one or two phases – cannot benefit more than integrated firms from particular market trends, but this does not apply in the long run (unless protective barriers are in operation). Figure 1.4. shows the relative advantages that normally come into play in the trend of unit costs between the transport of crude oil, its refining and the transport of the refined products, as the size of the plants (and the market) increases.

A typical L-shaped curve of transport of the crude oil and refining – which shows an advantage in increasing the dimensions of the quantities transported and treated at least up to a certain minimum optimal dimension (MOD) – is countered by the cost of transport of products that grows unitarily and to such an extent as to counterbalance, beyond a certain market dimension, the advantages of scale that are to be had at the upstream phases. The result of the three cost curves identifies an *optimal dimension* of the refining plants in relation to their location with respect both to the sources of oil supply and to the markets of final consumption. The optimal solution will be given by the combination of the three variables (scale of plants, transport of crude oil, transport of products) and will not necessarily be the same as the sum (vertical) of the single MODs. In certain situations it could thus be advantageous to operate with refining capacities below the optimum – when for example the market is limited and the distance between other refining centres is great – while in other cases the dimension might be even 4-5 times greater than the MOD.

How are the company policies affected by the existence of operative interdependencies? We may answer this question by referring to the analysis of the *interdependent planning uncertainty* proposed by Malmgren in 1961 following the work of Coase in 1937. In his opinion, the *combined* planning of activities with a close degree of interdependence gives – when faced with a highly variable and uncertain context (certainly the case with petroleum) – cost advantages and makes it easier to reach conditions of equilibrium of the markets. The internal organisation of the companies, in a better position to control the circulation of information, set up compatible planning for complementary activities, and co-ordinate the expectations of the single units, will thus be preferred to the intermediation of the market, also because the prices are not efficient indicators of future events. "Interdependence does not by itself cause difficulties" wrote March and Simon, "if the pattern of interdependence is stable and fixed. For in this case each sub-program can be designed to take account of all the other sub-programs with which it interacts. Difficulties arise only if program rests on contingencies that cannot be predicted perfectly in advance. In case, co-ordinating activity is required to secure agreement about the relevant activities of the others" (March and Simon, 1958, p. 159). Vertical integration makes it possible not only to better plan the entity of the investments with respect to the expected market demands, but also to ensure a combined exploitation close to the optimal one and in any case better than that attainable through contracts which are inevitably incomplete. The company is also able to be more accurate in planning the capacity level used which is needed to give stability to prices and production in the long term. This is, in turn, what helps to reduce the entity of transaction costs.

The same conclusion – the preference for hierarchy to the market – is reached if we make the analysis in terms of *external economies*. These originate when investments set up by two or more companies are profitable only if they do not operate independently of each other. "If the firms in this interdependent group were integrated, the pecuniary external economies would become internal economies, with a resulting gain in profits to the investors of the integrated firm" (Malmgren, 1961, p. 412). The technical and operative complementarities mean that certain investments are profitable only if other investments are made at the same time, and if the decisions in their management are adopted coherently with the maximisation of the combined profit. This is possible only if the circulation of information is efficient, in other words if the market prices are able to consider this type of complementarity when the decisions are adopted independently by the companies. "Where output and profitability of various production units are closely interdependent, a company is formed to undertake decisions concerning all or some of the production units simultaneously, so as to maximise the joint profit and total output. What would have been 'external economies' and therefore not registered in the price system become internalised" (idem).

The close technical interdependence which runs between the subsequent phases of the petroleum industry – associated with a highly variable and uncertain environment – has brought out *economies of co-ordination* which have strongly favoured the integrated structures of the major companies.

1.6.1. The case of refining

Since the very beginning of the petroleum industry, the size of the refineries has been one of the strongest reasons for the competitive advantage of the companies. In 1870 when John D. Rockefeller founded the Standard Oil Company together with Samuel Andrew and other minor partners, the refineries that he had built not long before in Cleveland were the largest in the world, with a capacity exceeding 50 thousand tons/year. This enabled him to almost halve the unit costs from over 5 cents to 3 cents per gallon, and to overcome competition.[4]

In the century which followed, the constant improvement in processing technologies – to optimise the yields to the dynamics of consumption, to make the processes more continuous, and to increase safety in the plants – was to be accompanied by an uninterrupted expansion in the size of the plants, stimulated by an exponential increase in consumption. The increase in the quantities processed and marketed made it possible to reduce transport costs, thus even overcoming competition in the distribution field.

In the years following the Second World War, the *average dimension* of refineries in industrialised countries increased by a factor of 20 – from about 250 thousand to about 5 million tons/year – over a range of unitary values between minimums of 0.5 to peaks exceeding 30 million tons: showing the varieties of situations and different factors that determined the dimensional choice. For an initial approximation, it may be said that the *minimum optimal dimension* is around 4-5 million tons per year. Beyond this threshold the curve of economies of scale tends to flatten out until it becomes asymptotic from 7-8 mill. ton/year upwards. From unit costs of 7 doll./bbl for plants up to 1 mill. ton/year there is a drop to values around 2.0-2.5 doll./ bbl for dimensions equal to or exceeding 5 mill. ton/year. Between the Fifties and the Eighties the distribution of the plants changed drastically. Initially 70% concentrated in plants smaller than 1 mill. ton/year capacity. Now only 10% is in this category, 50% between 1 and 5 mill. ton/year, and the remaining 40% in that exceeding 5 mill. ton/year.

A typical process industry, refining has represented a schoolbook case of reference for the study of economies of scale.

Table 1.3. *Producing capacity, size, localisation of oil refineries (mill. tonn.)*

	1940	1950	1960	1970	1980	1990	1995		
								No.	Size[1]
United States	222	334	511	628	883	809	763	169	4.51
Europe	21	43	198	693	1,011	709	709	109	6.50
Middle East	18	46	71	114	151	222	262	40	6.55
Northern Africa	1	2	6	39	83	140	141	43	3.28
Eastern Europe[2]	30	57	161	349	689	767	640	81	7.90
Asia	13	14	73	292	597	607	718	129	5.57
Others	51	84	187	324	546	458	460	134	3.53
Total	356	580	1,207	2,439	3,960	3,712	3,693	705	5.24

1. mill. tonn./plant; 2. Ex-USSR included.

Source: *International Petroleum Encyclopedia*

There are two main reasons for this: (a) the *relative indivisibility of the production factors*, for which the various plant units into which a refinery is divided could only be used fully for production levels higher than certain thresholds; (b) the *non proportionality between scale of plants and their costs*, which may be approximated by the ratio between volume (production capacity) and surface (cost) of the individual sites.

If C is the investment corresponding to a plant of capacity Q, the following ratio applies:

$$\frac{C}{C_0} = \left(\frac{Q}{Q_0}\right)^b$$

where C is the cost of the larger plant, C_0 that of the smaller; Q and Q_0 are the respective capacities while b is indicated as "scale factor". The lower its value, the greater the advantages in unit costs obtained by expanding the dimensions of the plants. Estimates carried out by various scholars (Pratten, 1965 and 1971, Bruni, 1964) relative to the Sixties have given b values generally of between 0.6 and 0.8 (for the different plant units in a refinery or also for the whole).

In addition to the technical economies of scale, there has been a reduction in the costs of refining due to the great advances made in engineering and processing techniques. The layout of the refineries is now entirely centralised, where each process unit was once a separate plant with its own control room and its own laboratories. The area of a refinery has been considerably reduced, as have the costs of the infrastructures within. Other developments have been made in automatic control, optimising the operations through computers and in the automatic mixing of the refined products, with subsequent reduction in the storage requirements. These innovations considered as a whole may be estimated as having reduced refining costs during the Sixties by about 30% (Frankel and Newton, 1968, p. 29)

The combination of high economies of scale with a low absolute value of variable operating costs (basically fuel and electricity) and their invariance for unit of output, has led to three conditions which should be taken into account in evaluating company behaviour: (a) a continuously decreasing trend in unit costs; (b) a clear predominance, in their structure, of fixed over variable costs: a ratio of 4 to 1 (for a full use of the capacity); (c) direct or marginal costs much lower than the mean costs (0.20-0.60 doll./bbl compared with 2.0-7.0 according to the different types of refinery). Considering the first rather than the second has always been a temptation and a hazard for the companies. The President of Standard Oil stated very succinctly that "the 'incremental throughput' philosophy has spread like a virus throughout our industry" (Frankel, 1962, p. 95). In fact, an increase in processing implies additional costs, insignificant compared with the sunken costs already incurred, while allowing the margin of contribution to their covering to be increased at least until prices fall below the low marginal costs. This tends to lead to oversupply, especially in times of low demand, and subsequent and frequent price wars. This risk, fairly typical in heavy industries, is increased in the case of the refining industry by the great number of product-markets in which it operates: each featuring a different user value (price), different elasticity, and different competition between the companies. The result is that at any time one of these products is always able to guarantee prices equal to or higher than the respective average cost, so that the refiner has an incentive to optimise the incremental output on this product, placing the residual products at prices lower than the average cost.

The oil crises of the Seventies upset the conditions which had motivated the uninterrupted race to expand the scale of the refineries. *First of all*, there was a drastic curtailment in the profiles of growth in consumption, just when the expansive cycle of investment started in the early Seventies was drawing to a close. The world refining capacity in 1980 reached an all-time high of 4 bill. tonn., about 70% higher than that of a decade before. The stream factor – equal to the ratio between quantity of crude oil processed and nominal capacity – fell almost everywhere below the break-even point, in some European countries (including Italy) dropping to below 50%. Scale became a penalising factor. *Secondly*, lower demand was accompanied by a drastic change in its quality structure: to the advantage of the medium-light fractions but with the large conventional refineries unable to cope with it without activating costly policies of technological qualification. *Thirdly*, the new capacity tended to de-localise towards the production areas (especially the Middle East) or towards the new consumer markets of Pacific Asia, modifying the traditional trade patterns.

The refineries all registered serious losses – also because of the dynamics of the prices of finished products which only in part reflected the increases of the raw material – which led to severe policies of rationalisation. A number of plants, and not only the marginal ones, closed down, with processing concentrated on the more efficient and larger centres.[5] Instead of increasing the scale of single plants – as always happened before – the capacity tended to expand (and improve the quality) by repeating plant units at the level of MOD: in order to acquire greater *flexibility*, given the uncertain and unpredictable dynamics of the markets both of raw material and of finished products.

So from the end of the Seventies the size of the refineries ceased to be the *discriminating* element in competition between companies. However this factor did not lose importance overall, since indeed the average scale continued to increase, also due to the arithmetic effect of the expulsion of the marginal refineries. Besides this, growing importance was given to the following: *the degree of technological complexity of the plants*, in other words the capacity to make the best evaluation of the processed barrel optimising its yield in light and medium products (with greater value of use); *the productivity of the refineries*, linked to the progress made in their technological design, in the criteria of construction, and in the construction of the facilities of distribution of the products; their *level of environmental quality* as regards both the impact on the environment of their operations and the capacity to produce refined products with a low pollution content.

These factors have conditioned both the production costs and the specific nature of the investments. The unit costs increase with the level of technological complexity of the refineries: normally classified in three types as the extracted proportion of medium-light fractions increases. The first is the so-called *simple* refinery: with separation through atmospheric distillation of the light fractions (motor spirit), middle fractions (diesel oils), and heavy fractions (fuel oil). Through processes of catalytic reforming or hydrosulphuration, complementary units improve the quality of the products (respectively by increasing the number of octanes of the motor spirit or reducing the sulphur level in fuel oil). The second type of refinery is defined as *complex*, due to the addition of units for the conversion of heavy fractions into light fractions through processes of catalytic cracking, hydrocracking and the reduction of viscosity. More sophisticated equipment, allowing an even more radical conversion of the heavy fractions (through processes of hydroconversion, cokefaction, visbreaking, etc.) identifies the third type known as *deep conversion*.

The various technological solutions for the upgrading of refineries may lead to configurations which may even be very different from these, with the relative costs of investment and of production. Highly complex plants are also inevitably associated with large dimensions, so that it is difficult to distinguish clearly between the costs of the scale effect and the upgrading effect. For a simple refinery of 5 mill. ton./year, Masseron (1992) calculates a processing cost of around 2.0-2.5 doll./bbl which rises to 4 for a complex refinery of 8 mill. tonn./year and to about 7 for a deep conversion refinery of the same capacity.

The negative effect of the great complexity of the refineries on the costs is compensated by the advantages that this gives the company in terms of: (a) better *exploitation of the processed barrel*, guaranteeing middle-light products unit profits exceeding by 50%-70% those of heavy products; (b) greater *operative flexibility* both in the range of crude oils that can be used and in the range of products which can be obtained from their processing.

Table 1.4. *Estimated plant scale economies (I) and totals (T)*

Capacity	Pratten (1965)		Pratten (1969)		Masseron (1981)	
(mill./tonn./y)	*I*	*T*	*I*	*T*	*I*	*T*
1	100	100	100	100	100	100
2	85	96	75	92	88	99
5	66	91	56	86	76	98
10	52	87	44	82	69	97

N.B. The average cost (of plant or total) equal to 100 for a capacity of 1 mill./tonn./year.
Source: Goldoni (1987).

Therefore in the competitive strategies of oil companies the upgrading of the refineries has provided a response to an expansion in the types of crude oil marketed; to the trend towards the deterioration of their quality (the opposite to what was observed in the structure of final consumption); to the variability and uncertainty of markets both upstream and downstream.

The last effects of refining upgrading are found in the *increase of the specificity of these investments*, in the sense that Williamson gives this term in his analysis of transaction costs. This is true in three of the five distinctions with which he analyses the phenomenon: (a) *site specificity*: with the tendency to concentrate crude oil processing in certain sites, in order to optimise transport costs, storage costs and the overall logistic layout with respect to the distribution units; (b) *physical asset specificity*: with plants destined for the production of qualities which may not be used for other transactions; (c) *dedicated assets*: when the investment is strictly functional to a sole buyer. The increased specificity, in an environment which has become much more uncertain in the dynamics of price and demand, has boosted incentives towards vertical integration or towards long-term contracts with complex structures of management and control of bilateral trading.

1.6.2. The case of transport

Reality and dynamics not unlike that of refining may be described for the *transport of crude oil*. The ease with which petroleum can be shifted and the asymmetry between areas of production and areas of consumption have led to an enormous development in the international trading of this source of energy since its industrial origins. Even in the mid-Thirties 40% of the world production of petroleum and refined products was transported by sea from Venezuela to North America or from the Middle East to Europe, exceeding in quantity *all* the international trading of coal, which was then the main source in the energy balances of every country. In 40 years these flows have increased 18 times to reach a peak of 1.8 bill. tonn. in 1979; equal to 10% of all world trade *in value* to 60% of sea-going trade *in quantity* (Masseron, 1982, p. 164). Likewise, the capacity of the world fleet of oil tankers rose from 16 million dead weight ton (DWT) in 1939 to 23 in 1950, 63 in 1960, 131 in 1970, 328 in 1980 – a more than 20-fold increase in 40 years, with the number of ships multiplying to as many as 7,000 units. The increase in petroleum prices, the subsequent drop in the demand and the increase of production within the large consumer areas have led to the emergence of a huge surplus of shipping and to the abandoning in the Eighties of a capacity of 150 mill. DWT, technologically not yet obsolete. This was an enormous destruction of wealth which was followed by a structural crisis in the entire international shipbuilding industry.

With the exponential increase of the trade and extension of routes, the unit dimensions of ships saw a huge expansion, from *average* values (for ships in construction) of 21 thousand DWT in 1953 to about 50 thousand in 1963, reaching 164 thousand in 1973, while the *maximum* dimensions in the early Eighties exceeded 550 thousand DWT. As a consequence, the structure of the fleet, by dimensional class – at the end of the Fifties 98% centred on ships with a capacity lower than 50 thousand DWT – experienced a gradual shift toward the higher categories (Tab. 1.5.). At the end of the Eighties about 60% of the tonnage exceeded 125 thousand DWT and 45% exceeded 200 thousand DWT.

Table 1.5. *Oil tanker fleet by size (%)('000 dwt)*

	< 50	50-100	100-150	150-200	>200
1957	98	2			
1960	96	4			
1965	72	26	2		
1970	50	35	6	2	7
1975	16	25	8	43	7
1980	12	21	10	43	14
1985	15	24	12	11	38
1990	15	23	11	11	40
1995	14	21	10	12	43

Source: 1957-85: Jenkins (1986); 1990-95: Arthur Andersen, CERA (1995).

If we look at the curves of economies of scale, we can see that in 1958 the curve of unit costs in the long term was "flattened" to a size of 75 thousand DWT, in 1966 to 250 thousand DWT and in 1979, when gigantism reached an all-time high, to 475 thousand DWT (Zannetos, 1987, p. 248). This shifting of the cost curves was due to great improvements in the technologies of welding, propulsion, loading and unloading, navigation and safety.

However there were also determining changes happening at the same time in the stages upstream and downstream: increases in the dimensions of the refineries and the capacities of storage, and expansion in the receptive capacities of the ports of departure and arrival. As a result, in the early Eighties the long-term transport cost of crude oil in a ship of 475 thousand DWT, presumed to be operating fully loaded, was just 40% of the cost of a ship of optimal dimension of 75 thousand DWT in 1958 and only 20% of the cost of a ship of 30 thousand DWT.

As shown in Figure 1.5., the unit transport costs (for the shipowner) registered decreases as the gross capacity of the ship increased for all the cost items. In particular:

- *technical economies*: the cost of the vessel is a little less than proportional to the ship's tonnage while it is proportional to its power in the propulsive apparatus. The technical economies are particularly significant up to 200-300 thousand DWT, after which they flatten or show turning points.[6]
- *management economies*: due to the huge progresses made in the automation of ships, the number of people needed to operate them has been drastically reduced. A ship of 100 thousand DWT (and more) requires about thirty people compared with the 50 usually working on ships of 15-30 thousand DWT. Their level of professional skill and the difficulty (also psychological) of working on modern oil tankers however involves a much higher unit cost, though this is, in practical terms, negligible in the structure of overall costs. The economies of scale, which can be associated with the use of the fuel needed to power a ship, have become much more significant, especially when the price of this fuel has gone up. At full speed a ship of 200-300 thousand DWT has a daily fuel consumption from 4.5 to 6 times greater than that of a ship of 20 thousand DWT. The consumption per ton transported will thus be 56% lower, assuming the ship is fully loaded. Instead, other items in the management costs are unfavourable to the dimension, thus taking on an increasingly important role: for example, the costs of insurance of the risk regarding the ship itself and marine

Figure 1.5. *Transport costs and oil tanker size*

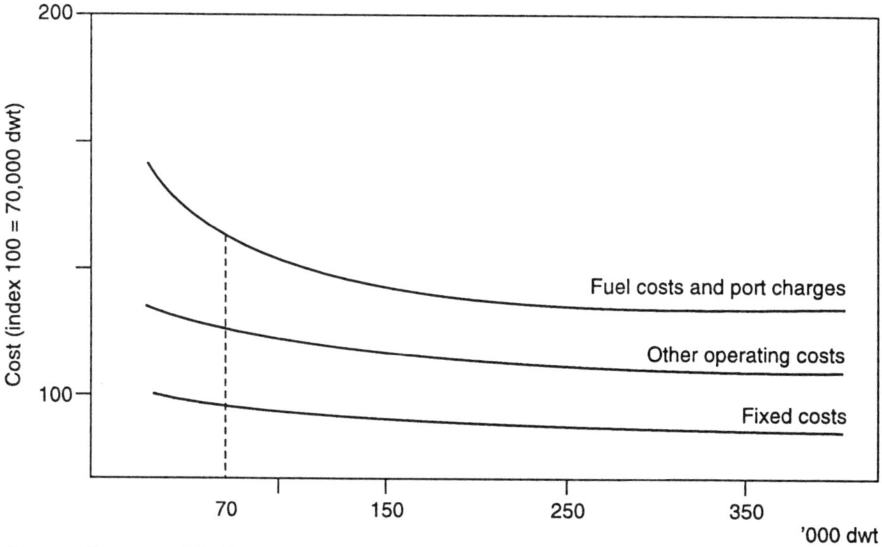

Source: Masseron (1982).

pollution. For the former the insurance premiums are 1%-2% of the insured value for ships up to 100 thousand DWT and 2% and more for ships from 250 thousand DWT.

Thus, as we have seen for refining, it may also be said that in the transport stage there are no longer those conditions – as regards the market and technology – which have favoured the economies of scale for half a century. These certainly maintain an important role in the entire economy of transport, but other variables combine with them in determining the strategies of the companies and the sectorial dynamics.

Innovations in the technology of petroleum transport and the reduction of relative costs – in absolute terms and in relation to the final price of the products consumed – have played a major role in the growth of the absolute extension of markets and the overcoming of barriers which once split the whole petroleum market into many regional or sub-regional markets. The great flexibility provided by super-tankers, able to reach any distance and which may be diverted at short notice towards the most advantageous outlets, has greatly favoured the competition between suppliers and the convergence of prices in commercial transactions, both on the physical market and in paper transactions. Finally, what has increased is the degree of *competition* in the petroleum markets (both of crude oil and of products), given the possibility of accessing them immediately and without sunken costs, one-off extra profit situations are created, brought about for example by price differentials in the various areas of export.

Of the overall transport capacity of oil tankers, 40% is controlled by the oil companies, while the remaining 60% is owned by shipping companies which offer transport services with timecharter contracts for limited periods, for consecutive voyages or for single spot voyages. While for company-owned ships, it is the transport cost analysed above which is high, for the transport services acquired externally the prices on the hire market are high. This is a strongly competitive market, highly unstable even in its day to day quotations.

After sea-borne transport by ship, the other form of transport which has become very important in the petroleum industry is through mainly overland *oil pipelines*. When they were first constructed in the United States at the end of the 1870s, the Standard Oil Company gained such a competitive advantage – in terms of costs and area of influence – that it was able to overcome the competition of the other companies forced to rely on the more costly rail transport. Like sea-borne transport, pipeline transport is characterised by strong economies of scale. Its flow capacity is in fact fundamentally determined by the diameter of the pipeline and since this increases less than proportionally to the capacity, the construction costs – and thus the total costs, since management costs are very low – decrease as the capacity increases. The capacity is defined both by the flow capacity of the oil pipelines and by the number and power of the stations pumping the liquid flowing through (this latter is an item which has more influence on the management costs). For each ton transported there is more than one optimal dimension for the pipeline according to its length, which however encounters specific technical and economic limits, of construction costs and pumping costs. Thus over relatively short distances (or in the absence of alternatives), the pipeline is competitive with tankers. Limits to the economies of scale are encountered therefore on handling much lower volumes in pipelines than in oil tankers, so that the former is generally preferred when the quantities are not excessively high and above all when the distance by land and by sea is very different.

The high economies of scale in the transport of crude oil and those relatively lower in the transport of refined products have gradually influenced the siting policies of the companies, their commercial policies, and the nature and geographical distribution of the traffic flows. The most significant change has been seen in the de-localisation of refineries (outside the American market) from the areas of production of crude oil to the areas of consumption. At the end of the Thirties about 70% of the refining capacity was located in the Middle East, the Far East and in the Caribbean; only 30% in the markets of consumption. Three decades later the situation is more than reversed: 80% of a capacity which has increased four-fold in the meantime was located in the markets of consumption; 7% in an intermediate position between production and consumption areas; and just 13% close to the oil fields. Several reasons have brought about this change: (a) *the economic advantage of transporting crude oil rather than oil products*. While in fact the curve of the unit costs of transport of crude oil decreases given an increase in the dimension of the market and also of the distances covered, the products curve goes up sharply. The distribution of oil products must therefore be restricted to relatively limited geographical areas; (b) the greater possibility of *adapting the technical design of the refineries to the structure of the final consumption* of the area to be served: thus optimising the investment cost and the working processes; (c) the interest of many consumer-importer countries in *having their own refining facilities*, with the additional aim of reducing the political vulnerability of the exporting countries (Frankel and Newton, 1961).

1.7. THE "TIME FACTOR"

There are basic conditions whose relevance is not always averted in its entirety both because they are secondary with respect to other conditions and because of the strategic-operative responses that have been provided to them by the companies. In the petroleum case, one of these conditions is unquestionably the "time factor": the period which runs in the *current* activity (of every phase) between the time of the decision and the time in which its effects are felt. More specifically, this period involves the number of days needed to vary the level of crude oil production (when this is permitted by capacity), send it to the gathering points and load it; transport it to the refining centres (when they are not nearby); refine it and finally distribute it to the various categories of final users. Any specific quantification of this time is hindered by the enormous diversity of individual operating conditions, and is in any case irrelevant for our purpose here. What should be stressed here is that the *entity* of the "time factor" throughout the different phases may even be quite large, thus complicating the optimisation of the entire petroleum cycle from both a quantitative and economic point of view. Situations of *stability* in the markets and prices do not generate anxiety; this is aroused conditions of uncertainty prevail. The acquisition of a load of crude oil, at the time t_0, at a price considered remunerative on the basis of the current quotations for its refined products, may turn out to be completely disastrous if in the meantime the quotations have plunged. Similar situations have become commonplace since the Eighties, when prices (absolute and relative for crude oil and oil products) were subject to increasing variation with differences of up to 10%-20%, even from one day to another. Market risks have become central in the activity of every operator – at any stage – and they are all the greater the more extensive the "time factor" (*ceteris paribus*). Without considering the oil extraction stage (where indeed it is the time necessary to vary the production flow of a certain entity which is important), different periods may apply to its transport: between a minimum of 3-5 days for short routes (North Africa-Genoa), to 13-15 days for ships over medium distances (Persian Gulf-Naples via Suez) up to 30-35 days for long distances (Persian Gulf-Naples via Cape). Once the ship has reached a port for unloading the petroleum, this operation itself requires 3-5 days, plus another 6-12 days for the crude oil to go through the entire process of refining to become available for sale and to enter the final distribution circuit. A total of between 15 and 60 days may be required between the time of the decision to buy a load of crude oil and the time of sale of

Figure 1.6. *Oil planning*

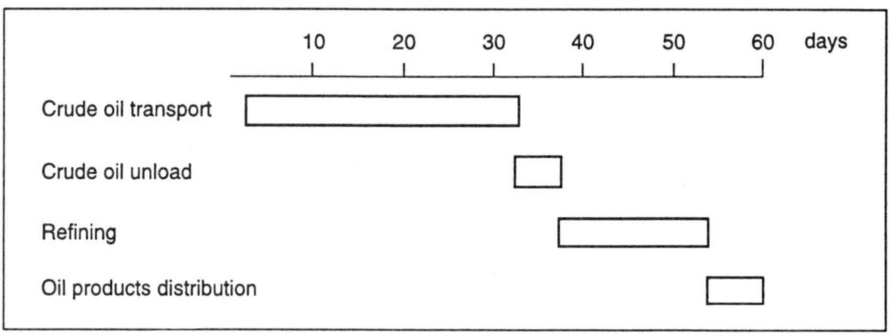

the finished product (Fig. 1.6.). We certainly cannot imagine that everything happens according to the rigid sequence of times described, since the refineries prepare themselves in seasonal advance for the expected requirements of their clients, but there is no doubt that, since it is impossible to rely exclusively on the policy of stocks, the times of current decisions are not short and the operators are constantly exposed to the risk of the market.

One consequence of this which is particularly worth highlighting is that the overall equilibrium of the petroleum system (balancing of demand and supply) requires the operators of each stage to have a *suitable capacity for predicting the market dynamics* as well as a similarly timely capacity for response when the actual situations diverge from the expected trends. In other words, for the system to operate with equilibrium, the producers upstream must be able to predict, with a time difference of 30-60 days, the possible volume of demand from the refiners or the commercial traders, and so on downwards throughout the stages of the cycle. The possibility for the system as a whole to manifest capacities of response prediction to market dynamics, in order to avoid situations of pathological unbalance, is closely linked to the *operating co-ordination* which constantly intervenes between companies of the same stage and those of the stages upstream or downstream. But this is part of the company policy which we will be dealing with later.

The importance of the "time factor" also varies in relation to the *predictability of the rates of growth of demand* for crude oil and derivative products (since this influences the quality of the crude oil and therefore their geographical origins). The more certain the market dynamics, the less important the time lags, and vice versa. Throughout the Fifties and Sixties, the demand for petroleum registered a strong and very stable growth from year to year in particular regional areas of consumption (United States, Europe and Japan). The differences in the relative dynamics between areas were constant, as were the differences in the growth trends of the various products. The system expanded stably and uniformly in terms of both geography and commodities. This enabled each producer to plan his own production and commercial activity in time and without error on an annual and multi-annual basis.

This synchrony changed radically in the Seventies and Eighties for two reasons: (a) there were no longer the instruments which in the past had regulated the inter-company operational co-ordination (which we will be dealing with in the following chapters); (b) the petroleum price shocks. The trends in growth of the demand for petroleum and its particular products thus were not only reduced considerably but, more importantly for the purposes of our subject, they also saw a considerable *diversification* from area to area and from year to year, due to:

- the different effect of the price shocks on economies, on their rates of expansion and thus on the growth of total energy consumption;
- the different energy policies followed by individual countries, with differentiated impacts on the processes of substitution of petroleum products.

A phase of strong and stable growth in consumption was followed by one of low and differentiated growth. The combination of this change with the absence of traditional mechanisms of inter-company co-operation led to an enormous increase in the *planning difficulties* for the operators, and the role of the "time factor" on these. For the producers and exporters (thus mainly the Middle East and Africa, covering about half the international trade), the lack of any informational feedback from the companies downstream (which had previously been guaranteed by vertical integration) forced them to "work in the dark", relying solely on the validity of predictions formulated by analysts, international organisations and research centres. Over years in which these predictions have more often than not been revealed as incorrect "prophecies" it is easy to understand why the market registered constant unbalances – even on a seasonal basis – which may also be attributed in part to the long-term problems of intersectorial planning described in this paragraph.

1.8. THE DEMAND FUNCTION

The basic conditions which we have examined so far as regards the petroleum supply would not be so important if their effects did not combine with those induced by the characteristics of demand, mainly in relation to three aspects: (a) degrees of reactivity of demand to its main determinants (*elasticity effect*); (b) relation of competition-substitution between the various sources in the different use functions of petroleum (*substitution effect*); (c) trend of consumption over the year (*seasonal effect*). Each of these aspects has some specific consequences on the operativity of the petroleum companies, on their strategic choices, and on their market power. In order to deal with them analytically, we should first of all give a brief description of the salient features of the energy demand function before looking at petroleum demand.

1.8.1. The energy demand: qualitative specificities

Energy demand is considered a *derived* demand, since it is not requested in itself but to satisfy the various needs or *final consumers*. What is required is therefore mechanical energy (to power electrical motors or internal combustion engines); thermal energy (to heat homes or produce heat in industrial boilers); electrical energy (for lighting, industrial use or rail transport); chemical energy (in obligate industrial processes such as the steel or chemical industry). Each use may be satisfied by one or more *energy sources*: a term we use to refer to the materials or systems which through the effect of transformation are able to provide energy in forms which can be harnessed. They may be divided into renewable sources (hydraulic, solar, biomasses, wind, etc.), fossil sources (coal, petroleum, natural gas) geothermal and nuclear.

Starting from the energy sources and through the processes of transformation to which each of these can be subjected, we reach the *energy forms* – manifestations of energy through which our needs are satisfied – and through their transport, to the different segments of final demand. Each final use is distinguished by strong elements of *specificity*, and this is an element of enormous importance in conditioning the choices of consumers. This concerns the forms of energy required, place and time of demand, and dimensions of use. Energy – in other words the energy sources through which it is expressed – should thus be assessed, first of all, *for what it is used for*. The suitability of each source for satisfying the specific requirements for each use (in each historical period) may not be judged solely from observation of its relative abundance nor from its relative prices (if this were so, coal would always have prevailed), but it *must first and foremost take into account the quality characteristics of each source*. In particular: location with respect to use, seasonal nature and degree of reliability, thermodynamic quality, concentration, ease of use, and possibility of storage. All elements that cannot be expressed fully in the price factor.

"The qualitative specificities of each source of energy and the technological innovations which are associated with them determine their prevalence over the others. This coherence between specificity and innovation explains why, in any historical period, a single source tended to become *dominant* over the others regardless of their availability and relative price – and because the period for a source to become dominant (to pass from 1% to 10% of world consumption) has always been between 50 and 60 years" (Clô, 1993). In current economic terminology, we could say that this link between "specificity and innovation" determines and is translated into a *growing inelasticity to the prices of the demand of the dominating source*, to the extent that it consolidates its affirmation over other sources. In the case of petroleum, this process began in the Forties (although its consumption dates back to the second half of the last century) and has its peak in the late Sixties, when petroleum became the dominating source in the energy balance of industrialised countries, replacing coal which had dominated for about a century (Figs. 1.7. and 1.8.). The analysis of the processes of substitution between sources of energy cannot therefore – as normally happens – be stylised into a *linear* passage from one source to another due to the effect of suitable mechanisms of competition founded on the relative costs. This should be borne in mind in the theoretic analysis of the formation of petroleum prices.

Figure 1.7. *World energy consumption by source*

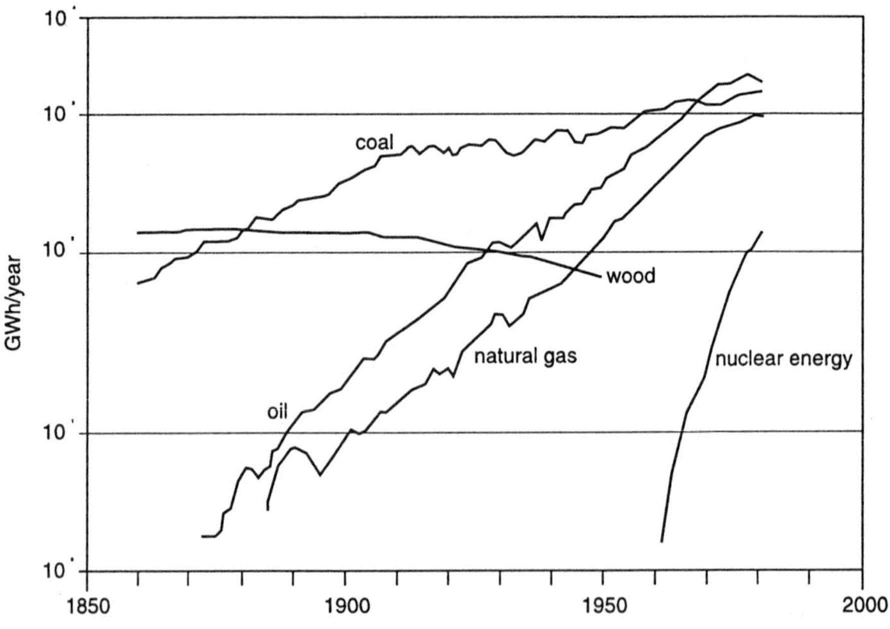

Figure 1.8. *Energy sources market share on total consumption (%)*

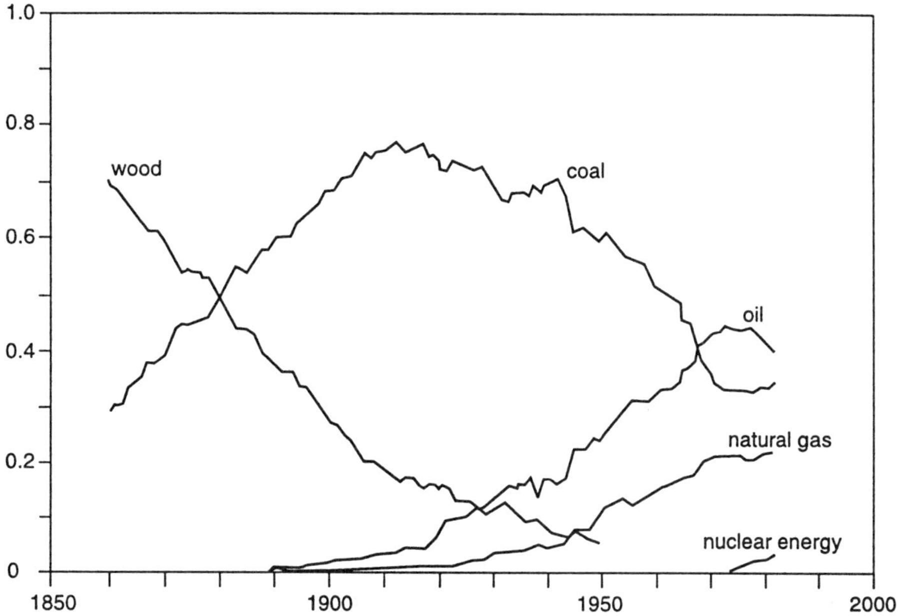

1.8.2. The energy demand: main determinant

We will be returning later to the relationship of competition-substitution between the sources, after dealing with the theme of determinants of energy demand function. Extensive literature on the subject (Girod, 1977) has identified hundreds of these features, though they can mainly be attributed (directly or indirectly) to economic production.

By studying the long-term trends of energy consumption – multiplied by a factor of 15 from the start of the century to the end of the Eighties – we may see that 80%-90% of their increase can be attributed to the growth of economies (as expressed by gross domestic product). The reasons for this may be: the intensity and continuity that this has manifested, especially just after the Second World War; the favourable trend of real energy prices; the changes in the composition of the income (per sector of economic activity) and in the adoption of consumer durables (from cars to electrical household appliances, etc.); the processes of urbanisation etc. All these elements helped to raise global energy consumption in all countries and they were unable to be offset in full by those technological improvements in the production processes which led to reductions in the coefficients of energy use (per unit of output). Until the early Seventies, the *income elasticity* – the ratio between the percentage variation of energy consumption and the percentage variation of gross domestic product – was basically the only term of reference in evaluating past and future trends of energy consumption. And indeed, its value manifested a strong continuity in time and a correspondingly strong spatial similarity (at least in relatively homogenous economic situations) with levels close to or exceeding unity. The drastic changes to be seen later placed the spotlight on the other, apparently dormant, determinants of the demand function – real prices, technology, institutional aspects, other non-price factors – whose role tended to increase in relation to the huge variations that each of them was to witness, but also in relation to the concomitant huge drop in the rates of growth of the economies. In any case, this remains the first determinant of energy consumption. In very general terms, the energy demand function may be expressed as follows:

$$C = f(Y, P, T, Z)$$

where:
C = energy consumption
Y = gross domestic product
P = real prices
T = technology
Z = other important factors.

The empirical analyses give the following more significant conclusions:

(a) a *high partial energy income elasticity*, with values for the OECD area of between 0.8 and 1.0. These values are much higher in the more recently industrialised countries, assuming an increasingly declining trend when levels of full industrial maturity are reached.

Table 1.6. *Energy demand: short and long term price elasticity in the most industrialised countries*

Sector	Short-term elasticity[1]	Asset turnover[2] years	Long-term elasticity[1] range	Long-term elasticity[1] mean	Implicit primary elasticity[3]
Residential	–0.12	15-20	–0.71/–1.14	–0.9	–0.3
Transport	–0.24	8-10	–0.36/–1.28	–0.8	–0.2
Industrial	–0.16	5-25	–0.30/–0.52	–0.7	–0.4
Total			–0.66/–1.5	–0.8	–0.3

1. In reference to final demand. 2. Approximate estimates referring for residential use to heating systems; for transport to motor vehicle fleet renewal; for industry to small and large boilers or electrical systems. 3. Obtained by dividing the mean values of the long-term elasticity by the ratio between consumer prices and crude prices.

Source: first and second column our estimates; third and fourth column: Nordhaus (1980).

(b) a *low partial price elasticity*, with values in the short term of between 0.12 and 0.24 and in the long term between 0.7 and 0.9. Thus even in the long term, the energy demand remains inelastic to prices, with values in any case considerably differentiated between sectors of consumption, income classes of consumers and income levels of countries;

(c) the third set of determinants, which may be identified as the *structural factors* linked to technological progress or to the composition of the production mix, with relevance in economies with high industrialisation above all.

1.8.3. The demand for petroleum: general lines

Previous analysis has examined *all energy* sources consumed. Other factors determine the specific demand for *each energy* source, including in particular:
- the *quality characteristics* of each source, which we have said influence their demand given their relations of functionality with technology and consumer preference;
- the *relative prices of the sources*, in the areas of substitution allowed by the existing plants and technique;
- the *stock of plant*, which in certain cases may allow an energy source to be replaced even in the short term (for example in electricity power stations),
- the *security of supplies*, which may induce a country to choose to use a source even though it is not strictly convenient in economic terms;
- the *environmental effects* connected to the production and use of each source.

There is no doubt that these factors as a whole have favoured the penetration of petroleum in the world energy balances, rising from 21 mill. tons in 1900 to about 300 on the outbreak of the Second World War, and to about 3,000 mill. tons in 1990 (Tab. 1.7.). The conditions which favour this penetration (Clô, 1993) show why consumers widely prefer this source in terms of both *income elasticity* and of *price elasticity*. All the empirical evaluations, which refer to the period between the Forties and the Sixties, show a high income elasticity of petroleum demand (higher than the overall demand for energy), with a low price elasticity (less than the overall one).

Both coefficients would clearly have given a high *market power* to the sellers, allowing them, once petroleum had overcome the other sources and asserted its leadership, to practice manoeuvres of price raising without this bringing about negative effects as regards demand in the *immediate future*.

In order to look closer at these aspects it is useful to divide petroleum consumption into *function of use* and thus into categories of final consumer. Thus *gasolines* are directed almost entirely at the transport sector (families and industries) in an almost monopolistic situation, since there is basically no substitute for its use in road transport (LPG and methane are in fact completely marginal fractions, despite their being much more economical); *virgin naphtha* is used in petrochemistry; here too in a situation of almost obligate technology, at least as regards the European and Japanese chemical industries (while American industry has always used natural gas for raw material). As regards the *middle distillates*, these are destined mainly to families, for heating, for which a large proportion of diesel oils and kerosene is used; also families and industries as regards transport, with diesel fuel used in diesel engines. While in this second case the competition between sources is almost entirely within the petroleum system (the alternatives being between diesel and gasoline), as regards heating, competition is both between petroleum alternatives (diesel, kerosene, LPG and fuel oil) and, outside the petroleum industry, with other sources (natural gas, coal, electricity). *Heavy distillates* have historically had the most influence in the dominating role of petroleum, since they have superseded all other sources in electricity generating plants as well as in industrial usage. So in this case, the elasticity of substitution (given by the percentage variation of the demand of a product in relation to the percentage variation of the price of the alternative product) is relatively higher with respect to that of middle distillates and particularly higher than that of light distillates.

Table 1.7. *World consumption of traded energy sources by primary source: 1880-1998*[1]

Year	Coal[2]		Oil[3]	Natural gas		Hydroelectricity		Nuclear energy		Total
	mill. tce	mill. toe	mill. toe	bill. cubic metre	mill. toe	mld. KWh	mill. toe	mld. KWh	mill. toe	mill. toe
1880	326	228	4	–	–	11	2	–	–	234
1900	737	516	21	7	6	16	4	–	–	547
1920	1,272	890	100	24	20	64	14	–	–	1,024
1940	1,523	1,066	299	82	68	193	42	–	–	1,475
1950	1,473	1,031	446	187	154	296	65	–	–	1,696
1960	1,981	1,387	901	456	376	609	134	1	–	2,798
1970	2,184	1,529	1,923	994	820	1,080	238	75	17	4,527
1980	2,632	1,837	2,590	1,557	1,284	1,768	389	681	150	6,250
1990	3,238	2,267	2,808	2,175	1,794	2,249	495	1,985	437	7,801
1995	3,153	2,207	3,235	2,318	1,912	2,630	579	2,727	600	8,533
1998	3,170	2,219	3,389	2,444	2,016	2,755	606	2,850	627	8,857

1. In the conversion into tons oil equivalent (toe) the following factors have been used: for coal 7,000 kcal/kg; for gas 8,250 kcal/cubic metres; for hydroelectricity and nuclear energy 2,200 kcal/kWh. The toe values are rounded up. 2. Includes lignite and coal. 3. Includes natural gas liquids.

Source: up to 1990 United Nation's, "Statistical Review of World Energy", and Doe (various years).

Figure 1.9. *Monthly oil consumption (% of total consumption)*

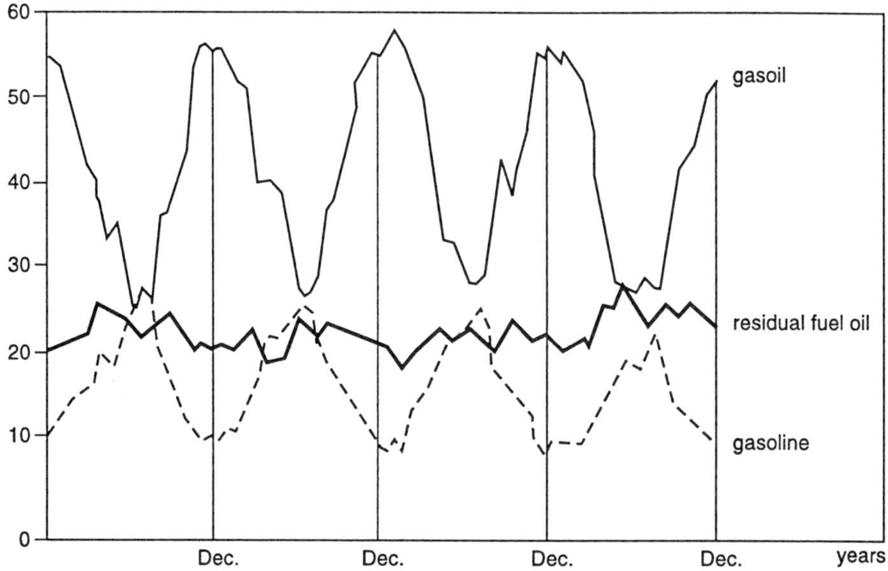

1.8.4. The seasonal nature of petroleum consumption

One final condition should also be taken into account, although it is certainly of lesser significance than those so far examined. This is the seasonal nature of petroleum consumption, which is extremely *marked* in some products and *differentiated* between them, in relation to the multiple factors which condition their dynamics of an economic, sociological and climatic nature. Figure 1.9. shows the mean seasonal trend of petroleum consumption on a monthly basis from which we can observe: (a) the relative stability of the consumption of *fuel oil* (at one time conditioned both by the seasonal nature of production and by that of climate, but today influenced by the former alone since for heating it has been substituted by other products); (b) the marked seasonal nature of *gasoline* (transport) and *diesel* (transport and heating), with trends which are complementary but not such as to cancel each other out.

The seasonal factor produces two consequences: one affects the operativity of the companies and the other the trend of the primary crude oil demand (more specifically, the demand aimed directly at the international markets or at the extracting stage). The *logistic layout* that the refiners-distributors must have (to store the single products) is influenced by *company operativity*, as is the *cycle of stocks* both of crude oil and of products (to balance the different seasonal factors against a continuity in the working flows of crude oil and to predict the single flows of demand adequately). Both effects involve the commitment of investments and financial costs which become higher the greater the volumes of output affected, petroleum prices, and interest rates.

The second effect worthy of note regards the supply: in other words the trend within the year of the *primary crude oil acquisitions* on the international market.

According to long-term trends, these go through a *destoring* stage in the winter, with a total of 2 mill. bbl/d and a *storing* stage during the spring-summer season of a similar amount. In the Eighties these values represented about 5% of the petroleum production in countries with market economies and 8% of the quantities traded on the international markets.

1.9. SHORT AND LONG TERM

In Table 1.6. we have seen that demand/price elasticity (both of energy and petroleum) is higher in the long term than in the short, though remaining at relatively *low* values. This means that in the majority of activities, industrial or domestic, the energy value – in other words the cost opportunity not to have any – is *higher* than other values involved. This applies both to energy as a whole and to the petroleum source, at least as regards recent history. It is only over a span of several years that variations in prices lead to changes in consumption. Indeed, in the majority of cases, expense planning is required for new installations that require varying lengths of time for setting up and putting into operation. Elasticity is in some way conditioned by "sluggishness" linked to the "time factor". Using a dynamic series of demand curves (with constant elasticity), as shown in Figure 1.10., on consolidation of the process of adjustment of the consumers to the drop in prices from P_1 to P_2, the demand curve tends to shift to the right with a gradually decreasing inclination and greater elasticity (at every point) until it reaches a new position of equilibrium.

First of all, the drop in prices generates an *effect of impact* on consumption, which goes from $0Q_0$ to $0Q_1$, while the *cumulative effect*, which takes into account the entire process of adjustment, sees consumption increase to $0Q_*$, a new quantity of equilibrium. Therefore the dimension of the response (elasticity) increases with time, and this considerably complicates its calculation, since the wider the interval, the more phenomena of other types (for example economic growth) tend to interfere in consumption, also influencing its dynamics.

On the basis of what has been said, we can define as *short term* the span of time in which the reaction of the consumers to the variation of prices takes place with the *same* stock of plants (whether electricity plants, vehicles, or heating plants), since they are allowed either to give up using them (which as we have seen would involve more costs than benefits) or alter their *reaction* so that they would be more attentive to the efficiency of yield of the plants, or enough in any case to reduce energy consumption while still satisfying the essential needs (by guaranteeing for example the better maintenance of plants, or reducing the average temperature of the rooms). The global effect of these decisions, which we have called "effect of impact" remains however very slight.[7]

The *long term* we use instead to mean the span of time necessary to allow a full adjustment to the new structure of prices relative to the stock of plants. Its *actual duration* will depend on several factors: the amount of stock; the speed of its *turnover* (in turn connected to the trend of the income); the improvements in the energy efficiency incorporated in the new machines; their times of penetration. The complexity and diversity of these factors show: (a) the difficulty of having an exact perception of the effects of consistent variations in prices on consumption; (b) the different type of response which is given on different markets (also according to the structure of uses).

Figure 1.10. *Short and long term demand curves*

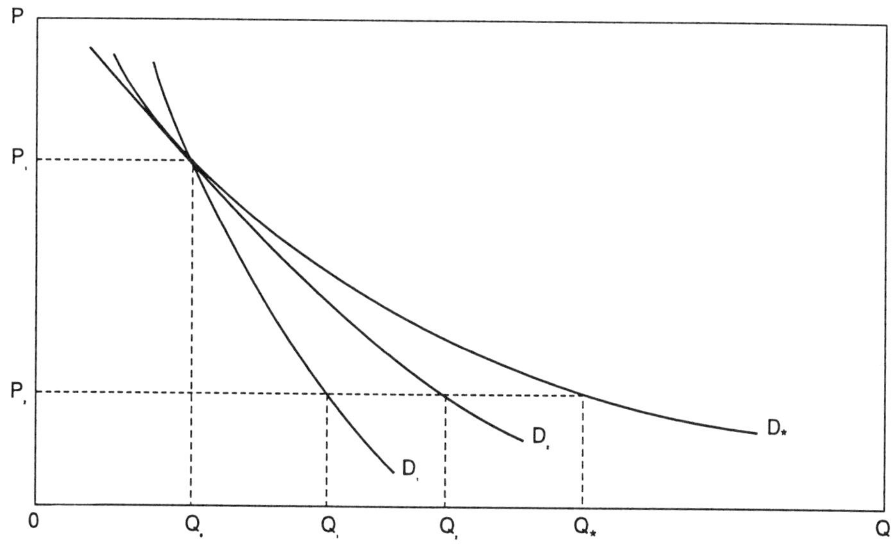

In spite of this, two conclusions seem to be fairly certain. The *first* is that the long-term adjustments are located in time spans which we can roughly place between a minimum of 3-6 years (domestic plant or renewal of motor vehicle fleets in some countries) and a maximum of up to 20-30 years or more (construction, industrial plants, etc.). The *second* is that the degree of response still remains, on the whole, fairly limited.

We have so far concentrated on energy in its *globality*: seen as a *production factor* of industrial activities and as such, to a certain extent replaceable with other factors as its relative prices vary, or as a consumer associated with the use of other goods. However every *single* source of energy is another matter. In the case of petroleum, variations in price are added, in their effects on the quantities consumed, to the general factors examined above, those *details* linked to the competition-substitution between sources. And this is when the variation in prices is not homogenous between all the sources. But we will have the opportunity to look at this later on.

To conclude this examination of the demand/price elasticity, these points in particular should be highlighted:

(a) in the petroleum industry, decisions also regarding the demand generate effects which are *much delayed* in time, as has been seen for the supply;

(b) a superficial evaluation of these effects – which are indeed complicated by the time lags themselves, and by the lack of homogeneity of the market situations – may bring results which do not comply with (or may be completely opposite to) the expected results.

Notes

1. A new find of a significant dimension does not today cost less than 500-1,000 million dollars.

2. In the period 1983-87 the average rates of success in drilling activities exceeded 40% in OPEC countries; 35%-40% in some non-OPEC countries where investments have been directed only recently (including: Norway, Oman, Angola, Columbia, Thailand) and 25%-35% in other non-OPEC countries (Great Britain, Holland, Brazil, Egypt, Malaysia, India, Australia). In evaluating the significance of this data – sometimes (apparently) more comforting than those of the past – we must bear in mind that they exclude extensive areas where the results have been clearly negative (such as North America) and especially the fact that they refer to *absolute* levels of investment which fell sharply in the period examined, due to the slump in crude oil prices, with a number of explored wells 30%-40% lower in Africa and Latin America; 11% in Europe, 3%-4% in the Middle and Far East. Forced to reduce their spending plans, because of fewer available funds and the lower profitability expected from the investments, the firms have concentrated on relatively low-risk projects and lower costs, thus sometimes leading to an improvement in the average success rate.

3. If for example a country wishes to maintain a level of reserves equal to 15 years of production, with a rate in the growth of production of 3% per year, extracting a ton of crude oil in the year N, in the subsequent year it needs to discover N + 1 : 1.03 + (15 × 1.03 – 15) = 1.48 tons of petroleum (Masseron, 1982, p. 17).

4. Just one of those refineries was equal to the sum of the three largest that followed it on the scale.

5. In the Eighties in Europe alone about 50 plants closed down, with a cut in capacity of about 40%.

6. According to Masseron, in 1981 a ship of 50 thousand DWT had a construction cost of 500-550 dollars/DWT compared with the 400 of a ship of 100 thousand DWT and 300-320 doll. of a ship of 200 thousand DWT. With an expected lifetime of 15 years and given the usual conditions of financing, the annual cost of the amortisation cost for a new ship of 250-300 thousand DWT was 45%-55% lower than for one of 50 thousand DWT.

7. The increase of around 80% in the (real) prices of energy between 1973 and 1976 in Japan and in Italy led to reductions in consumption, over the same span of time, of 0.3% and 3.4%, thus with an elasticity of just -0.004 and -0.04. There was much more reaction from consumption in Canada and in the United States mainly due to their different distribution according to sectors of use (much more concentrated in the sector of transports and in domestic use).

2 KEY PLAYERS AND STRUCTURE

2.1. THE *ECONOMICS* AND *POLITICS* OF OIL

It would be impossible to get a clear idea of how things work in the oil industry by referring merely to its basic *economics*. This would exclude the *political* aspect, which has always been vital to the workings of the oil market. With the arrival of oil the development of modern economies, at first centred around the national coal industries, became detached from direct or part ownership of its essential energy supplies. The countries which previously dominated the industrial economy became progressively *dependent* on other countries for their energy supplies, causing both economic and political vulnerability.

Winston Churchill, speaking as First Lord of the Admiralty, was already well aware of this danger when, during a debate concerning the British Admiralty held at the House of Commons on 17 July 1913, he strongly emphasised the extremely high priority that the oil issue had acquired for the British Navy and for national survival.

He stated that "If we cannot get oil, we cannot get corn, we cannot get cotton and we cannot get a thousand and one commodities necessary for the preservation of the economic energies of Great Britain" (Ferrier, 1982, p. 182). Considering that (even in those days) Churchill claimed that the "oil open market is becoming an open mockery", he proposed that the Admiralty took on a direct role by: (a) building up a strategic reserve, capable of meeting basic needs in times of war and overriding price fluctuations in peacetime; (b) buying crude oil directly when market prices were particularly low; (c) acquiring control of oil companies. This speech had an enormous impact on the daily newspapers and public opinion, and had the effect of accelerating the implementation of an (already drafted) plan for the government to take an active interest in the oil industry. This was to be put into practice a year later, in May 1914, when the British Treasury acquired a majority shareholding of the Anglo-Persian Oil Company (changing to British Petroleum in 1952) for the sum of two million pounds sterling. This company, founded in 1909 by the rich English lawyer William Knox D'Arcy, had been granted exclusive rights for exploration, extraction and exportation of crude oil over almost the entire Persian Empire, in a licence lasting for 60 years. The agreement with the Treasury gave the company a huge injection of liquid assets and a long term outlet supplying fuel to the Royal Navy. On its part, the British government, as well as ensuring a constant supply of fuel for its fleet, also acquired a crucial role in the Mediterranean area, guaranteeing full political protection for the expansion of the Anglo-Persian Oil Co. The only condition it imposed was the right to appoint two members to the Board of Directors, while guaranteeing that it would not interfere in any way with company policy-making. This principle was to be confirmed and respected by successive governments. Until 1964, the British Treasury kept an absolute shareholding majority in British Petroleum, and until 1983 it had a controlling majority, and only in 1987 was the company completely privatised (while the government still maintained some quite considerable special powers).

The reasons which prompted the British government to take an interest in the oil industry were to influence the behaviour of almost all the other major industrial nations, though in different ways. With oil, energy ceased to be a purely economic matter and became a *motive* and *arena* for political *conflicts*, whether between importing countries, between importing and exporting countries, or between companies and governments. There was a lot at stake: acquiring direct control of oil resources; increasing the security of supplies; consolidating one's own position with respect to other countries; and making a healthy profit.

In order to fully understand the developments that were to occur in the oil market and the substitution of oil in the place of other energy sources, it would be impossible not to take into account their *political value* (although it would be wrong to find political explanations for every single event). It is clear that the combination of *economics* and *politics* in the oil industry increases the level of uncertainty in our appreciation and understanding of reality. This makes it almost impossible to try to isolate relevant *causal factors* from events, as well as making it difficult to formulate theoretical generalisations from past trends and predicting what could happen in the future. These uncertainties and difficulties help to explain the numerous failings of economic theories when applied to the oil industry, although many economists would still refute this.

However, due importance must be given to the interaction between economics and politics in any historical assessment of the subject, as will be attempted in the course of this book, in the conviction that history often helps us to make sense of the present and throw light on the future. Over the years this interaction has differed greatly, according to the specific state of the international relations involved, to the geopolitical areas to which they refer, and to the parties involved.

In our analysis we can define three separate types of negotiations which have proved crucial to the development of the world oil industry. The first are those between *oil producing countries* and *oil companies*. Institutions, organisations and property rights have, over time, defined the incentives for companies operating in the sector, their competitive business relations and production costs. Profit-sharing in the oil industry has (at least until the late Eighties) always been the subject of negotiation between conflicting spheres of interest, in which purely political considerations interact and often overlap with other typically economic ones. Intergovernmental relations, sometimes with the companies themselves as mediators, have been a dominant characteristic of past controversies over the issuing of oil concessions, the definition of their terms, or the selection of companies allowed to compete for them. The results, and the very nature of the negotiating process between each individual country and the oil companies, depend not so much on direct contractual relations, but on the political leverage that was derived from a broader system of international relations and the system of ties connected with it.

The second type of negotiation (separately identified, although closely connected to the first) is that between the major *consumer countries* (oil importers) for the control of the oil resources in the Middle East and North Africa. From the Twenties to the end of the Forties, conflicts were all within the bounds of *Western* diplomacy – particularly between Britain and America – and involved the sharing out of geographical areas, and thus political and economic influence, among the major oil companies that had sprung up on both sides of the Atlantic from the end of the nineteenth century. The peculiarly multinational features of these negotiations actually derived from the mutually beneficial relationship which the companies always maintained with their governments of origin. Relatively secure oil supplies were ensured for the governments, whilst in return, the companies received support for the expansion and protection of their economic interests abroad. This happened, for example, in 1951, when the Iranian government of Mossadeq nationalised the Anglo-Iranian Oil Company. Two years later the government fell (following a *coup d'état* that was to bring Reza Pahlevi to power) because they were unable to sell a single barrel of oil on foreign markets as a result of boycotting by the West (Odell, 1986).

The conflict between American and British diplomacy came to a head at the end of the Second World War. The leaders of the two countries at the time, President Franklin Delano Roosevelt and Winston Churchill were directly involved. The Americans proposed dividing up the Middle East between the United States and Great Britain, with Saudi Arabia in the American zone of influence. Formal negotiations began in spring 1944, leading to the signing of the *Anglo-American Petroleum Agreement* on 8 August of that year. This agreement defined their respective areas of influence and, through the *International Petroleum Commission*, the ways of managing supplies and distributing them among the various importing nations. The agreement was to take various factors into account, including "available reserves, appropriate technical methods, intrinsic economic factors and the interests of producer and consumer countries, in order to satisfy increasing demand" (Yergin, 1993, page 345). In February 1945, Roosevelt met the powerful Saudi Arabian king, Ibn Saud, on the American ship Quincy, anchored in the Great Bitter Lake, and obtained his consent. The American president's death only a few months later and his successor, Harry Truman's refusal to comply with the suggested intergovernmental cartel with Britain, effectively put an end to the possibility of an agreement. However, this did not prevent the USA from obtaining a definitive *supremacy* over the global oil industry and over the whole Middle East area. In 1948 the Americans produced 275 million tons of oil, out of a total global production of 435 million tons (excluding the USSR). Five American companies controlled over 50% of the oil reserves in the Middle East, compared with only 10% in 1940. In Europe, the arrival of American companies was facilitated by the Marshall Plan, even in countries that were keen on protecting their own newly-founded national oil companies (France, Italy, Spain and Belgium). Esso, Mobil and Gulf even managed to enter the protected Japanese market. Anglo-American companies were also given the primary task of managing the allocation of petroleum to Western countries as economically as possible, but one must take into account that these companies were instruments of American foreign policy and that their interests were substantially identical to the national interests of the United States – as stated by the Church Commission of the American Senate in its 1975 *Multinational Oil Corporations and U.S. Foreign Policy* Report (p. 14).

A third type of negotiation is that between *producer* and *consumer* countries which was to get firmly underway with the entry of the former on the international political scene. The number of conflicting interests and potential disputes was to increase, but, above all, the negotiations expanded to involve matters that went well beyond the sphere of the oil industry itself, including such areas as North/South world relations, and relations between First and Third World countries.

2.2. MARKET FAILURES AND PUBLIC POLICIES

Modern states have always considered the energy sector as an *exceptional* case within the general panorama of industrial activities. Its importance in the development of their economies, in their currency balance, in the distribution of wealth *within* and *among* them, and in their status in international relations, was considered so great that states could not be expected to let the markets alone achieve goals of general interest that (it was thought) would not otherwise have been followed spontaneously.

Among the various reasons for *market failure*, security of foreign oil supplies has historically proven to be the most frequent. From an economic point of view, this can be expressed as a reduction in wealth caused by such factors as: a drop in supply, with respect to current or future levels of demand; a strong price volatility with negative effects on investment growth; the exercising of strong "market power" by those who control oil production and trade; and the risk of international political crises with shock effects on prices and quantities.

These reasons point to an equally large number of *negative externalities* connected with the oil industry (especially when importation is involved), which have traditionally been cited as the rational base of public policies. The term *externalities* is used in economics to indicate costs and benefits that are not reflected in prices and therefore not transmitted into market decisions. This leads to a loss of efficiency for the economy, in the sense that a re-allocation of resources would lead to greater wealth. In particular, an avoided cost means that the private sector *underestimates* the total social cost connected with petroleum use, leading to excessive demand, a minor production of alternative energy sources and higher import levels. The opposite happens for benefits reflected in prices, to the detriment of the use of alternative energy supplies.

The presence of externalities justifies public intervention to improve the allocation of resources, even though it does not, in itself, ensure that the benefits obtained will necessarily outweigh the avoided costs. In the case of security of supply, public (and, as we shall see, also company) policies have tended to pursue the aim of a greater *diversification* in sources of oil supplies (and energy supplies in general), than would have been dictated simply by their relative costs or physical availability. As Helm points out, "there are good theoretical and practical reasons for believing that a private market will not provide an appropriate diversity of fuel-mix" (Helm, 1993, p. 420).

A perfectly competitive economy would not allow us to resolve problems arising from the need – due to levels of demand and the physical and political limits of resources – to keep energy supplies or oil production supplies with completely different cost levels, cost structures, production and investment rates on the same market. In a competitive market (not affected by extra-economic considerations) companies would tend to invest in areas with the best resources and the lowest production risks and costs. If this were the case, the Middle East – containing 2/3 of the world's oil proven reserves with extraordinarily low technical costs – would have production levels over a third higher than those at present. The same observation could be made concerning the relation between petroleum and alternative energy sources (to the detriment of the latter).

Another aspect to be considered when assessing failures in the energy market is the *long term view,* which has to be taken into account in order to assess the ability of markets to cope with expected levels of consumption. From a competitive and privatistic point of view it is clear that investment choices would tend to favour technological solutions with relatively low capital investment, that are rapidly implemented, and provide more immediate returns. Company behaviour would be conditioned by *short-termism* and the resulting investment would probably not coincide with what was socially desirable.

This conclusion is reinforced by the high price instability that has historically characterised the oil market and, therefore, the dynamic of the markets for alternative energy supplies. The continual price and structural fluctuations of a competitive economy have led to substitutions in production supplies often bearing no relation to past investments, or to the relative availability of supplies or their long term development costs, which are also extremely difficult to predict because of the high risk factor. In an ideal world sale prices should accurately reflect production costs and the relative scarcity of every energy source, in order to level the long term marginal development costs. Rather than being the lowest price, the best price should actually be able to cover the investments and costs needed to explore and produce energy in sufficient quantities to satisfy long term consumption with a socially desirable variety of energy sources. However, as we shall see, the oil sector is anything but ideal.

Whatever the case, the perception of real and presumed market failures has led to *pervasive* public intervention in the energy industry, which is found in relatively few other sectors. This intervention involves many countries, even those with a strong liberal tradition, and involves the use of different instruments according to their specific economic, social and institutional context. Market controls, exclusive rights, public property, protection and subsidies, and fiscal policies are some of the many forms this intervention can take. All this has led to the strong conditioning of company policies and their degree of freedom in the decision-making process. Market structures have developed patterns that are far from competitive. The possibility of understanding market performance using traditional theoretical models is therefore limited, even though some aspects cannot be attributed to political reasoning and motivation.

The "*philosophy of public intervention*" can be said to have dominated energy policies in all countries from the beginning of the century until the end of the Seventies (and for many countries still today) with varying intensity, but always greater whenever a traumatic international crisis occurred. It was only with the arrival of the Eighties that a "*free-market philosophy*" began to take hold and spread, both in the theoretical debate and in government policies. This was designed to dismantle the whole system of public intervention in national markets, allowing them to develop competitiveness and rivalry. This included sectors such as electricity and gas, which had naturally tended to be structured as monopolies. This philosophy is founded on the theoretical assumption (with considerable empirical evidence) that the reasons for failure – in terms of a reduction in efficiency – can more often be traced to public intervention and regulation than to market causes. This led to the conclusion that the task of energy policies was to substantially "retract from" their traditional way of thinking. In the words of the British Secretary of State for Energy, Nigel Lawson, "Our task is to set up a framework which will ensure that markets operate in the energy sectors with a minimum of distortion and that energy is produced and consumed efficiently" (Lawson, 1985, p.23).

The idea was, therefore, to dismantle every form of public intervention and reinforce the efficiency of the markets and the workings of competition as much as possible. Energy, in this case, is no different from any other product or *commodity*. We do not intend in this book to analyse this new "*market philosophy*", but it seems useful to point out the limits and hypocrisy associated with it. The main limit is in considering, out of the whole set of objectives used to measure the effectiveness of an energy policy, only those relating to the efficiency of national markets, while completely ignoring the aspect of the *security* of international supplies, which was always a prime concern of previous energy policies. This aspect should only be ignored when the following assumptions are true: (a) that the motives behind the political and economic instability that have caused no less than three separate international oil crises in the last years (the Kippur crisis in 1973-74, the Iranian crisis in 1979-80 and the Gulf crisis in 1990) are no longer valid; (b) that the structures of the oil market and its refined products have reached such efficiency that every risk or possible event can be taken into account in determining contracts and prices (both current and future). Both these assumptions are difficult to sustain on the basis of mere observation of the data. The *hypocrisy* lies in claiming that the level of oil prices on international markets is now only determined by free trade. There is no longer any doubt that the extent of the markets, the large number and variety of operators, the complexity of transactions, and the contract types enable market forces to play a much more important role today than ever before, as we will see. But, it would be very wrong to underestimate the role of politics. Nobody, in fact, can realistically say what the oil price would be today if, on 28 February 1991, the American army had not been able to return the State of Kuwait to full sovereignty, after it had been invaded and annexed by the Iraqi troops of Saddam Hussein during the night of 1-2 August 1990. It took the greatest mobilisation of men and military force ever observed since the end of the Second World War to recreate conditions of international legality in the Persian Gulf. But it is also true that this noble motivation – which did not lead to the same show of force in other similar situations – was also accompanied by the important intention of preventing Iraq from acquiring control, with the annexation of Kuwait, of over 1/5 of the world oil reserves and a position of absolute political-military leadership in the Gulf area. So, if it is true that market forces now count more in the oil sector, it is equally true that political (and military) forces have not ceased to count!

2.3. THE KEY PLAYERS

In an attempt to categorise the many interests acting within the oil market – on the basis of differing roles – we have grouped them into four main sectors, the "key players": the oil companies, the governments of producing/exporting countries, the governments of importing countries, and international organisations.

The structure of the oil industry and its market performance are the result of complex interaction between these interests, the prevailing air of conflict or co-operation, the contractual relations that evolved as a result, and the chain and counter-reactions that these provoked. Rivalry between producer countries, the strategic and political interests of governments, and the political and economic conditions in oil-rich areas are only a few of the aspects that have marked the development of the relations between the key players in the oil arena. This interactive process is, by its very nature, highly complex and cannot therefore be analysed or provide significant results in the short term. This complexity is further compounded by the diversity, or lack of overlap, of interests (and therefore of objective functions) within each category or group. Therefore, we need to be more specific in our examination and definition of the key players themselves.

2.3.1. The oil companies

The traditional distinction that is usually made within the category of oil companies is between the *major companies*, who could be said to have shaped the development of the industry from the beginning (for better or worse depending on one's point of view), and the *smaller companies*, which entered the market in the Fifties and Sixties. This distinction, still substantially correct from an economic point of view, has lost much of its significance as regards existing *power relations* in the market, ever since producer countries have been able to act independently. However, it is useful to remember this distinction in order to understand how things worked in the past.

The term *international majors* (or *seven sisters*, as defined by Enrico Mattei, president of ENI) refers to the following companies:

- *Exxon*: previously named Standard Oil of New Jersey, which itself derived from the dismantling of the Standard Oil Trust, founded by John D. Rockefeller in 1870;
- *Royal Dutch-Shell*: founded in 1907 from the merger of the Royal Dutch Petr. Co. with the British Shell Transport and Trading Co., which held 60% and 40% of the capital, respectively;
- *British Petroleum* (BP): previously the Anglo-Persian Oil Co., founded in 1909 by William Knox D'Arcy, and completely British owned;
- *Gulf Oil*: created in 1907 by the American Mellon family; sold in 1984 to Chevron;
- *Chevron*: previously named Standard Oil of California, itself derived from the dismantling in 1911 of the Standard Oil Trust;
- *Mobil Oil*: previously named Socony Mobil and also deriving from the dismantling of the Standard Oil Trust;
- *Texaco*: American, founded at the beginning of the century.

The *Compagnie Française des Pétroles* (CFP) is normally associated with these seven majors, because of its similar role and historical importance. It was founded by the French government in 1924 to replace German interests in the Ottoman Empire after the end of the First World War. Some common characteristics of all these companies were: the long tradition of professional experience that they had all acquired; their large size (in absolute and relative terms); their high level of geographical and production diversification (in both *non-oil* and *non-energy* sectors); their high and balanced degree of vertical integration, starting from the essential and highly remunerative production of crude oil; their high level of control over low-cost oil fields; and their fundamentally private management structure (which was also true of BP and CFP, even though the majority of their capital was actually state-owned). All these characteristics gave them a *differential advantage* over other companies, which again led to better performance, especially as far as higher long term profitability levels were concerned. Until the beginning of the Seventies the majors maintained almost total control over the international oil sector. In fact, just under 70% of the entire oil production in market economies was in their hands. However, as we shall see, things have changed radically since then, their percentage dropping sharply to nearly a fifth of that figure, barely reaching 14% (Tab. 2.1.). Despite their more limited role on the international crude oil market, the oil majors still maintain a position of absolute and relative dominance within the overall global industrial panorama. R.D. Shell and Exxon have always been among the top ten industrial giants. In 1995 they made revenues of 150 and 124 billion dollars, respectively, and net profits of 7 billion dollars each. Altogether the seven majors had a turnover of about 830,000 billion Italian Lire or, to get a clearer idea, the equivalent of a little less than half the total Italian GNP.

The second category of companies are those that could be defined as *independent companies*, because they are not part of the dominant club of the majors. They are generally much smaller companies; they have entered the market using typical *newcomer* strategies; they often have a vertically integrated structure, which is usually more concentrated downstream; and they show much more limited geographical diversification and often have a more national outlook. This category of companies – which can be considered as a whole because of the policies they have adopted rather than their management structures – includes most of the American companies, such as: Phillips Petroleum, Occidental, Conoco, Atlantic Richfield, Amerada Hess, Standard Oil of Indiana, Amoco, Union Oil, Getty Oil, etc., some European companies, such as Petrofina, and others, such as the Japanese Arabian Oil Company.

Table 2.1. *Majors' up and down: oil production[1] (mill. bbl/d)*

Companies	1950	1960	1970[2]	1980[3]	1990	1995	1998
Exxon	1.3	2.5	6.1	4.0	1.7	1.7	1.6
Gulf Oil	0.5	1.6	3.2	1.2	[5]	[5]	[5]
Chevron	0.4	1.0	2.6	3.0	0.9	1.0	1.1
Texaco	0.5	1.4	3.2	3.3	0.8	0.8	0.9
Mobil Oil	0.3	0.8	2.1	2.0	0.9	0.8	0.9
Royal Dutch-Shell	0.9	2.0	5.1	3.7	1.9	2.2	2.4
BP	0.7	1.5	4.0	2.4	1.3	1.2	2.0[6]
Majors total	4.7	10.9	26.4	19.6	7.5	7.7	8.9
World total[4]	8.5	18.7	40.0	47.9	51.3	56.4	62.5
Quota of majors (%)	55	58	66	41	15	14	14

1. Gross crude production (including condensates and natural gas liquids). 2. Includes long-term purchases and overlifting on participation agreements, equal in total to 2 mill. bbl/d. 3. Includes equity crude and buy-back crude. 4. Excluding countries with state-controlled economy. 5. Purchased in 1984 by Chevron. 6. Includes Amoco.

Source: 1950: FTC (1952); 1960 and 1970: company balance sheets; 1980, 1990 and 1995: "Energia"; 1998: "Petroleum Intelligence Weekly"; world total: BP, "Statistical Review of World Energy".

A third category consists of *State companies*, oil companies with equity controlled by consumer or producer states. In both cases, these companies constituted the prime instrument through which public policies follow their national interests. In the case of *consumer countries* these were: to acquire control over oil resources; to challenge the dominion of the majors; to consolidate the national industry to enable it to involve the whole oil cycle. This was the task given to companies such as ENI, founded in Italy in 1953, ELF-Erap, founded in France in 1964, Hispanoil in Spain, Petrobras in Brazil, and so on. The importance of these companies lay, above all, in the type of competitive strategy that they were able to adopt towards the majors and in the publicist attitude that governed their behaviour. This aspect characterised the strategy of State companies during the first two or three decades of their activity, while, today, there are no longer any substantial differences between their strategies and those of privately-owned or other categories of companies. This even applies to companies that have remained under public control.

However, the public vocation has not lessened in any way for companies controlled by *producing-exporting countries*, whose domestic economies, as we shall also see in the oil industry as a whole, have tended to become more centralised, as they have acquired greater control over crude oil supplies. They were founded as a result of a number of different requirements: (a) to gain direct experience in the oil sector in order to gradually take over from foreign companies; (b) to diversify the structure of the national economy, centred around oil export, by setting up a process of industrialisation in the basic sectors of refining, petrochemicals and chemicals; (c) to develop their national distribution of gasoline and other refined products. The first major company of this sort, the National Iranian Oil Company (NIOC), was founded in Iran in 1951. This was followed in 1960 by the Venezolana de Petroleo Corporation in Venezuela; in 1962 by Permina, Pertamina and Parmigan in Indonesia; also in 1962 by the Kuwait Petroleum Company in Kuwait; and in 1964 by Petromina in Saudi Arabia. Since the Eighties, these companies have become the new majors of crude oil production, managing to overtake and substitute the former majors.

Finally, there is another category of companies, which were once paid very little attention partly because they were considered marginal to the other companies, but also because they appeared to have little effect on the performance of the market. Here I am mainly referring to: (a) a large number of *minor companies*, operating in individual periods and/or in single countries, occupying small market niches and locally well-established; and (b)*traders* and *brokers*, who only operate in the trading of crude oil and its products. The vertical disintegration that characterised the Eighties has given them powers which are anything but marginal – as intermediaries, in the control of information about the market and, ultimately, in the fixing of oil prices.

2.3.2. The states

It would not be necessary to give much attention to this category – the states involved – if their role were limited to the part they normally play in other industrial sectors, where they have important effects on the national economy, foreign affairs, production costs, price dynamics, and so on. The situation changes, however, when – as in the case of the oil industry – these areas of economic interest are supplemented by other political interests that together have forced states to adopt behaviour that has strongly conditioned company strategy and performance, sometimes crucially. This has happened, as far as *consuming countries* are concerned, firstly by altering market incentives in order to redirect consumer choices or producer investments in a particular direction;

secondly, by facilitating the introduction of national industries and their interests into foreign markets, and giving them protection; thirdly, by directing the flow of trade towards countries that are considered more reliable, or with whom they have special political and trade relations; and, finally, by regulating company trade in order to prevent the exploitation of their dominant position over consumers, or to ensure the maximum security of supplies (Murat, 1969).

As regards *producing countries* policies have, on the other hand, been designed to sustain the development of national industries; to impose restraints on foreign companies; to capture most of the producer's profit; to intervene with companies and sometimes in trade flows in order to exert political pressure on the company's countries of origin or on consuming countries. We will be returning to this later in this book.

Here I wish to emphasise that: (a) *over long periods* the action taken by different states has varied so considerably – both in intensity and in direction – that it cannot always be considered a decisive factor for market performance; (b) apart from certain general intentions, strongly diverging interests have always existed between the two main categories of countries (consumers and producers); (c) all this makes it almost impossible to consider their respective actions as a *decisional whole*, apart from particularly limited and rare occasions. These diverse, if not conflicting, interests have often characterised and strongly influenced relations within consumer and producer countries.

The reasons behind these conflicting interests can be traced directly to the *variety* of situations to be found in the different countries as regards the oil industry. Let us now examine the main differences. As regards consumer countries the first distinction must surely be between those who are forced to import crude oil (or other energy sources) because they are substantially lacking in any domestic production (this is particularly the case for Italy and Japan) and those who are still forced to import, but who can boost their imports with ample domestic production, or potential production (this is the case of the United States, in particular). While the first are mainly interested in lowering the price of oil imports as much as possible, the second are anxious to avoid allowing low international prices to undermine the competitiveness (and, therefore, supply) of their production.[1] The second distinction involving consumer countries separates rich countries, who are able to import oil at any price, from poor countries, who are forced to cut imports when prices reach levels that are incompatible with the country's own expenditure and burden of debt capacity.[2]

As far as producing countries are concerned, there are also two distinctions to be made. The *first* is between net producer-exporters, who place most of their production on foreign markets, and producer-consumers, for whom almost all of their production is destined for the internal market. These two situations clearly produce conflicting interests in terms of the dynamic of international prices. The *second* distinction only affects the producer-exporter group, and is between those who possess considerable oil reserves (with a low exploitation rate) and those who have limited reserves nearing depletion. To aggravate this distinction, in both cases, nature has additionally created a very different absorption rate of oil profits. In the first case, this is very low (because of the small population and low rate of development), whereas it can even be exorbitantly high for the second group. These situations lead to very different objectives, and therefore to different preferences as far as *short* and *long term* price-quantity combinations are concerned.

As well as this form of differentiation *within* the two main groups of countries, there is also a geopolitical imbalance that has developed *between* them as oil has increased in importance world-wide. From a substantially balanced phase before the Second World War with an approximately equal proportion of production and consumption areas – when the United States was still a net exporter – the situation has changed leading to a long phase in which the centre of additional production has tended to move further and further towards the Middle East and North African countries, while consumption remained firmly attached to the most industrialised areas of the world (United States, Europe and Japan).

2.3.3. International organisations

Finally we come to a third category of key players, international organisations, whose role has certainly been less evident than that of the oil companies and countries involved, but who have nevertheless acquired an importance, in certain situations, that should not be undervalued. This role cannot be traced to a decision-making capacity on the part of these organisations that is *distinct* from that of their member countries. It is rather an expression of the willingness of these countries to arrive at common policies and decisions. The more co-operation manages to overcome internal divisions, the more essential the role of international organisations becomes, while the opposite occurs when reasons for conflict get the upper hand.

Among the international organisations that have proved most influential, within the group of consumer countries, we find the former *European Economic Community* (EEC), founded by the six European countries who signed the Treaty of Rome on 25 March 1957. This expanded, in various stages, to include twelve and then fifteen countries. Energy has always been a field of particular interest to the EEC. In 1950 the European Community of Coal and Steel (ECCS) was founded, even though recognition of the *strategic* nature of this raw material was eventually to force it to come under decision-making processes which were subject to the full national sovereignty of each member nation. The *Organisation for Economic Co-operation and Development* (OCSE), which was founded in 1961 by the industrialised nations, has become another point of reference for consumer nations, especially since 1974, when it was joined by a special *International Agency for Energy* (IAE). This was designed to represent, in the intention of its founders and especially of the American Secretary of State of the time, Henry Kissinger, an organisation counterbalancing the cartel of producer countries. This intention (though perhaps not entirely fruitless) has remained only on paper.

The main organisation of producer countries is the *Organisation of Petroleum Exporting Countries* (OPEC), founded in Baghdad on 14 September 1960 by five countries: Iraq, Iran, Venezuela, Saudi Arabia and Kuwait. There are four main objectives indicated in the Statute of OPEC:

(a) to co-ordinate and unify the oil policies of the member states;
(b) to determine the best means of safeguarding the interests of member states, both individually and collectively;
(c) to ensure price stability on international markets by eliminating damaging price fluctuations;
(d) to ensure a constant income for producer countries; an efficient, economic and regular oil supply to consumer nations; and a balanced revenue for the oil industry.

The creation of OPEC – to which the Western world reacted with absolute indifference (only "The Economist" gave it any mention at all) – by gathering together and homogenising the policies of the major producer countries, was to *accelerate* the process of modification of contractual relations at the time, towards a progressive reduction in the dominant position that foreign capital had acquired. The five founder nations were later joined by eight others: Qatar (1961), Libya (1962), Indonesia (1962), the United Arab Emirates (1967), Algeria (1969), Nigeria (1971), Gabon (1973) and Ecuador (1973). At the beginning of the Seventies, OPEC and all its members represented over half of world oil production. The Western world quite suddenly and unexpectedly was to realise what had happened.

2.4. MONOPOLY OR COMPETITION?

Let us now return to the discussion begun in the first chapter, concerning an analysis of basic oil *economics* and company policies, to evaluate their effects on the structure of the industry. These issues are complicated by the varying *significance* that can be attributed to each single economic *issue* and combination of issues; by the various assessments that can be made of them (as we have seen in terms of cost); by variations that can occur over time, and so on. This complexity – and how, in economics, "the same events have been used by different authors to support opposing theories", as A. Marshall wrote (Becattini, 1981, p. 145) – became the subject of an interesting debate that arose in the Sixties between Paul Frankel, an academic and highly acclaimed consultant for companies and governments, Morris Adelman, an academic from MIT, and Edith Penrose, a British company economist who summarised and developed this debate brilliantly (Penrose, 1964 and 1968). The discussion basically revolved around three key issues:

(a) whether the high level of market control practised by the major international companies – with both producers and consumers paradoxically accusing them of "monopolistic exploitation" – should be considered an inevitable result, so to speak, of the *economics* of the oil industry;

(b) whether this result should be considered economically and socially significant, or whether (more) competitive structural configurations would have been able to ensure a better allocation of resources and better conditions for the development of the energy markets;

(c) whether the "monopolistic" structure of the sector effectively corresponded to "monopolistic" behaviour, or whether competitive attitudes and risks had actually always existed on the market, even among the majors.

Some of the points made during this debate should be discussed here, since they can help us to understand events that have happened on the markets, and events that could occur in the future.

2.4.1. A "not self-adjusting" industry

The title of this paragraph contains the conclusion of Frankel's argument, which can be summarised as follows: because of the high exploration risk, the high capital intensity in all phases of the industrial process, and the high inelasticity of prices, in respect of both demand and supply, the industry cannot spontaneously reach an equilibrium – through the "servo-mechanism" of prices – in the generally accepted sense for economic theory. So, in the absence of any strict control over supplies, the industry is bound to be subject to inevitable crises of *oversupply* or *deficit*, leading to a consequent instability of markets and prices, which is to no-one's advantage. "The problem of oil is that there is always too much or too little". The need for companies to recuperate their investments as soon as possible, and the continuing steady *decrease* in production costs (with extremely low variable costs), have prompted companies to produce as much as possible in the shortest possible time, in an attempt to recuperate their *sunk costs* and to at least partially cover their *overheads*. Even a small surplus in supply can lead to a drop in price below production costs, without being counterbalanced by an expansion in demand. This leads to a need and preference for major international companies, capable, thanks to their integrated vertical structure, of ensuring a continuous supply flow, which is also efficiently administered with respect to expected levels of demand. This guarantees conditions of price stability, in both the short and long term, and, therefore, a high development potential for the whole industry. In Frankel's opinion, only through an "almost monopolistic" structure can the oil industry "adjust itself", even though the economics alone cannot be expected to resolve all its problems (Frankel, 1946, p. 98).

2.4.2. An opposite point of view

Adelman supports exactly the opposite point of view – both in his analysis and conclusions. In his opinion, the fact that the dominance of the major international companies and their integrated organisation has characterised the history of the oil industry does not provide any proof or guarantee that the industry would not have operated in an equally efficient way with a very different organisation. This could have been through a system of inter-company contracts and, more importantly, through a freer market. As a mechanism for the co-ordination of decisions and the behaviour of a large number of individual firms, the market and free competition would have been able to ensure greater efficiency and, above all, in what was to become an obsessive concern for Adelman, lower price levels. As for the empirical studies on which Frankel based his conclusions, in Adelman's opinion these had no foundation in fact. "The international oil industry is beset by myths... The myth of oil as an industry of decreasing cost (and) that vertical integration is, somehow, the 'right' way for the industry to be" (Adelman, 1972a, p. 3). Indeed, he believed that "the industry is at all stages subject to increasing costs except where inappropriate policies have led to the development of large excess capacity" (ibid. p.32). Therefore the conditions that would produce unlimited economies of scale and that would allow the oil industry to be considered a natural monopoly, unable to survive in competitive environments, do not exist. By introducing and analysing the development costs of discovered reserves and taking into account that "the basic fact about oil and gas production is the *production decline curve*", he concludes that, *ex ante*, in considering the development and production programmes, we need to consider the fact that both the total production and the total cost are dependent on time. In general, if the costs are fixed per time unit (and if they are increasing *a posteriori*, as they really are) and if production per time unit falls, *then production costs for the additional units increase with the increase in total production*

(Adelman, 1964a). If this is true, in a competitive situation the producer will be forced to produce until his increasing marginal costs equal his prices. In his opinion, the crude oil industry is inherently self-adjusting. In other words, it has a strong adjustment mechanism for determining the level of output and its division among various sources of supply, with the price acting upon the cost which must be incurred to bring up more output (ibid., p. 32). Therefore, according to Adelman, the oil industry could actually exploit the mechanism of competition as an excellent means of self-regulation and adjustment, helped by rising development costs that should prevent the non-profitable increases in supply described by Frankel from occurring.

2.4.3. Theory and performance

Adelman counters Frankel's analysis, which was based on historical experience and contained theoretical generalisations perhaps not universally valid, with theoretical arguments that are more convincing, but without sufficient evidence to explain why the mechanisms of competition – in the radical sense envisaged by him – have never actually prevailed and functioned correctly. In fact, Adelman himself admits that the mechanism of competitive adjustment could be subject to *destabilising* tensions, because of unexpected changes in the balance of supply and demand. This was exactly the basis of Frankel's main arguments. Large-size oil discoveries, compared to the actual market supply, that could be exploited at a lower cost than most of the current supply, could move the supply curve sharply to the right. As Penrose writes "such discoveries necessitate substantial readjustments among producers and sources of production which may be unacceptable to governments as well as to the industry, and may lead to some form of public or private intervention to mitigate the effects" (1968, p. 169). Although she disagrees with Frankel's analysis, which would lead to an inevitable series of oil surpluses and oil deficits without the dominant influence of the major international companies, Penrose comes to the same conclusions as regards the need for and inevitability of some form of control on supply. In fact, Adelman's analysis ignores the dynamic aspects of the changes observed in the oil industry, as well as the complications deriving from the consequent adjustment processes. In order to demonstrate that the oil industry could operate efficiently in competitive conditions, it is not enough to prove that this would be desirable. One also needs to understand how the forces of supply and demand could operate in practice, and why the behaviour of oil companies has continued to be far removed from a purely competitive mechanism. Adelman's argument therefore seems to be an obsessive attempt to view co-operation as triumphing over the constantly re-emerging reasons for conflict. In order to fully understand this issue, an examination of oligopoly theory could be more enlightening than any further analysis of the reasoning behind the Frankel's and Adelman's positions.

2.4.4. Some comments about oligopoly

Oligopoly can initially be defined as a market structure characterised by the presence of a small number of supplying firms, that are large enough to be able to determine price variations through changes in their volume of production. In other words, we can speak of oligopoly when each of the major companies is individually able, through its own decisions, to influence the results and decisions of another company and, therefore, of the whole group.

Connotations of non-competitiveness are often erroneously attributed to this kind of market. In fact, the presence of a small number of firms with a strong market power does not, in itself, imply an absence of competition. On the contrary, in an oligopoly competition and competitiveness abound in all their various forms and possibilities. For an oligopolist, "competition consists not in a quiescent stalemate of perfect competition, where there is no battle, because there is never anyone strong enough to disturb the peace. Rather, to him, true competition consists of the life of constant struggle" (Baumol, 1968, p. 328).

The aspect of oligopoly which most theoretical analyses have examined in the past, and which has proved central to oligopoly models is the *decisional interdependence of firms*, which actually distinguishes an oligopoly from other forms of market. The less consideration is given to this aspect, the more indeterminate the theoretical models become, and it becomes impossible to identify a situation of stable equilibrium in the market. Fundamentally, according to the theory of perfect competition, a firm operates on the basis of data fixed by the market, obtained by means of predetermined relations independent from their decisions. Therefore, according to this hypothesis, a situation of equilibrium was easy to define. On the other hand, the behaviour of oligopolistic companies is not *unequivocally* determined by the characteristics of the market. The more complex the behavioural criteria become, the more problematic the possibility of identifying a position of equilibrium.

This emphasis placed on the aspect of interdependence is actually a reflection of the fact that a company's decision-making process is never actually finalised, but rather consists of a constant readjustment of decisions within a continuous process of reaction and counter-reaction. As Sylos Labini observes "the path of conjectural variations (I think that he thought that I would have thought) is never-ending" (1964, p. 46). And there are many authors who have stressed the impossibility, in an oligopoly, of identifying set patterns of behaviour corresponding to given market situations. The actions of firms in an oligopolistic market can better be studied through Von Clausewitz's theories of strategy and tactics, than on the basis of the economic theories of Smith and Marshall.

In other words, traditional theoretical models *did not* show the aspect of interdependence, and this made it possible to construct precise cost and demand functions and, therefore, to reach positions of well-defined and stable equilibrium. The first formulations of the theory of the firm assumed that a firm was fully aware of its own cost and demand functions, as well as of the environment in which it operated, and tended not to consider the way in which this knowledge was acquired. They merely stated that the firm proceeded to maximise profits, after processing all the information relating to costs and revenues. Using this approach, no attention was given to uncertainty, risk aversion, and the time horizon of entrepreneurs. Using the traditional theory of value one could arrive at positions of well-defined and stable equilibrium on the basis of relatively simple assumptions. It was assumed that the company was in a position to define its demand and cost (and therefore also supply) functions *ex ante*, and then reach an equilibrium through the application of the principle of equality between marginal cost and marginal revenue.

Oligopoly models, however, recognise that firms are not totally aware of their own costs, their own revenues or the environment in which they are operating, given the uncertainties governing the real world. What is more, these models stress that the decision-making process is continually evolving. The conduct of every company is, in fact, dominated by its concern with the reactions of its rivals and, in this sense, studies of oligopoly represent a further approximation of reality (after Chamberlain's and Robinson's rejection of the traditional formulation of the theory of the firm). In oligopoly, the firm is not in a position to determine *ex ante,* with any certainty, its exact supply and demand functions, precisely because its choices have been materially conditioned by those of its competitors (and not only its *actual* competitors, but also *potential* ones). The quantity that an oligopolist is able to sell at every price range (and therefore its supply function) depends on the prices offered by its competitors, that are conditioned, in their turn, by the price that he intends to set. The same is true of the demand function. The firm is not able to predict a demand function only by referring to consumer preferences, since even these depend on the price/quantity decisions of the firm and its competitors. This difficulty in determining supply and demand functions then prevents a position of (quantity/price) equilibrium from being calculated. We must therefore conclude that there is no marginalistic solution to the problem of oligopoly. The equality between marginal costs and marginal revenue does not provide a solution for oligopoly, in which the main analytical difficulties lie in the reactions of rival firms (which worry every entrepreneur much more than the possible reactions of consumers).

In these conditions the solution is indeterminate or, more precisely, "pluri-determinate". In general, there could be any number of price/quantity combinations leading to a situation of equilibrium, in the sense that, *ceteris paribus*, each single oligopolist tends to try to remain in that position. However, in actual fact, it is impossible to definitively establish which of these situations will occur.

As a result, the various models of oligopoly reach positions of equilibrium only because they impose restrictions on competitors' behaviour. Fellner writes that, in oligopoly, a determinate equilibrium seems impossible to maintain and instability impossible to eliminate, unless firms arrive at some implicit or explicit form of agreement (Fellner, 1949). In Fellner's analysis, the importance of decisional interdependence on the industry's performance finds its strongest expression. This level of awareness then forms the basis for the establishment of a more or less explicit process of agreements between the firms themselves. This "conjectural interdependence" (or awareness of interdependence) forms the basis of a process of negotiation and agreement between firms (although development does not necessarily follow: a company's willingness to take the behaviour of its competitors into account does not always lead to an agreement between them). However, on the whole, an awareness of interdependence is often the starting point for a collusive process. This could result in a *cartel* (a real agreement between firms, aiming to anticipate competitors behaviour and therefore eliminate sources of indetermination) or in a *quasi-agreement* (making it possible to discount the behaviour of others to some extent) or, possibly, in a spontaneous co-ordination of interests.

In general, the fact that an oligopolistic system is incapable of creating a stable equilibrium within the system produces negative effects, and it is precisely this awareness of negative effects that induces firms to adopt a collusive behaviour. An unstable market, incapable of reaching an equilibrium, becomes socially unacceptable, and this can prompt compulsory state intervention and regulation in order to solve conflicts within the sector. On the other hand, the firms themselves are well aware that isolation is the worst situation for them, and that contact with other firms, even as part of a moderately conflicting relationship, can help to exclude any open and strong competition (which is always extremely costly).

Fellner concludes that collusive agreements are a way of reducing the uncertainty triggered off by the interdependence of the behaviour of oligopolistic firms, even if it must be understood that the situations of equilibrium reached are not necessarily stable. One only needs to think of the difficulties involved in predicting all the factors contributing towards a particular result and all the relevant economic variables which create uncertainty when taking on any reciprocal commitment, or the fact that the greater the number of firms in the sector, the more complex it is to co-ordinate their interests. Therefore, all things considered, any form of co-ordination between firms is inevitably imperfect, and there will always be a residual "competitive space" that will not allow competition to reach insignificant levels. Even when an agreement has been made, a "latent competitiveness" still exists that is destined to explode whenever diverging interests or technological disparities take the upper hand.

2.4.5. Market structure and price dynamics: some preliminary issues

The unique characteristics of the oil industry have, as we have seen, created the necessary conditions to form the large-scale "economic-technical" groups that dominated the oil business until the end of the Sixties. Under these circumstances, the use of price as a direct competitive force was not likely to occur. The limited number of *suppliers*, their size and relative strength, together with substantial product uniformity, made (and still make) the oil industry a typical case of strictly controlled oligopoly (*competition among the few*) in which the potential price cutter has to consider all the probable means of retaliation by his rivals and understand that price reduction, since it can easily be imitated, is not a particularly advantageous move. As Hartshorn (1966) argued, international groups, in an oligopolistic market structure, have no incentive to compete with each other using methods that would have led to a general lowering of price levels in the consumer markets. In an oligopoly, in fact, any direct action aiming to modify price levels or the output level on the part of a major, would certainly have altered the general market equilibrium for all the others. Each international group had to consider that, if they cut their crude oil prices, their competitors would certainly have done the same, and probably with an even greater price reduction. It was therefore necessary to decide how to proceed when confronted with the type of reaction provoked by this initial move. What is more, every major sold its own products on very different markets. It could therefore exert pressure, but would still be vulnerable on many fronts, in the sense that a price-cut designed to put pressure on a competitor in a particular market could generate "backlashing" pressure in some other part of the world, where the competitor was better placed. And Penrose observed that "each company may endeavour to maintain or improve its own position *vis-à-vis* the others, but each accepts the restraint that arises from the recognition that some actions, perhaps profitable in the short run, would in the long run be defeated by the retaliation of others. The particular restraint most often exercised is in respect of prices, although at the same time non-price competition can be alarmingly vigorous. It is this latter type of competition, and the sense of rivalry behind it, together with the occasional price-competition and the numerous limitations on their freedom of action, that makes the companies feel that the charge of monopoly is absurd and that they are essentially competitive. It is the restraint each exercises in the common interest of all, together with the high degree of positive collaboration in many markets and producing areas, that makes large numbers of outsiders feel that the industry is essentially monopolistic" (1964, p. 180).

In short, *instability is a structural characteristic* of the oil industry, that can be traced to the basic conditions that we have analysed above, thus being fully in agreement with Frankel's ideas. Every time this occurs, operators are strongly tempted to initiate some kind of collusive action. Rockefeller himself maintained that the industry was not capable of self-regulation. In his opinion, the theory of free competition worked fairly well when an industry consisted of a large number of small firms, but ceased to function as soon as a few multinational companies entered the sector. If competition were to push prices below production costs, large companies would not be able to interrupt production temporarily. Overall costs, interest on invested capital, maintenance costs etc., would necessarily have to continue, sometimes with dramatic results. In his opinion, the only solution was some form of agreement. Frankel, on his part, pointed out that, in the oil sector, associations between producers have always existed in every country of Europe, and that "untrammelled competition by oil companies in the field took place only during the aftermath of the breakdown of one organisation, and represented no more than a period of preparation for a new agreement" (Frankel, 1946, p. 81). The entire history of the oil industry has been marked by an obsessive and continuous effort to reduce the uneconomic aspects and costs of free competition, through centralised control over the industry, for the sake of market stability. Taking this point of view, there are five different phases in the history of the oil industry that can be identified (see Tab. 2.2.). Each of these is characterised, on the one hand, by a different degree of competitive conflict between firms and, on the other, by a consequently different degree of price stability. This would be low in phases in which competitive conditions prevailed, and medium or high whenever market control became oligopolistic or quasi-monopolistic. This functional relation between market structure and price performance is the working hypothesis that we will be developing in the course of this book.

Table 2.2. *Phases of the oil industry*

Phases	Period	Life cycle phase	Market structure	Price stability
I	1859-1900	setting-up	fromcompetition to quasi-monopoly	low-medium
II	1900-1940	development	competition	low
III	1940-1970	growth	oligopolistic co-ordination	high
IV	1970-1985	maturity	transition	low-medium
V	1985-?	maturity	competition	low

Notes

1. This conflict of interests has occurred many times (both in the past and today) and, in the mid-70s, for example, it prompted the American government to propose the idea of a *floor price* for petroleum to its European partners, ostensibly in order to defend "common interests", but actually in order to protect its domestic production. This proposal, not having any counterpart, was not successful.

2. This distinction was particularly evident during the second oil crisis (1978-80), when Germany and Japan openly declared that they would take no steps to stop rising oil prices, so that the market would be left to "ration out" demand.

3 POLICIES AND TACTICS OF OLIGOPOLISTIC CO-ORDINATION

3.1. AMERICAN ADVENTURERS AND THE BIRTH OF THE INDUSTRY

In the historical development of the oil industry, we can usefully distinguish a *take-off phase* – roughly between 1859, the year oil was first discovered at Titusville (Pennsylvania), and the end of the nineteenth century – which was a phenomenon almost unique to the United States, and marginal even in the context of the energy resources of that country. Until 1900, kerosene was the main product, followed by lubricating oils, naphtha and oils for medical purposes. Demand for gasoline was to become significant only later. Even so, this first phase is extremely important for many reasons.

The great issues that still characterise oil markets – in particular how to reconcile competition, stability and growth – became evident in that period. It was then that John David Rockefeller's Standard Oil Company, symbol of American big business, first became established. In the opinion of Alfred Chandler Standard Oil was the "first mover" not only in the oil industry, but in the whole of modern industry (Chandler, 1994). It was the first company to have based its efficiency and its competitive advantage on a triple, interrelated investment pattern, in production, distribution and management. Finally, it was in those years that the technological dynamics (economies of scale) and organisational strategies formed which were to distinguish the structure of the sector from the beginning.

In 1870 oil consumption still accounted for less than 1% of the American energy balance, and in 1900 just 4%, in absolute terms at a little less than 9 million tons. What had halted growth was not only the limited technology of the time, but also the high transport (rail) costs, and the great variation in selling price: from 16 dollars a barrel in 1859 to 9.25 in 1860, 0.49 in 1861, and 1 in 1862, before jumping to 11 dollars in December 1864. Faced with such extreme variation, it is clear why an "*Association of Oil Producers*" aiming to protect the well owners' interests already existed in Pennsylvania by 1869. Its aim was not so much to impose higher prices on buyers, as to contain production as far as possible within the limits of actual demand. The start of production in *new richer oil fields*, which upset the precarious equilibrium of the markets, was, in general, the driving force behind the formation of agreements between competing producers. The owners of the older wells, whose productivity was already falling, and who could not survive if prices went down too far, made urgent calls to limit drilling, and even to partially close down operations at some wells.

Frankel argues that the main reason for the continuous difficulties of the oil industry can be found in the rapid expansion of the market. If demand had been relatively stable, supply would have found its own natural level of equilibrium; but awareness of the strong and continuous expansion of the market caused an over-estimation of the opportunities for growth on the part of the entrepreneurs, in part caused by the unpredictability of future supply levels. This situation led to over-investment, which could only have been eliminated by either severe competition, or by an *agreement* that restricted the rate of production capacity and allowed the formation of an acceptable price. The first instances of collusive behaviour were thus inspired by the desire to stabilise the market, rather than an attempt to push prices up (although it is hard to trace a clear line between the two policies, since the two results often coincide).

The outcome of these agreements was, however, almost always unsatisfactory; the higher the price, the greater was the temptation to increase supply (and one's own earnings) by going above the assigned quota; and those that were the first to cheat gained the dual advantage of high prices and huge sales. Once this practice spread, it did not take long for those who had previously lost out to also break the agreements. The weakness of arrangements of this kind lies precisely in the fact that while they require the consensus of a vast majority to be signed, they can be destroyed by the defection of a small minority. Frankel's opinion is that an association of direct competitors, with equal opportunities, could survive for a while, but not for long, because – however paradoxical this may sound – the participants interests are so *identical* that in the long term they are *incompatible* (Frankel, 1946).

The impossibility of reaching any form of co-ordination of supply can first of all be explained by the large number of companies, which in turn is the consequence of the property rights for the exploitation of mineral resources. Two of these are of particular importance. *Firstly*, the right to extract oil had to be negotiated with the individual owners of the land, thus giving rise to a large number of small and medium sized companies, given the extremely scattered nature of the reserves. *Secondly*, the so-called "law of capture" recognised that the owner of the land where the oil was brought to the surface had ownership of the oil extracted, independently of where oil was actually located underground. Each field could therefore belong to more than one owner, and each producer had a great incentive to get the oil out as soon as possible to prevent others from seizing the chance to "capture" it first.

Through time the number of companies involved gradually decreased, although volume of the oil supply still remained at a low level compared to elsewhere. By the mid Eighteen Sixties there were 16,500 companies in the oil and gas mining sector, of which 9,500 were in the oil sector alone.

Given the rigidity of the institutional framework, it was not possible to think of gaining a dominant position in the industry via the control of the mining *upstream*, and Rockefeller was well aware of this. He believed that rather than aiming for indiscriminate control of all stages, he should go for those that were strategically much more important. So he concentrated his energies first on refining and then on transport, reaching a quasi-monopolistic control of the industry by the end of the century.

Rockefeller gained a dominant position in refining in the 1870s, being the first to exploit the great competitive advantages associated with economies of scale. With processing costs cut by half, he was able to sweep away competitors and give himself a powerful negotiating weapon against the railways to get transport rates reduced. "The greater the volume of traffic the greater the discount. It is precisely Standard Oil's unprecedented throughput that led to the reduction of rates – and not vice versa. In April 1868 it was the railway companies who sought out the 28 year-old refining entrepreneur" (Chandler, 1994, p. 152). Low transport costs, low refining costs, and high productivity gave an increasing competitive advantage to the Standard Oil Alliance – a "federation" of five oil magnates. Standard exchanged the shares of the individual members – which remained independent from the legal point of view – with those of more than 30 refining companies. The result was an interdependent financial structure that was the force behind the regulation of prices. Price volatility seemed to diminish in some years, although in the 1870s it continued to oscillate yearly by over 30%.

The reduction of competitive pressures and the rising dominance of a relatively small number of companies was to be the result of another major technological innovation: the introduction of long-distance oil pipelines in 1878. Though first constructed by the Tide Water Oil Company (an association of crude oil producers), Standard Oil was the first to fully appreciate their potential, investing 30 million dollars (the company assets then stood at three million) in the construction of its own pipeline network. This allowed it not only to cut transport costs by three times (from 1.5 to 0.45 cents a gallon), but also to greatly increase its storage capacity, thus ensuring a greater and more constant flow of oil to the refineries.

Standard Oil – which was made an Oil Trust by its constituent members at the time – followed up this technological revolution with one of another kind, concerning its internal organisation. It created a huge distribution and marketing network on a national and international level, so that its sales could keep up with the unprecedented refinery output. It centralised the oil supplies mainly from external producers, and it invested in human resources not only for better control of broadened administrative structures, but also to co-ordinate and control its new functional activities. The competitive advantage of Standard Oil was no longer based on lower costs alone, but equally on greater functional and strategic efficiency. That organisational model was to become a reference model for the entire international oligopoly.

Towards the end of the century – when the companies of the Standard Oil Trust passed under the control of the new Standard Oil Company of New Jersey (1899) – Standard Oil dominated the American market almost completely. Prices became relatively stable at between 0.5 and 1.0 doll./bbl. This allowed the industry to consolidate its growth inside and outside the United States, transforming itself from "a sector of adventurers into a methodically organised industry".

3.2. FROM AN AMERICAN INDUSTRY TO A GLOBAL INDUSTRY

A *second phase* in the history of the oil industry, lasting roughly from the beginning of the century to the end of the Thirties, was marked by: the progressive increase in oil consumption outside the United States; the entry into the market of new areas of production in the Middle East; the setting up of all the international majors; and the rise of an international market organised into extensive regional markets (the United States, Europe, Russia, Asia). This phase marked the passage from the *take-off* to the *development* of the oil industry, as well as from being an *American* phenomenon to being a *global* one. There were many reasons behind the take-off, but above all it was due to the extraordinary sequence of inventions and innovations from the end of the nineteenth century: the first cars with the internal combustion engine (Benz, 1885); the birth of the Ford Company (1902); the Wright brothers' first flight (1903); the introduction of the Model T Ford (1907), with production quickly rising to 650,000 a year by 1911; and the invention of synthetic rubber (Hofman, 1909). The onset of globalisation came with the first great discoveries of oil outside the United States, the spread of the innovations in technology into Europe, and the entry on to the stage of most of the protagonists listed above.

Fostering and then accelerating this process of transformation was an event of traumatic significance occurring in the United States in 1911. In the case of the Standard Oil Company versus the Government of the United States, the Supreme Court – applying the 1890 Sherman Act for the first time – ordered the dissolution of the New Jersey holding because "through intimidatory action it had acquired control of the shares of over 70 oil companies with the aim of monopolising trade". After this injunction, Standard Oil distributed its shares proportionally to its own shareholders, splitting up into 33 separate companies (of which one remained the Standard Oil of New Jersey, together with Standard Oil of California, Socony-Mobil, Standard Oil of Indiana, etc.).

Partly because of this decision, the second decade of the twentieth century saw the transformation of the American oil industry from a quasi-monopoly to oligopoly. On the one hand, this made every collusive internal strategy problematic, because of the risk of falling into the clutches of the anti-trust forces. On the other hand it both old and new members of the United States oligopoly into new expansion strategies. Two of these strategies, *vertical integration* and *internationalisation*, were to prove especially successful. Since the Twenties, the American majors as well as the main independents were beginning to set up refineries and distribution networks in Europe, seeking to seize the areas of possible production in the Middle East. Operative, functional and strategic competition for taking control of production and consumption markets shifted to an international level, with many different problems from those experienced in America. Our attention from now onwards will be focused primarily on this international scene.

The whole history of the oil industry's second phase could be summed up in the long march of the majors to gain centralised control of the international markets. The objective was the same as that of the "first movers" in the first American phase: to ensure stable market conditions (especially in prices) and then proceed to an orderly, harmonious development of both investments and industry. The institutional basis from which such a strategy could get off the ground, as well as the infrastructure available, were very different. On an international scale it was neither sufficient nor possible to take control of transport and refining facilities, as Rockefeller had done in the United States, in order to take control of the entire market. The most important strategic sector of the industry, in this case, was the crude oil supply, which was spread around a growing number of countries within a huge geographical area. The possibility of gaining an appreciable measure of control was dependent on two factors. First, the emergence of *property rights* in the rich producer countries of the Middle East, allowed the companies fuller control of their reserves there and broader decision-making autonomy. Second, the adoption, between producer companies, of oligopolistic co-ordination which impeded all forms of dangerous competition and allowed a strict matching of supply to levels of demand, thus making direct price control implicitly possible. The first condition was dealt with through *concessions*, the second through *consortia*.

3.3. CONCESSIONS

Concessions were the juridical instrument that from the beginning regulated most of the relationships between the states with oil reserves and foreign companies. Once a concession was obtained, a company would enjoy a kind of *absolute right* over a certain surface area of the state which had conceded it, allowing them to search for, extract and sell any amount of oil (at any price), in exchange for compensation of financial nature. From a legal point of view, there is no doubt that the concessions took private, commercial and contractual forms (El Sayed, 1967), as well as combining elements of public law due to the specific nature of their subject and the collective interest involved. Hence the great controversy (Verdross, 1964; Cattan, 1967 a,b; Rouhani, 1970) over the juridical nature of the concessions: there were those who stressed their public character, thus subjecting it to the domestic laws of the host states, and those who on the other hand considered them almost *international agreements*, and as such subject to international law and juridical rule *pacta sunt servanda*.

The latter interpretation was certainly not disliked by the oil companies, given the great advantages and freedom of action the concessions guaranteed them. This was thanks to the politically subordinate role played by the Middle Eastern states in relation to the western powers, and thanks to their (almost) complete ignorance of every subject dealt with in the concessions. In civil law it would certainly have been possible to find the existence of invalid intent, since one of the contracting parties lacked any knowledge of the realities involved. However, from a historical perspective, it cannot be denied that concessions played a decisive role in guaranteeing the development of a resource that was to prove essential to the economies of the producing countries.

This does not change the fact that the more the desire for independence grew in those countries, the more concessions were increasingly considered the primary instruments in the "exploitation" that they considered themselves subjected to by western states and companies. As an Indonesian politician wrote, the concessions were increasingly seen as "transfer of national sovereignty" to third parties (Sutowo, 1962, p. 21). Indeed no one could deny that the early concessions from the first years of the century were "more imposed than negotiated" (El Sayed, 1967, p. 48) and that they created a reference point for those issued later, so that "their similarity was equalled only by their simplicity" (Cattan, 1967a, p. 2).

The concessions defined the *area* involved in company activities, in general covering most or even all the territory of the state (this was mainly to create an insuperable barrier to the entry of other companies); the *duration* of the agreement, usually between 60 and 75 years; and the *financial revenues* the state would receive as owner of the resources (a *royalty* in proportion to the amount sold) and as fiscal authority (an *income tax* in proportion to the profits earned). These revenues were also based solely on an annual company statement of accounts over which the states had absolutely no control (Mikdashi, 1966 and 1972). As regards the interests of the companies, the concessions fulfilled three essential tasks: (a) they maximised their *decision-making autonomy*, relegating the States to mere receivers of revenue; (b) they guaranteed *conditions of certainty* for the future, preventing the States from unilaterally revising the terms of the concessions; and (c) they blocked or strongly conditioned any potential *newcomers*.

3.4. CONSORTIA

While concessions served to regulate the relations with the host states, consortia were the main instrument for the regulation of relationships *between* companies. In general this happened in the more critical and strategic oil extraction phase, and did not necessarily lead to symmetrical collaboration in refining or marketing, where, on the contrary, partners were left with greater opportunities to compete with each other.

The first big consortium in which the American and European majors joined forces was the Iraq Petroleum Company (IPC), founded in 1928; it was formed from the Turkish Petroleum Company with the shareholder entry of the Standard Oil of New Jersey and Mobil, as well as BP, Royal Dutch-Shell, CFP and Mr Gulbenkian, an influential businessman. The IPC is important for several reasons. *First of all*, it marked the definitive establishment of an American presence in Middle Eastern oil producing areas, despite hitherto stiff English opposition. *Secondly*, it was a point of reference for later consortia formed by the majors elsewhere. *Thirdly*, it was the most striking illustration of "joint control through common ownership" (FTC, 1952, p. 109) set up by the majors to prevent any kind of competition between themselves and other companies. Analysis of the IPC allows us to detail the ways through which this essential instrument of co-ordination between the majors operated. Its principles were to be extended to all the other consortia set up at that time in Saudi Arabia, in Iran, and in Kuwait, covering almost the entire Middle East.

The IPC, like other consortia, did not operate as an independent business entity. "Its policies and its management were determined by and made to serve the mutual interests of the major international oil companies which jointly owned the majority of its shares" (idem, p. 47) a sort of "confraternity of oil merchants", as its managing director described it. Its function was essentially *the joint management of oil production and its pro rata distribution among the shareholders*. Its profits were kept at a very low level, due to the practice of fixing transfer prices very much lower than market prices. Another important function was that of controlling both the current as well as potential oil supply; i.e. what might derive from new concessions assigned to the same IPC partners or to other companies. To prevent this possible source of competition, the members of the IPC signed the *Red Line Agreement*, which prevented them from competing with each other to obtain concessions in an area including most of the former Ottoman Empire. In reality, the IPC was aiming to stop autonomous initiatives by the companies in the consortium (or any others) within the immense red line area, and to do this it tried to gain as many concessions as possible.[1]

No less interesting is the control the IPC exercised over its own supply, with the aim of preventing even a minimal oil surplus. Although the original 1925 concession covered the whole of Iraq, the IPC limited its upstream activity to oil fields that amounted to less than 1% of the territory's surface area. The reasons behind this behaviour can be understood in the light of the fact that during the Great Depression of 1929 oil supply significantly exceeded demand, and the companies were afraid that higher production in Iraq would have pushed down prices even further. Dilatory tactics were thus applied not only in drilling and exploitation activities, but also in the negotiations on the rights of passage of the pipelines. Right from the start of its activity, the IPC more or less *suffocated* production in Iraq and in neighbouring regions, being careful to conceal its behaviour from the government, in fear of a reduction in the size of the concession or its transfer to independent companies, hungry for crude oil and anxious to maximise production.[2] To sum up, the foundation of the IPC, as of other consortia, allowed: the regulation of the entry of new sources of production into the markets in line with the expected demand growth; the reduction of competition that would otherwise have occurred, with negative effects on prices; the limitation of potential competition from new entrants; and the solution in a structural fashion of conflicts of interest not otherwise solvable.

3.5. THE CARTELS: THE EVENTS OF 1928

Although the assignment of the concessions and establishment of the consortia date from the Twenties, it is only much later that their role was to be fully realised. Both strategies (together with others that we shall be examining later) make up the oil majors' ingenious long-term structural reply to the market dynamics which were becoming evident in what we have called the "second phase". These dynamics were marked by continuous rivalry between producers, with momentary alliances and truces, intrigues of every kind, continuous interference by governments, and frequent price wars.

The most important price war exploded with amazing virulence between the two giants, Royal Dutch-Shell and Standard Oil of New Jersey, in India in 1926-1927. Having been expropriated of some properties by the Russian government in 1918, Dutch-Shell had continued to purchase crude oil from that country for the Indian market, which Standard Oil also supplied. In 1926, in order to force the Soviet government to pay compensation, Shell decided to stop purchasing Russian oil, asking Standard Oil to do the same out of solidarity, although it was unable to make alternative supplies available immediately. Faced with a choice between strengthening Dutch-Shell's position *vis-à-vis* the USSR (underlining principles of international law that other countries might be tempted to violate), and therefore recognising it as the leader *de facto* of the Indian market (thus weakening its own competitive position), Standard Oil did not hesitate. It continued to buy Soviet crude oil, provoking the violent reaction of Royal Dutch-Shell on September 10th 1927.

In the space of a few days a price war began which quickly spread, in a chain reaction, throughout all the consumer markets. To Standard's tough reply on the Indian market, Shell answered with price reductions on the American market, involving the entire business system in that country. Standard Oil replied with exactly comparable cuts on the European market. The violence of the clash showed everyone how the growth of demand and the entry of new companies risked compromising an *orderly* growth in the markets in the presence of inevitable competitive processes, where the equality of the forces in the field would have prevented either of the leaders coming out on top. The moment had arrived to take measures to stop this happening again.

The first and still most famous *cartel agreement* was signed on September 27th 1928 in Achnacarry Castle, where Sir John Cadman (chairman of the Anglo-Persian Company) and Mr. Walter C. Teagle (chairman of Standard Oil Company) were the guests of Lady and Sir Deterning (chairman of Royal Dutch-Shell). In a 17-page document, they fixed the general principles and practical means of co-operation between the oil producing companies to end "destructive rather than constructive competition, resulting in much higher operating costs. The petroleum industry has not of late years earned a return on its investment sufficient to enable it to continue to carry in the future the burden and the responsibilities placed upon it in the public's interest... Recognising this, economies must be effected, waste must be eliminated, the expensive duplication of facilities curtailed" (FTC, 1952, p. 200).

The controversy that years afterwards would divide Frankel and Adelman is summed up in these words, together with the criticism of the traditional economic theory of the dichotomy between competition and monopoly. According to this criticism there were restrictive practices and monopoly situations (or very nearly) that ought to be stopped, but in reality there is no reason at all why monopolies should be fought against *in all cases*, even where it does not offer a natural solution. The indiscriminate struggle against it is thus based on ideological prejudices rather than economic principles or realities. Indeed, the theory itself demonstrates that monopolies do not always find it suits them to raise their prices since this depends on the elasticity of the stretch of the demand curve from which the variation moves, and also on the movement of the cost curve. Price rises may suit only if they correspond to a variation less than proportional to the amount in demand (hence with elasticity from 0 to 1), with a resulting increase in total profits, and at the same time the total costs increase proportionally less than the profits. Thus, for instance, while a shift upwards of the demand curve generates in competition (in the short term) a new equilibrium characterised by greater quantity and higher price, in monopoly the new equilibrium will result in greater quantity, and a price that may be higher, lower or the same as before (Koutsoyannis, 1981, p. 287).

Just like a monopoly, cartels suffer from false myths: for example, that they are a way of imposing higher prices and gaining fatter profits, although this may naturally be their legitimate intention. But there are other objectives that do not coincide with this: the stabilisation of prices, production and profits; the protection of invested capital; and defence from the risks of destructive competition, whose negative effects would fall not only on the producers. When an industry is going through a phase of recession or oversupply, with prices going down steeply, the necessity of cartel agreements becomes all the more evident. However the opposite is certainly not true, i.e. that higher prices can be attributed to cartel manoeuvres, even where forms of association exist that could be assimilated to practices of this kind. If we look closer, when the market situation justifies a steep rise in prices, an intervention by a cartel appears to be quite futile when aiming to slow down or even halt price movements. This should be remembered when we are analysing events of the recent past, and responsibilities wrongly attributed to the OPEC cartel.

As regards the role and effects of the 1928 cartel in the international market, there is plenty of room for uncertainty, notwithstanding the stream of literature devoted to it (especially the very detailed examination carried out by the Federal Trade Commission in 1952). Though most of the majors joined, a few potential competitors escaped its net; these were a minority overall, but a significant presence in individual markets. Because of this limitation, the importance of the Achnacarry agreement is to be considered mainly from the point of view of its general principles, of its aims, and of the procedures adopted to stabilise the market, more than in the results it obtained. As far as the principles go, the following requirements were upheld: (a) to accept and maintain the market share of each member at the time of the formation of the cartel, giving rise to the phrase "as is agreement"; (b) to make the existing facilities available to competitors on favourable conditions, but not at less than the actual cost; (c) to improve facilities only insofar as new consumption made it necessary; (d) for each production area, to maintain the economic advantages deriving from its location; (e) to stop all surplus supply.

The declared *aims* of the cartel, to be implemented through a set of *procedures* regulating intra-cartel trading and interrelations, were: to administer rather than ration the supply, especially if additional; to rationalise the costs of investment; to improve the system overall; and to stabilise and harmonise the price structure. As regards the price system, a multiple basing point system was adopted, in which each loading port was nominated a *base point*, with the Gulf price used as a reference, to which was added the cost of transport to the port of arrival.

On the basis of the principles, aims, and procedures of the "as is agreement", specific organisations were set up and further agreements reached to describe in greater detail the functions of the local cartels in individual countries or areas.[3] Despite the great determination shown in pursuing the sought-for objectives, and despite the substantial support of the majors, the agreement did not fulfil its intentions, as the continuing price instability throughout the Thirties demonstrates.

There were several reasons for this, but the inability to control the small fringe of independent producers and refiners in the American market, where they were structurally more numerous, as well as in the international market, was especially relevant. An essential fact of life of the cartel economy is that even small outside operators can render the efforts of the cartel members futile, since they are able to benefit precisely from those efforts. Even the acquisition of independent companies by the majors, as the "as is agreement" recommended, failed to reduce the threat from that quarter. Behaviour towards them was conditioned by the awareness that if they were not admitted into the agreement they could reply by cutting prices, while at the same time it was recognised that if they were admitted, they would have to be allocated a slice of the market, which was thus lost to the majors. Precise rules were then fixed to discipline the provision of crude oil to the independent refiners by the majors, so as to prevent them constituting a market outlet for the small oil producers, and to put a stop to other potentially destabilising situations.[4]

The experience of the Achnacarry cartel was certainly not a good one, but it was not entirely worthless. It showed clearly that market stabilisation through cartel agreements, in the context of an expanding market, was actually impossible, even with a high level of support and loyalty on the part of the biggest operators. Other paths, more structural and therefore less risky, would have to be followed. And it was in this direction that the majors had started to tread, with attention directed to their internal organisation and the structural framework of the global international industry.

3.6. DOMINATION AND CO-ORDINATION

From the end of the Thirties the practice of formal cartel agreements was gradually abandoned. And yet it was from that moment and right up to the end of the Sixties, in what may be termed its *third phase*, that the oil industry went through a period of greater price stability. In 22 of the 30 years from 1940 to 1970, there was substantial *nominal price* stability (with variations of 5%). Substantial oscillations only occurred during exceptional times (the Second World War, the Suez crisis of 1956-1957, and the Six Days War of 1967). For around three decades the oil price on the international market stayed between 1.5 and 2 dollars a barrel: quite an extraordinary feat, when one remembers that during that time total demand increased by a factor of 8, to around 3 billion tons, while

Figure 3.1. *Joint ownerships by international oil companies of subsidiary and affiliated companies in the Middle East*[1]

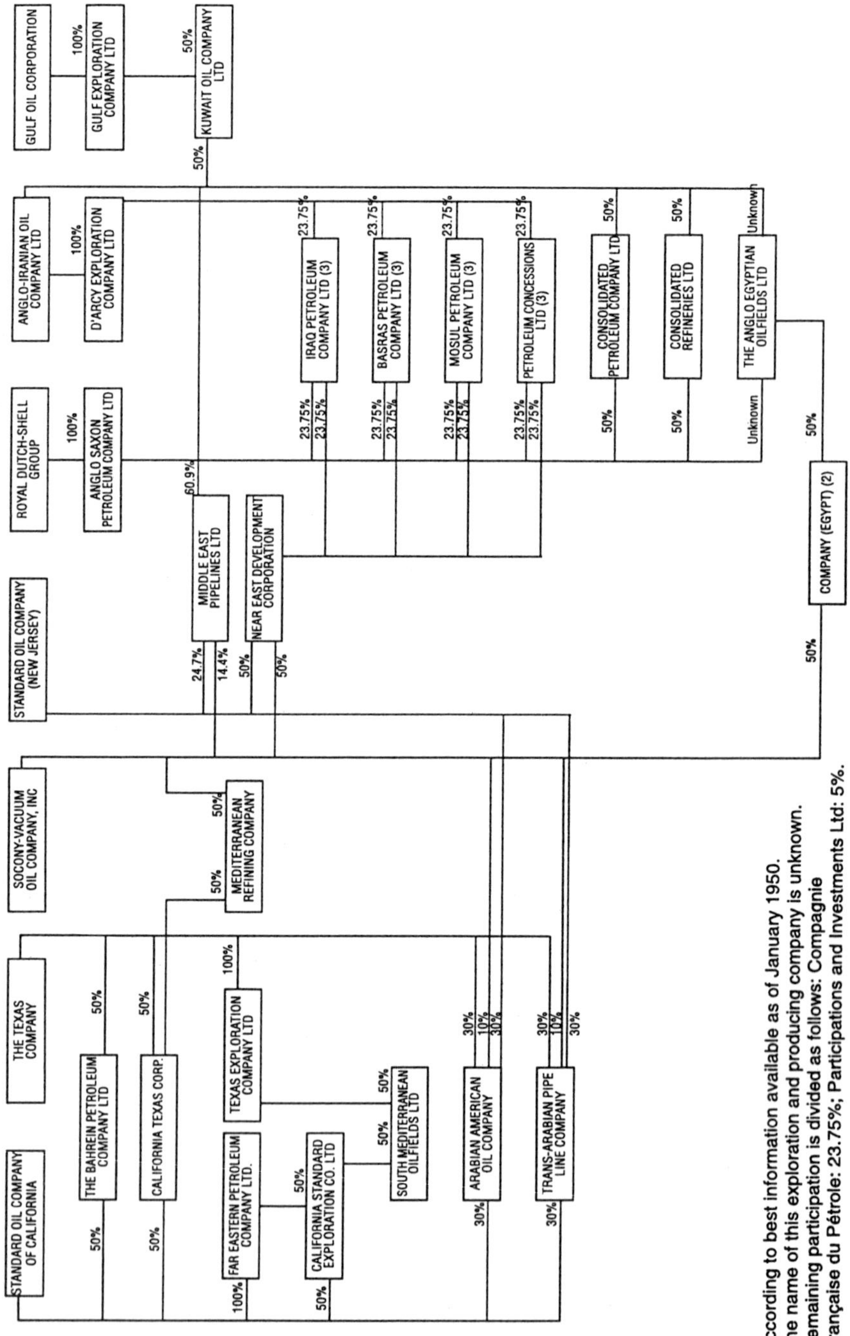

1. According to best information available as of January 1950.
2. The name of this exploration and producing company is unknown.
3. Remaining participation is divided as follows: Compagnie Française du Pétrole: 23.75%; Participations and Investments Ltd: 5%.

the flow of Middle East supply multiplied 70 times to over 1 billion tons.

How can this be explained, after eighty years of chronic price instability? The answer lies in the *structure* that the international oil industry had developed, with a situation of near monopoly created by the majors, and in the *organisational structure* they had formed internally and in their relationships (Fig. 3.1.). At the end of the Forties they controlled (for the most part jointly) 82% of all proven oil reserves outside the United States and the Soviet Bloc, 86% of current production, and 77% of refining capacity (Tab. 3.1.).

This domination of the market does not in itself explain what happened in those years, unless attention is paid to other structural aspects. In particular:

- the *strongly integrated structure* of the majors: both vertical and horizontal;
- *close co-ordination* of their operative and strategic choices especially upstream;
- *restriction of competition and of the market* with a price fixing system.
- *administrative predetermination of quotas*, both upstream (crude oil) and downstream.

On these foundations an organisation was built which no other industry could match, and that, regardless of any ideological prejudice, has in actual fact allowed the western economy to utilise an essential input in times and ways that may otherwise have been unthinkable.

Table 3.1. *The power of the majors: 1949*[1]

	Reserves		Production		Refining Capacity	
	bill.bbl	%total	bill.bbl/d	%total	bill.bbl/d	%total
United States	9.4	33.6	1.6	31.5	3.0	44.4
Rest of the world[2]	41.3	82.1	3.0	85.7	3.1	77.3
World[2]	50.7	65.0	4.6	54.6	6.1	56.7

1. The absolute data refer to the seven majors; the percentages represent the ratio on the total of each area. 2. Excluding countries with state-controlled economy.

Source: FTC (1952).

3.6.1. Vertical integration

For several reasons the oil industry has historically been one of the main reference points for the theory of vertical integration. Firstly, it was one of the first industries to follow this kind of organisation (in 1910 the "first movers" of Texas and California were completely integrated). Secondly, there are a wide range of sectors involved (from those strictly petroleum-based to chemicals, transport, engineering etc.), and companies are very large, and typically multinational. And thirdly, because the long life cycle of the industry allows us to better calculate the efficiency of integration with respect to the various phases of its evolution.

Economists have always been divided between those who stressed the *improvements in efficiency* that integration encouraged, with reduction of costs compared to those elsewhere, and those who complained about the *restriction of competition* it led to, with the worsening of the conditions for the optimal allocation of resources (prices equal to marginal costs). It is certainly not possible to isolate *reasons* for this choice of organisation that are valid for all times and in all situations, nor the *outcomes* necessarily deriving from it, in terms of efficiency, competitiveness and performance. There are many variables involved, for example: the operative and institutional context within which the companies worked, government policies at home and abroad, the strategies of competitors, and the life cycle of the industry. Examination of the evolution of integration in the oil industry shows the models of Stigler (1951) and more recently of Harrigan (1983) are still valid. High risk and the high level of variables in the take-off and development phase of the industry – when the restriction of the markets did not allow investments to be utilised to the full – made integration advantageous, fostering quasi-monopolistic arrangements. In later phases, product specialisation, in a less uncertain context, may make it advantageous to quit sectors of higher productivity and reduce the sunk costs, substituting contractual for hierarchical relationships. From this, more competitive market situations arise. According to this theory, the degree of integration can only be placed within a historical process that leads from integration to specialisation as more efficient alternatives are developed – or as we shall see in the case of oil, alternatives are made necessary by the obsolescence of the basic premise on which integration depended (the control of the upstream). Having said this, we shall later be showing in a schematic way the various kinds of reasons that have sustained strategies of integration over time, giving separate consideration to those aiming to improve efficiency and to those aiming to restrict competition, although in reality the two often coincided.

3.6.2. Improvements in production efficiency

Two aspects are of particular importance here: uncertainty and operative interdependence.

Uncertainty, risk levels, investments.

The combination of these basic, inescapable conditions of oil economics (section 1.3) is such that only those companies managing to produce high revenue flows, reasonably stable over time, are able to provide flows of investment sufficient to ensure adequate replacements for exhausted oil fields, or the expansion of capacity according to the requirements of demand. Only in conditions of relative certainty over the sales of oil and the opportunities for profits can immense amounts of capital be risked. Especially in their first and second phases, oil companies thus tended to enter into and operate all the various stages of the oil cycle, so that: (a) risk levels at the source of production remained high and relevant, but not decisive for the whole business; (b) positive results from research and development activities would not be delayed by problems with selling the product either on the primary or final market. The oil-producing companies thus set about building the refineries and distribution networks necessary to take their business straight to the final consumers. In reality, in the take-off and development phase of the industry, the alternatives between hierarchy and market, between internal co-ordination costs and costs of market co-ordination, were not on the agenda, since the market had yet to develop.

Frankel stressed that "right from its earliest phase, the oil industry was dominated by those who were able to control the phases of production and marketing on a sufficiently broad scale to be able to make use of superior transport systems, thus improving their standing at the expense of the competitors." He also remarked that in the oil industry the relations between each phase and the next were particularly close, given the size and extent of the specialisation of the plants: "the refiner hopes to have a guaranteed market outlet, and the distributor needs to count on continuity of supplies. Therefore both need a constant flow for efficient operation, and this generates a high level of reciprocal interdependence, determined by the facts." So two substantial advantages derive from integration: "in the first place, guaranteed market outlets for the refiner and guaranteed supplies for the distributor; secondly, the advantage of being able to compensate the profitability of production, refining, and marketing: a company involved in each phase of the industry enjoys higher prospects of survival than those operating only at one or two levels" (Frankel, 1946). Blair likewise affirms that "the key to reducing uncertainty and attaining some profitability lay in an uninterrupted flow of crude oil, through the refining process to the final consumer. Where control over any of the downstream phases was to be found in other hands, the flow could be interrupted, and therefore the production of crude oil would find no outlet – with readily imaginable consequences on profits"(Blair, 1977).

Of course, these economic advantages were defended vigorously, especially by the oil companies themselves. The words of W.T. Slick, Senior Vice President of Exxon, before the U.S. Senate's Subcommittee on Antitrust and Monopoly in 1975 were particularly relevant: "Vertical integration is an efficient form of industrial organisation, commonly used by many companies in many business sectors. Indeed, it makes it easier to cope with the uncertainties of the economic environment, since it guarantees reliability of supplies, quality of the product, price and service. It also allows a better co-ordination of expectations and the activities of production stages with a high degree of interdependence, thus making it possible to cut down on stocks and circulating capital. Vertical integration also improves a company's access to the capital market: the cost of the capital is lower because of the lower risk, but also because the investors may be given a complete outline of the interdependent investments".

Operative interdependence and variability.

The high degree of operative interdependence in the oil industry (section 1.6), together with the high level of variables in the external environment, with the consequent need for frequent and unforeseen adaptation, makes the integrated form more efficient than separate stages of production (Viani, 1992). Interdependence alone is not a sufficient incentive to integration, since adequate contractual forms could be enough. In a highly variable environment, however, contracts become very complex and very expensive, involving transaction and renegotiation costs higher than those of integration (Williamson, 1971).

As we have seen in Chapter One, the cost levels that are obtained by considering the oil business as a whole are undoubtedly lower than those that would be obtained from adding the best results from individual phases of production. Vertical integration not only enables better planning of the amount, location and typology of investments according to long-term needs, but also ensures in the short term a better *joint exploitation* than would be possible with inevitably incomplete contracts, especially in cases where there is a high specificity of investments. To sum up, economies of co-ordination allowed higher levels of productive efficiency, greater competitiveness, better allocation of resources, and a more harmonious development of the industry (Cross, 1953; McLean and Haigh, 1954; Melvin et al., 1959).

3.6.3. The restriction of competition

According to Joe Bain, "integration which goes beyond mere technological reasons certainly has monopolist origins". This is undoubtedly the case with the oil industry. Independently of any actual intentions, the combined result of the high degree of vertical and horizontal integration was to: (a) raise *entry barriers* which could only be overcome with the greatest difficulty, especially concerning access to oil resources, resulting in a few large multinationals wielding a high degree of market power; (b) *greatly limit the amount of free commercial transactions* in primary and secondary markets, by actually substituting market exchanges (and prices) with internal transfers to the companies; (c) *reduce, overall, room for competition* within and between each phase of production.

The competitive advantages of integration for companies structured in this way meant that "once established, the concentrated and integrated structure of the market tended to perpetuate itself" (Adelman, 1983, p. 123). The relatively small size of intermediate markets failed to give guarantees, either in outlets or purchases, to companies who wanted to operate in only one of the production phases. Integration thus became inescapable, since it was itself a form of self-defence from the action of competitors already integrated (De Chazeau and Kahn, 1959). The effect was obviously to strengthen entry barriers, since entrance had to take place in several productive phases at once. "The vertically integrated firm" wrote Machlup and Taber (1960, p.116), "which has a monopoly in one stage of an industry may extend its monopoly power into other stages or subsidiary industries where competition might otherwise survive. A vertically integrated firm may be more able to discriminate between different purchasers and thus increase its profits, or strengthen its monopoly power, or both. Finally, it may, in periods of shortage, use its control of scarce material to eliminate or prevent the growth of independent producers".

After acknowledging the perfectly rational basis of integration (when large investments, factors of instability and high risk are considered), Frankel states that "integration, in itself, is a neutral force that may fulfil a variety of functions, both offensive and defensive, both monopolistic and competitive". But he then adds that "when the integration is combined with a substantial market power in one or more operating stages, it is destined to have a significant influence on price policies. In fact it restricts commercial channels, making the markets even more imperfect and reducing the practicable alternatives for any non-integrated companies; it enables integrated companies, with substantial market power at a certain stage, to extend their dominion upstream or downstream through the arbitrary fixing of prices not connected to the costs sustained in the other stages; it transfers price competition from the stages of production and refining to the product marketing stages, where the competition is more easily controlled; and it discourages new entrants, increasing the risk for non-integrated operators and forcing them to integrate themselves in a defensive way, which multiplies the minimum investment necessary" (Frankel, 1966).[5]

Figure 3.2. *Relations between primary, secondary and tertiary integration*

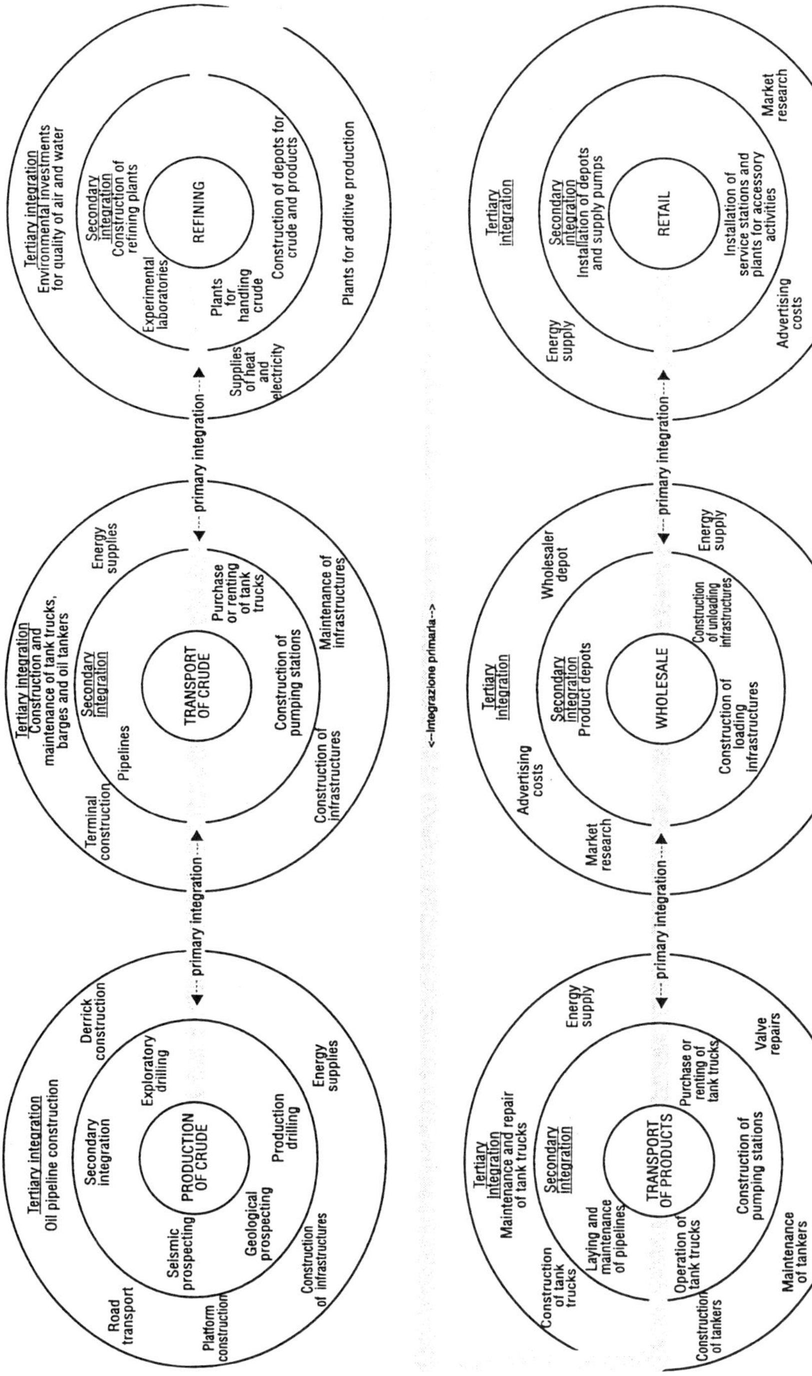

However, vertical integration would not have had such an effect if it had not been so closely interconnected to the strategy of *horizontal integration* pursued by the majors at the same time: through their internal growth, the acquisition of smaller companies, and the setting up of consortia. According to Edith Penrose, the most critical problem the majors had to face after the Second World War "was how to ensure that newly discovered reserves would be in 'strong hands', i.e. in the hands of companies who had market outlets and would not have to go after other companies' markets and cut prices to dispose of their oil. ....The obvious solution was to join in a somewhat polygamous wedlock – or at least in respectable long-term liaisons – ....through the joint ownership of producing companies or through long-term supply contracts" (Penrose, 1964, p. 159).

Horizontal integration made the controlling the supply of crude oil possible in several ways: the *joint and co-ordinated management of the crude oil supply* in areas subject to the power of the consortia; and the continuous *exchange of information* on the decisions to be taken in those phases not subject to joint control.

Both these factors helped to overcome the indeterminacy and instability that derived from oligopolistic interdependence by determining the *individual* demand and supply curves and therefore also the *aggregate* curves. In Figure 3.2., taken from De Chazeau and Kahn, we see the pattern of vertical integration that the majors were developing for and within each phase. If this pattern is combined with the one in Figure 3.1., we can have an exact idea of the overall structural form the oil industry was developing over those years.

3.6.4. Operative co-ordination

The central aspect of this configuration was the control of the oil supply. In an oligopolistic regime, the supply may in principle be controlled (and price competition eliminated) through an explicit collusion between the producers, or as a result of their conjectural interdependence. The limitations of the former solution have been widely commented upon, in analysing the unhappy experience of the 1928 cartel. As regards conjectural interdependence, it certainly had the basic premise of an oligopolistic structure, but a self-imposed limitation based only on awareness of reciprocal interests showed itself to be an even weaker solution, in operative terms, than the other one. The organisational structure carefully developed by the majors was far more effective, and based on two solid assumptions: (a) the rigid predetermination of the supply levels in line with expected demand levels; (b) concentration of decision-making in the hands of a small number of people.

3.6.5. Planning of supply growth

"If we had to exploit any oilfield at its capacity level, we'd be forced to cut down the use of the others. There would be no room for everyone, nor could we discharge the oil into the sea, there's a law against it". This was the ironic reasoning of William Page, Vice President of Standard Oil, before the Federal Trade Commission in 1952, in expressing the firm conviction of all the majors: that oil in excess of the amount the market could absorb at prevailing prices *should not* in any way reach that market. The system of control of supply developed by the majors aimed to govern the production rates of the various countries involved, so that a predetermined rate of growth of the total supply was not exceeded. The results obtained were truly surprising. Middle East production alone increased by around 40 times in the 1945-1970 period, without any noticeable price changes. An essential condition for such stability was the strict adherence of the increasing supply to expected levels of demand. This was all the more surprising if the continuous entry into the market of new producers is taken into account, as well as the great variation in production growth rates in individual countries.

Joint administration of supply was achieved, first of all, by predetermining the aggregate growth rate annually on the basis of demand estimates that flowed back from the integrated circuits of the majors, and later, sharing it out between the various countries controlled according to a variety of considerations. Basically, a mechanism of reallocation of production was applied, similar in its logic (though not in its outcomes) to the one proposed by economic theory in the case of a monopolist who produces homogeneous goods from several different plants. In this case there was one important difference: the criterion of reallocation was not only due to the relative marginal costs. Calculations of a political nature also influenced the consortia's decisions – especially the preference given to countries such as Iran and Saudi Arabia who were believed to be politically closer to the western powers and less subject to communist influence. With the growth rates of supply in these countries higher than those of international demand, itself high (on annual average, around 10%), an excess of supply could only be avoided by imposing relatively lower increases on other countries. This is what happened, for example, in Kuwait and Iraq, where production did not grow more than 5% a year on average.

3.6.6. Interlocking directorates

The majors' capacity to balance demand and supply (in both time and space) would not have been expressed so well unless it had been able to rely on another formidable operative instrument: the interlocking responsibilities and offices that they led and which linked together the activities of the various consortia. Each major was organised into a pyramid of subsidiaries and affiliates part-owned by at least one other major. These interlocking shareholdings were most important for the upstream phases, although there were also cases that stretched as far as the finished end-products (as in the case of Caltex: a joint-venture between Standard Oil of California and Texaco). This dense network of company relationships meant a strong concentration of decision-making powers in the hands of a small number of people, which facilitated the adoption of a common policy among the first-rate holdings, the fullest information on what the competitors intended to do, and the control of the adherence to the rules agreed to by the individual members of the 'club'.

This concentration of power and division of responsibilities, in day-to-day terms, eventually became the most important means of co-ordination between companies. As Figure 3.3. shows, in 1950 the decision-making powers of the entire Middle Eastern oil industry's activities were concentrated in the hands of just 28 people. Many of these held other directorates in the subsidiary companies in the downstream or in the upstream sector of other countries.

In examining interlocking shareholding over various areas, it is interesting to see that it also tended to involve companies outside the club of majors, according to the

Figure 3.3. *Interlocking directorates between international oil companies in the Middle East*

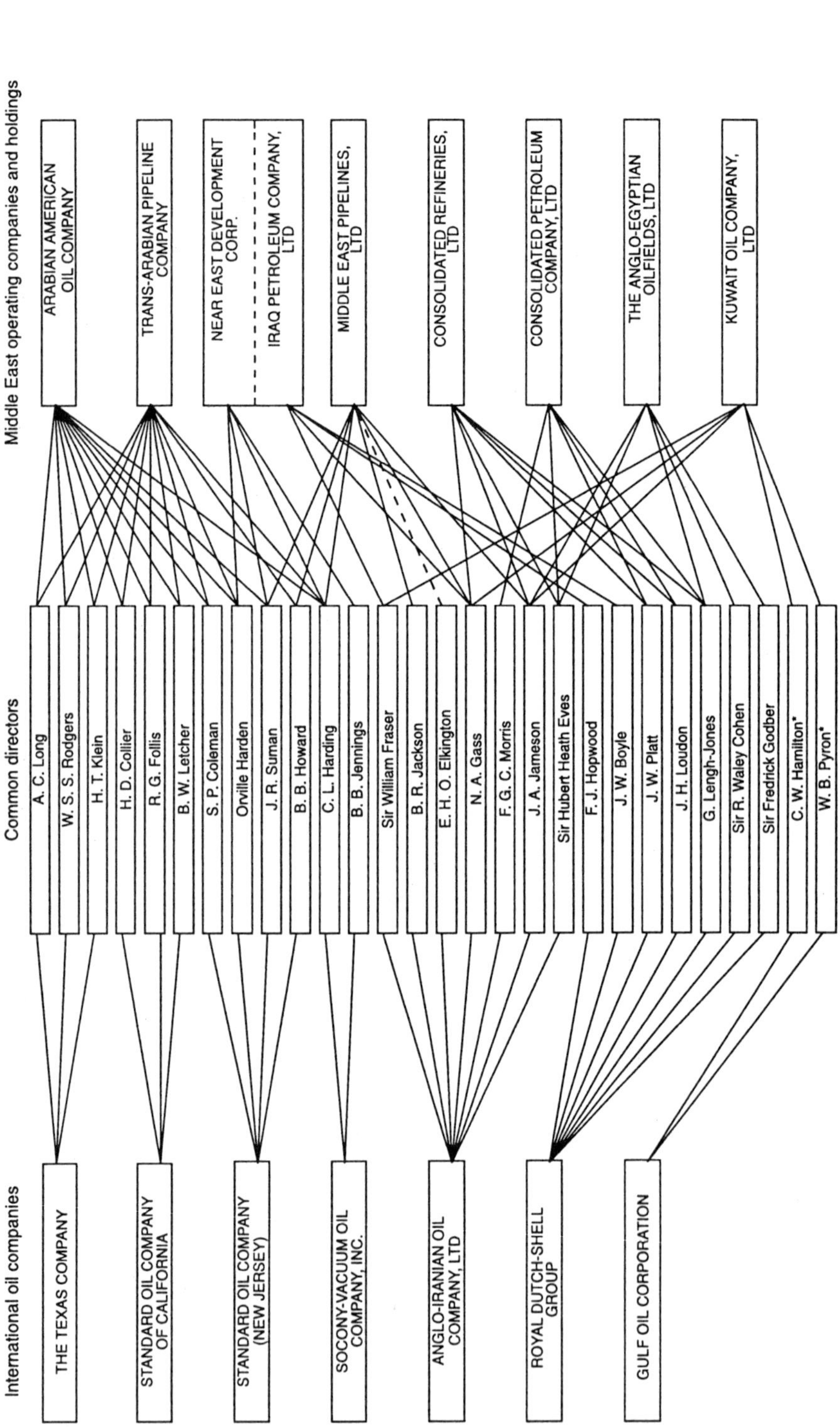

strategic role that each of them held in each particular market or in each particular phase of the cycle. Thus for instance in the United States the co-operation concerned companies operating in the strategic sector of oil pipelines; in the Middle East, the production of crude oil; in Europe, refining and distribution.

3.7. PUBLIC POLICY IN THE UNITED STATES

The structure of both the American and the international oil industry, the growth abroad of the American majors, the level of competition in the markets, and the movement of prices were all very much conditioned by the policies followed by the U.S. Administration. Through a stratification of federal and state regulation, aimed at different and often conflicting objectives, company decision-making autonomy was severely limited, the degree of competition falsified, and the allocation of resources compromised instead of expressing itself at optimum levels. It is by no means easy – and certainly beyond the scope of this book – to reconstruct the jumble of laws and rules through which intervention took place. Judging from the results obtained and the opinions of many scholars, it is more often cited as an example of the "failure of regulation" than as an efficacious model of reference.

In order to understand this rather harsh judgement, we must consider the ultimate aim aspired to by American policy: to guarantee the country a high degree of autonomy in its oil supplies. This has given rise to a dual and often inconsistent line of action: foreign policy and domestic economic policy. The latter can be split into four parts: *regulation* of production, *import quota restrictions*, *fiscal policy*, and the *control of domestic prices*.

3.7.1. The prorationing of domestic production

Right from the very beginnings of the oil industry, alarmed voices were heard in the United States concerning the predatory and savage nature of extraction activity, and thus a certain degree of control over domestic production had been recommended, to prevent the physical waste of resources. Public opinion of a Malthusian-environmentalist nature was already roused, and its influence was felt in government circles (Pinchot, 1910). Up to the end of the Nineteen-Twenties, the majors' attitude towards this "philosophy of conservation" of the resources had gone from indifference to open hostility; but these feelings soon gave way to the realisation that it could help them to stabilise the market.

After the Second World War, faced with an enormous increase in demand and hence in the exploitation of oilfields, there was increasing concern that the country was exhausting its oil resources. Without the intervention of some planned control, irreversible damage would be done to the wealth of the American nation. Although inspired by noble intentions, the philosophy of conservation eventually coincided exactly with the policy of oil proration and stabilisation of markets (and of prices) that the oil companies had previously pursued with mixed results. What was impossible to obtain on their own was gained instead through the law of the land. The American administration eventually fulfilled the function of co-ordination and control of supply elsewhere carried out by the majors.

The policy of production proration was enforced through a dense and extremely complicated mass of state and federal regulations, through recommendations formulated by committees from the integrated companies, and by the Bureau of Mines and the Department of the Interior. "In itself, none of these elements could be defined as a decisive instrument of control of the industry; as a whole, however, they formed a perfect model of monopolistic control of the production of petroleum, of its distribution and finally of the price paid by the public"(FTC, 1952). The control of production was structured in such a way as to avoid creating even a minimum oil surplus, thus avoiding pressures towards price lowering. This control was achieved through quota restrictions for the days in which production was allowed, altering them from year to year according to requirements. In Texas, one of the oil-richest areas, production was allowed in 1952 for 261 days, going down to 194 during the slight recession of 1954, to 122 in 1958 after the reopening of the Suez Canal, and to a minimum of 97 days in 1962. Control of production was also differentiated among the various oilfields, hitting mainly those with the lowest costs and the highest productivity, while those which were marginal (with a daily flow of less than ten barrels) were entirely unaffected.

The consequences of the control of production were, in essence, an increase in both prices and costs: in *prices*, as a result of the reduction of supply to levels strictly in line with those of current demand; in *costs*, because the reduction hit the relatively cheaper productions, creating a rise in the level of average costs (Kahn, 1964). However, no less important were the responses of the leading companies, especially as regards investment and the strategy of growth. Since it had become impossible to develop their reserves adequately in the United States, and attracted by the greater profitability of operations abroad, they drastically reduced their financial commitments at home, shifting the centre of their interests to the more promising areas of the Middle East. The foundations were thus laid for a progressive reduction of proven oil reserves in the United States, and therefore of their chance of keeping up with the inertial growth of domestic consumption. This result was exactly the opposite to the one the Administration had set out to achieve.

3.7.2. Import quota restrictions

Since 1932, the Federal Oil Conservation Board had recognised that import quota restrictions would be an essential part of an efficient mechanism for the control of supply. The discovery of new oilfields in Texas allowed the United States to continue to dominate world production for a while (exporting more they imported), and the problem was set aside for a couple of decades. At the beginning of the Fifties, however, the enormous resources available cheaply abroad, together with the restrictions on domestic production, were a powerful incentive for many companies to search for their crude oil abroad. They would then import it into the American market, where prices were held artificially high precisely because of the restrictions. Imports thus began to rise sharply. So-called voluntary controls were introduced as a result, with importers "voluntarily" agreeing to accept the quotas established by the Oil Import Administration, although the results were disappointing.[6] In 1959 the Eisenhower administration set up a mandatory program of import quota restrictions with the aim of "protecting national security", with the slogan "*drain American first*!". Its overall result, however, was to raise prices for the American consumer, without halting the rapid depletion of domestic reserves, because companies continued to favour more profitable production abroad.[7] The import quota had three effects:

a) it helped to intensify the exploitation of American reserves, with a consequent reduction of the reserves/production ratio;
b) it forced consumers to pay more;[8]
c) it encouraged the flow of cheap Middle Eastern oil into Europe and Japan.

3.7.3. Fiscal policy

A third aspect of US government policy was the favourable taxation of companies' activities. This policy was openly aimed at: (a) *strengthening national security* by encouraging upstream investments; (b) *reducing upstream risk* through compensatory taxation; (c) guaranteeing *conditions of neutrality* in the taxation of industries with high capital intensity like the oil industry; and (d) avoiding *double taxation* for companies operating abroad. Behind these policies lay the conviction that market mechanisms were inadequate for meeting objectives of this kind, although not so much attention was paid to the costs involved in terms of the inefficient allocation of resources and lower tax revenues. This policy took three basic forms: depletion allowance, deduction of intangible costs, and foreign tax credit.

Depletion allowance meant companies being allowed (from 1925) a fixed percentage of gross revenues from oil production (27.5% until 1965, then 22%) tax-free. This was recognised for the entire productive lifetime of the reservoir, and was hence in general higher than the technical depreciation of the assets, so it was in reality a tax concession. It was introduced to compensate producing companies for the risks undertaken and to support self-financing in the replacement of resources already exploited. "Mining research" stated the Revenue Act of 1918, "requires many years and the use of large sums of money in unsuccessful searches. When a productive area is identified and production begins, it is senseless and unjust to tax the profit at the maximum rate, as if it were income which could be attributed to normal operation in a single year".

What effects did the *depletion allowance* produce? It was essentially a negative *ad valorem* tax. An *ad valorem* tax is a tax not paid in a fixed amount per physical unit, but in a fixed percentage of the relative price. The higher the price, the greater the monetary amount paid. The *depletion allowance*, on the other hand, is the exact opposite: a subsidy proportional to the value of the oil produced.[9] Companies are therefore encouraged to produce, at every price level, more than would otherwise be good for them; with a slide to the right of the aggregate supply curve. This result, the opposite of the one obtained by the production quota restrictions, was aggravated by the companies being able to deduct from their gross revenues all the *intangible costs* involved in the drilling and exploitation of reservoirs, elsewhere written off as for tangible costs.[10]

The importance of the above measures was, however, obscured by *foreign tax credit*, already included in the 1918 Revenue Act tax regulations but not fully utilised by the companies until the beginning of the Fifties. This is basically the right conceded to those companies operating on an international level to consider the taxes they paid in a foreign country as a *tax credit* for those owed in their country of origin. Basically, the country in which the oil was produced collected the taxes first, while the company's country of origin collected taxes only where its tax rates exceed those of the producing country.

The potential offered by this arrangement was only fully exploited from the end of the Forties, when Saudi Arabia expressed dissatisfaction for the royalties it collected from Aramco (12% of the posted price). However, Aramco's partners were unwilling to concede an increase. The solution allowed Saudi Arabia greater profit sharing (up to 50%), without at the same time reducing the profit margins of Aramco. The essence of this simple but ingenious solution was that the difference between the 12% and 50% was provided by an income tax paid to Saudi Arabia. Thanks to the foreign tax credit, these sums were creditable against the taxes owed to the US Treasury. Instead of deducting these sums from revenue as operating expenses, as happened with royalties, Aramco was allowed to consider them as a tax credit that compensated for every tax otherwise owed to the United States Treasury. On December 30th 1950 Aramco "submitted" to this solution, and before long the principle was extended to all the other consortia operating in the Middle East. In this way, the companies satisfied the producing countries without altering their own position.[11]

3.8. PRICING POLICIES

As we mentioned at the beginning of this book, one of our aims was to gain an idea of the functional connections over time between the industry's structure and its prices. By comparing, for each of the industry's historical phases, specific structural aspects and the particular criteria of fixing prices with which they associated themselves, we can achieve this result. This methodology can help us to understand what actually happened, rather than adopting theoretical schemes which refer to optimisation models founded on the theory of exhaustible resources, or models of the Cournot-Nash type, or again on those founded on game theory (for a survey see Gately, 1984 and 1986b).

3.8.1. Basing point system

Until the outbreak of the Second World War, the United States, and especially Texas, was the world's greatest oil producer and exporter. It was clear in this context that the prices in the Gulf of Mexico would become the *marker price* for all other crude oil. The so-called *single basing point system* was thus adopted by most companies, first within the cartel agreement of 1928 and later in commercial practice. On the basis of this technique, "a certain place of production is accepted as the commonly agreed base point, and *all* prices are determined according to the official factory price of that place plus the cost of transport to destination" (Scherer, 1985, p. 369). Basically, all the sellers, obviously with a non-competitive background, behaved as if the product came from that single locality: in this particular instance, from the Gulf of Mexico, from which comes the system denomination, *Gulf Plus*. If the oil actually came from a different centre of production (for example the Middle East) the price at the port of delivery was equal to that of crude oil of the same quality coming from a port of the Gulf of Mexico. On the whole, the delivery prices were the same in every port wherever the oil came from, and entirely independent of the actual costs of production. For the consumer it thus made no difference which supplier one bought from, whereas the suppliers would have suffered losses (or gained profits) if the port of delivery had actually been further (or nearer) to the one calculated with respect to the Gulf of Mexico.

Table 3.2. *Basing point system of oil price in 1949 (East-USA parity) (doll./bbl)*

	United States West Texas (36° API)	Venezuela Oficina (36° API)	Saudi Arabia Ras Tanura (34° API)
FOB price	2.75	2.65	1.75
Freight	0.25	0.24	1.10
Quality differential	–	–	0.04
Customs duties	–	0.11	0.11
Basing point price	3.00	3.00	3.00

During the Second World War, oil became of crucial importance. All the belligerent governments were driven to intervene directly in the oil market. The protests of the British and American fleets against the high prices charged by the Anglo Iranian Oil Co. and Aramco led to the first significant modification of the *Gulf Plus* system: the fixing of a *second base point in the Persian Gulf.* The base price actually remained that of the Gulf of Mexico, while the only thing that changed was the transport cost charge, in which the actual location of the markets was taken into account. All countries east of Italy found it cheaper to be supplied by Middle Eastern countries, whereas those to the west found it more advantageous to go to the U.S. coast. In 1944 Italy was the *indifference line*, where the c.i.f. prices of the Mexican and Persian Gulfs were the same. With the increase in later years of Middle Eastern exports, and indeed to foster their outlet onto markets without price competition, this indifference line was gradually moved further west. In 1949 it was located on the East Coast of the United States, (more specifically, New York), because U.S. domestic production was becoming increasingly unable to meet the rise in consumption. The simple relation that fixed prices in every corner of the globe was then the following:

$$P+T=X+TR$$

where

P = f.o.b. selling price in Texas

T = transport cost from Texas to the East Coast (New York)

TR = transport cost from the East Coast to the point of delivery

X = f.o.b. sales price of the crude oil considered

The export price of Saudi Arabian crude oil, for example, was thus equal to the price of Texan crude oil in New York, less the cost of transport from there to the market of destination.

The price of Middle Eastern oil began to be officially posted only from 1950 onwards. From that year, the informal practice followed with the basing point system was made explicit through the fixing of posted prices, though remaining completely in line as regards the criteria adopted for their calculation. The official posting of prices was begun by Socony-Mobil for the sale of crude oil f.o.b. Ras Tanura; and before long all the other majors made public announcements of their posted prices. Mobil's decision was closely connected to the revision of the participation agreements with Saudi Arabia over Aramco's profits which, as we shall be seeing later, required prompt publication

Figure 3.4. *Crude oil prices in the United States and in the Middle East*

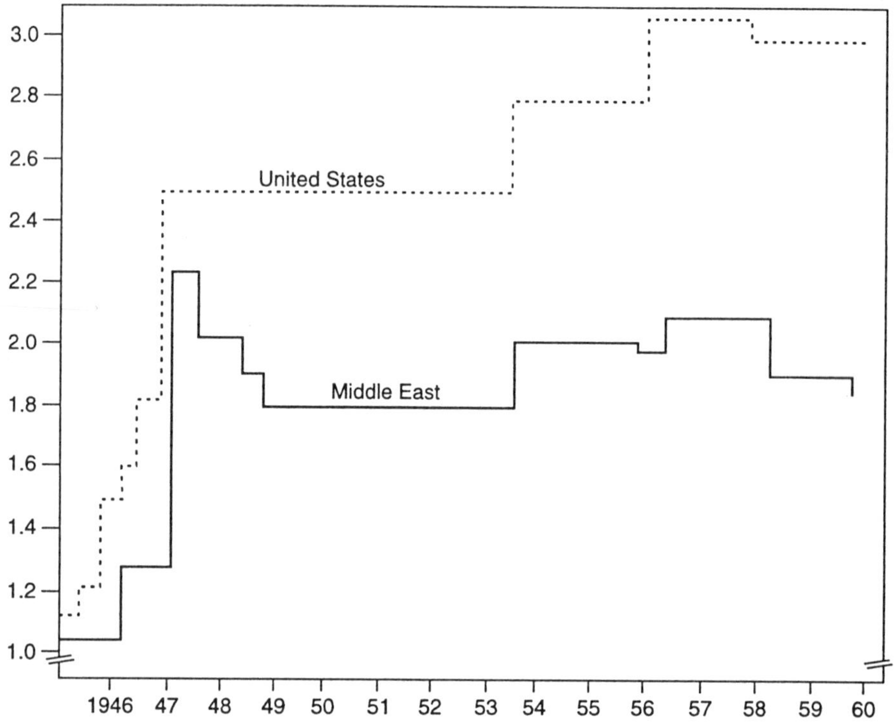

of the crude oil selling prices.

As Figure 3.4. shows, from 1946 to 1959 Middle Eastern oil prices, compared to those of the Eastern Hemisphere, at first moved along *separate curves*. Their differential widened from 0.30 doll./bbl in 1947 to 0.70 in 1950, when the indifference line shifted to New York. From then they ran *parallel*, until 1959 when the basing point system was eliminated with the closure of the American market.

The stability of oil prices in the years of the basing point system can certainly not be traced to the excellence of that method, but rather to the structural conditions of the industry and of the market that the method had generated. However, there is no doubt that the method did deserve a little of the credit. It enabled competition to be held in restraint, reducing the formation of the price to the application of a very simple formula, that had no relation to the production costs of the different crude oils. It was then able to solve the difficulties arising with the *qualitative differentiation* of the crude oils, and above all their *spatial differentiation*: because of the steady increase in the geographical points of production and exportation. Scherer's reflections, on the general effects of this technique of fixing prices, are worth remembering: "If each producer independently and unsystematically quoted prices to the thousands of destinations it might serve, it would almost certainly undercut rivals on some orders, sparking off retaliatory price cuts. But a common adherence to basing point formulas in effect eliminates discretion and uncertainty and, if each firm plays the game and sticks to the formulas, price competition is avoided" (Sherer, 1985, p. 372). So, it is clear that it is the latter which is the real pre-condition ensuring the success of the formula, since the aim of establishing stability could not have been reached if the producers had not chosen to submit to the rules of the game they had agreed to.

The great problem that the oil companies next had to face was how to reconcile the aim of stabilising prices with increasing flows of supply at costs significantly lower than those of the United States (in an average ratio of 1:10 and 1:20 in the most unfavourable cases). To this was added the fact that the rationing of American production kept domestic prices at artificially high levels. Without an adequate control of the system of fixing international prices, which would keep them artificially up to the levels of American domestic prices, American production would have been completely displaced by that of low cost areas. "Rather than accept the prospect of tariff barriers to protect the American market, it was considered preferable to establish an acceptable *modus vivendi* with the competitors on the foreign markets: the result was a policy which based the world oil prices on the Texas quotations. The price level in the United States was maintained with the assistance – or collusion, if you like – of the producers with lower costs, who at the same time acted in line with their interests. The result was a sort of invisible fence around the American market, strengthened by the policy of the major producers abroad. If there had been no international agreement, oil prices would almost certainly have been lower." (Blair, 1977).

3.8.2. The significance of posted prices

Crude oil posted prices have played an important role in the oil industry, in a different way from the commonly understood meaning of price as "the value of a good expressed in monetary terms". We should now, in the light of the specific structural characteristics of the oil industry, define the real meaning that ought to be given to them, and clarify their nature. Edith Penrose's analysis is especially penetrating here. She dismantles the case of those who have argued that the high profits of Middle East oil were revenues inevitably earned by the producers at a low cost; or else as profits from a monopoly deliberately set up by the majors, who kept the prices of crude oil high to protect their investments in the western hemisphere from an over-rapid expansion of Middle East oil. She is also critical of those who have interpreted the relative lowering of Middle East prices as proof of real competition between high-cost and low-cost sources of supply.

If there is some truth in either of these arguments, it is also true that "they give too much importance to the price of crude as a source of profit in itself for the companies. In actual fact, the market price of crude oil (or the posted price) had relatively little importance for the profitability of the majors, simply because they sold very little oil on the free market". So, although the posted price of crude oil was defined as "the official quotation of the price at which an oil producer is willing to sell", in reality most of the crude oil produced was used directly by the companies themselves. In actual fact, the sales were transfers from one affiliate to another of the same group, or else to other companies integrated on the basis of long-term contracts, at special prices. The most important factor, when a company was deciding whether to produce oil in Venezuela, the Middle

East or the United States, was the *cost of production* and transport. It made no sense for a company to decide on the basis of relative prices, when it neither obtained the crude oil at those prices nor sold it at those prices. For the same reason, the profit made by companies directly on the sale of crude oil to third parties was very small. It is true that the companies showed very high profits for their activities in the Middle East, but this happened because they had expressly chosen to evaluate the oil they utilised *as if* they had actually sold it to their refineries. How did the majors justify this behaviour, apart from the obvious fact that posting prices was also a way to ensure that all the oil companies followed a common price policy in their operations? Why did they attribute the largest share of their profits to the production of crude oil?

In general, in an integrated company the total cost of any amount of the finished product is equal to the sum of the costs sustained at each stage of production, and profit is formed only when the total revenues from sales goes over this cost. If, at every stage of the activity, the product is transferred to the succeeding stage at the relative cost, no extra-profit appears until the final sales. A profit may appear only if the product is transferred to the successive stages at an accounted price higher than its production cost.

From a strictly economic perspective, there are circumstances that can lead an integrated company to declare a profit (or a loss) corresponding to one specific stage of its operations. If the company operates entirely within just one country, and under the jurisdiction of just one tax authority, it makes relatively little difference *where* it decides to declare its profits; but it certainly does make a difference if the company operates on an international level (not just for the company, but also for the countries where it operates). This is the case of the oil industry: the companies' Middle Eastern subsidiaries sold virtually all the crude oil they produced to their shareholders, the majors, pricing the oil at the posted price: a cost considerably higher than the production cost. This allowed the companies themselves to declare a *profit* at the stage of crude oil production. In short: whether the integrated companies utilised the oil they produced directly, or whether they sold it outside on the basis of long-term supply agreements, it was assumed that all the transfers would take place at the posted price and all the profits were attributed to the crude oil production stage. There are several reasons for this choice of action:

a) to sell the crude oil to outsiders at the highest possible price, thus reducing their competitive potential on the product market. Given the relatively slight importance of the outside sales of crude oil, this factor contributed little to the increase of profits, whereas it did contribute a great deal to reducing competition;
b) for American companies, a second reason is to be found in the depletion allowance that enabled them to detract from their US taxation 27.5% of gross revenue earned at the extraction stage; this made income maximisation at that stage very advantageous;
c) a third reason came from the profit participation requested in an increasingly urgent manner by producing countries. Unless the companies wanted to give them a share of the integrated profits, it was necessary to find a way to attribute a profit to the crude oil production stage alone. The system was found in evaluating all the oil transfers at the posted price.[12]

By combining the rule of 50%-50% profit sharing, the depletion allowance, and the high posted prices of the crude oil, the tax liabilities of American companies towards their country of origin were virtually eliminated, and the majors placed in a much better position than they would have been if their profits had been declared in a later stage of their operations. In fact, some companies managed in this way to avoid paying any taxes at all to the US Treasury; and tax payments to producing countries involved no additional cost for them. On the other hand, the system used to allocate producing countries a share of profits was such that it gave those countries' governments a direct interest in the variation of crude oil prices. From their point of view, the system described was acceptable as long as the posted prices did not go down. Precisely because of this, the system introduced a powerful political incentive to keep the price level high.

To sum up, the high profitability attributed to oil production was the result of the price policy adopted by the majors, to the advantage both of the companies and the producing countries. This results of this policy were to: create an advantageous system to share the profits of the industry with the producing countries without costs for the companies; create a situation where an increase of the posted prices would mean that more taxes would be paid to the governments of the producing countries and less to the country of origin; and also to improve political relationships with the producer countries – at least initially – with the concession of sizeable profits. In addition this policy allowed a high price of crude oil to be maintained for independent refiners, reducing their profitability and hence potential competition. At the same time, high posted prices reduced the impact of the growing supply of Middle East oil compared to the production at higher cost in the western hemisphere.

The option of falling back on posted prices was therefore an arbitrary, but economically attractive choice, which allowed some of the political and economic problems of the international companies to be solved. Obviously, the profits calculated in this way had an insignificant economic meaning, since they had no relation to the real value of the crude oil utilised directly by the integrated companies, and after a few years not even to the price at which the majors sold crude oil to others. Given a certain profit from all the integrated activity, the higher the posted price, the lower the profits from refining and distribution.

The substance of this chapter can be summed up as follows: for nearly a decade after the end of the war, nothing happened that could disturb the structure of oil prices or alter the pattern of control of the industry significantly. The competing oil companies had established a *modus vivendi* that allowed them peaceful coexistence, and the relationships with the producing countries were relatively friendly, particularly considering the political differences between the regions involved. In general, the structure of the international oil industry, characterised by great concentration and vertical integration, had rendered price competition extremely improbable and had also provided convincing reasons for a policy of high oil price levels. It is for all these reasons that this period was called (with some exaggeration) the oil industry's *golden decade*.

Notes

1. Through the *Petroleum Concessions Ltd.* it controlled, IPC in 1944 had areas of around 300,000 km² outside Iraq.

2. Another example of this practice, indicated by the splendid expression *sitting on concessions*, is given by Qatar: fearing the area could fall into "outsider" hands, Anglo-Iranian in 1932

obtained an exclusive license until the area was absorbed into the *Red Line Agreement* in 1936. In actual fact, BP and Shell were not interested in developing greater production in the Persian Gulf, because of the consequences that this would have had on production in Iran; this explains why the first productive well was only completed in 1939 (Longrigg, 1968).

3. Such as the *Memorandum for European Markets* (1930), the *Heads of Agreement for Distribution* (1932), and the *Draft Memorandum of Principles* (1934).

4. Frankel wrote: "In the United States, the production of crude oil has never been able to be placed fully under control, and even on the world market, latent competition represented a constant threat for the majors. (...) Aware that their policies of price stabilisation were subject to destabilising interference from their less powerful competitors, the majors preferred to come to an agreement with them. It is obvious in fact that the market leaders will first try to crush a competitor, and then, when for some reason the other loses, they will accept him rapidly, with all the honours, as a sort of junior partner".

5. Kahn and De Chazeau have shown that a company with a "degree of integration" of 77% or more in crude oil production was able to increase its own profits by increasing the price of crude oil, even if this increase was not in any way transferred on to prices further downstream. The losses that the refining/sales branches would bear were more than compensated for, thanks to the rules of taxation, by the earnings of the branches of production/sales of crude oil.

6. Since in the voluntary program there was no limit on semi-manufactured oil and on refined products, the importers restricted themselves to substituting crude oil with the latter, and the overall imports continued to rise.

7. The mandatory program foresaw an overall ceiling to the importation of crude oil and oil products of 9% of the expected demand. The way round this by importing semi-manufactured products was eliminated with the rule that no company could import more than 10% of its quota in the form of semi-manufactured products. The "9% of demand" formula, devised in this way, allowed imports to increase more rapidly than domestic production; so from 1963 the program was altered, and the import ceiling was from that time calculated on the basis of domestic production and not on demand.

8. Estimated by Adelman, at the beginning of the Seventies, to be around four billion dollars a year.

9. The company could deduct 27.5% of the value of production up to the maximum limit of 50% of the net profit.

10. Intangible costs are those that do not have a recoverable value; labour, energy, transport, etc.

11. It was only with the *Tax Reduction Act* of 1975 that US Congress set out to reduce a little the long series of tax privileges the oil companies had enjoyed. 57 years after its introduction, the depletion allowance was abolished for the larger companies, and limits were introduced also for the smaller companies, while the depletion rate was gradually reduced to 15%. At the same time other measures were adopted to stop the intangible drilling expenses incurred abroad from compensating, for taxation purposes, for the revenue produced at home. Finally, the foreign tax credit was reduced slightly.

12. In working out the profit sharing agreements with the governments of the producer countries at the start of the Fifties, the companies did not actually have much choice. In order to give the governments a share of the profits, the hypothetical alternatives were some form of royalty payment per barrel, some form of profit sharing, or a combination of the two. The simple payment of a royalty, however, would no longer satisfy the governments, who would have always compared their revenues with the overall profits of the companies. A profit sharing system was therefore chosen, in the form of a tax on revenues paid to the producer countries by the companies, because this system allowed the American companies to benefit from substantial tax relief.

4 INERTIAL STABILITY AND EVOLUTIONARY PROCESSES

4.1. AN EXCEPTIONAL PERIOD

The early years of the Fifties saw the completion of a grand strategic design started by the majors two decades earlier. Over the next twenty years the organisational structure which had been skilfully constructed on an international level was to guarantee a *stability in nominal oil prices* that had never been seen before (and was never to be seen again). It is certainly true that this price stability was considerably influenced by its state administered nature, and did not derive from the interaction of the free play of market forces. However this was exactly what the large firms wanted: to manage prices through control of the real flows of the market, so that they could guarantee themselves the stability considered necessary for the entire system to develop in harmony. The framework of certainties within which the firms were operating allowed them to fulfil the primary function of *joint planning of the investments* of the entire chain of industry, a state which they considered not easily reachable in a competitive context.

This price stability, after eighty years of great instability, was even more surprising if we take into account the main phenomena observed between the Fifties and the Seventies: first of all, this period saw an *approximately fivefold increase in oil demand*, subsequent to the doubling of the individual income; to the substitution of other energy sources by oil; to the rise in importance of consumer durable goods, starting with the motor car, whose annual production rose from 8 to 22 million units; and to the similarly rapid development of the petrochemical industry. As the oil market expanded further, other effects were felt. In particular: (a) the demand curve shifted towards the right, leading to pressure to raise prices, with a given supply curve; and (b) an incentive towards the entry of new firms, especially in consideration of the high profitability registered by those already in the industry.

As far as supply is concerned, the competitive pressure was reinforced by the second of the phenomena characteristic of that period. This was the *shifting of the focus of world oil production* to the low-cost areas of the Middle East and North Africa. In 1950 the American oil fields and those of the Caribbean still made up 70% of the entire world oil supply (about 380 mill. tons). The Middle Eastern production consisted of just 85 mill. tons with North Africa producing 2 mill. tons. In 1970 the proportions changed drastically, with an increase of 150 times in North African production over the period, an eightfold increase in Middle Eastern production, but just a twofold one in North American production. Every year significant quantities of new crude oil appeared on the market; Iran, Iraq, Saudi Arabia and Kuwait were joined in the Fifties by Algeria, Nigeria, Qatar and the Arab Emirates. This additional supply led to average costs twenty times lower than previous productions, and even lower marginal costs . The majors were faced with the challenge of managing this huge flow of "new oil" without it causing shockwaves on the market.

There were three forms of competitive pressure on prices which had to be kept down. The first, standard in the cost-price ratio, could be represented with a shifting to the right (downwards) of the supply curve, with a given demand curve and a subsequent reduction of the equilibrium prices. The contemporary shifting of the demand curve to the right (upwards) led to negative forces on prices, with final price not easy to predict. The second form of competition was *spatial,* due to the fragmentation of the geographical areas of both the origin of crude oil and the consumer location. This led to an objective difficulty in standardising quotations so that the area of purchase would be indifferent to the consumer. The third form concerned the quality of the crude oils, and thus their relative values. Given these phenomena, the fact that the absolute and relative prices of crude remained more or less stable between the Fifties and the Seventies takes on even more relevance, especially considering the micro-tensions in the economic trend that in other conditions would have immediately impacted upon them. However, this was an exceptional period.

Beneath that mask of stability *evolutionary processes* were in fact maturing which were gradually to destroy the institutional conditions on which it rested. Two processes in particular were of importance:

(a) the entry of independent newcomers, with a low or partial degree of vertical integration and small size;

(b) the reinforcement of actions by the producing States to increase the control of the exploitation of their resources and their mining yield.

The combination of these two processes was to gradually erode the organisational structure of the majors to its foundations, preparing the ground for it to be superseded during the Seventies.

4.2. NEWCOMERS ON THE OIL MARKET

In order to make a thorough evaluation of the effects on the international market of the entry of new firms, we should consider production separately from refining, since these sectors had different origins and consequences.

Although the firms that we have defined independent took a major part in the oil industry right from the beginning of the century, it was only from the Fifties that their presence began to be felt outside the United States.[1] This was stimulated by the extra profits which could be gained and by the increase in demand. They were aided by the political support given to the newcomers by their governments of origin; the allocation of new mining licences in old and new producing countries; and the abandonment of these areas (considered marginal) by the old licensees. Their entry was associated in various cases with extensions in the control group of already formed consortia. In many other cases, however, through the formation of new consortia to which the majors were completely extraneous, individual concessions were gained and joint-ventures were set up with local firms. In 1946 just 9 companies operated in the Middle East, the seven majors plus the French CFP and the Armenian oilman Colouste Gulbenkian. In 1956 there were 19, mainly due to the entry of independent Americans, while in 1970 the number had risen to 81. As always happens, the growing number of companies – although their size was initially irrelevant with respect to the dominant positions of the majors – sparked off a *competitive process* which in time was to become greatly significant. The increased competition between old and new companies mainly had the effect of changing the nature and the basic conditions of the *contractual process* between firms and host States, shifting the power ratio to the latter.

Table 4.1. *The control of the market (mill. tonn.)*

	1950	%	1973	%
Majors	280	95	1,550	84
Independent companies[1]	14	5	191	10
State companies[2]	2	..	110	6
Total	296	100	1,851	100

1. Amoco, Atlantic Richfield, Conoco, Phillips, Marathon, Sun, Unocal. 2. CFP, Elf, ENI, Fina.

Before dealing with these political aspects, we should examine the economic aspects which gave motivation and strength to the former. The most immediate effect of the entry of new firms on the international scenario was an acceleration in the development of new production areas and the appearance of new players. Huge oil fields were discovered and put into production in Algeria (from 1945-55 through the work of French firms), in Libya (Oasis Group in 1956), in Nigeria and in the Arab Emirates. In the Sixties the proved oil reserves (in the non-communist world) increased by 41 bill. tons. Of these 32 were located in the Middle East and North Africa and just 4.5 in the United States. The finding costs of the new reserves amounted to just 1.5 cents per ton in the Middle East and 13 cents in Africa, compared with 1.4 dollars in the United States and 3 in Europe.

A large part of this additional supply, effective or potential, was controlled by the newcomers, outside the integrated circuits of the majors. The difficulty in placing it on the American market (due to the import quota) forced them to direct sales towards the European and Japanese markets, generating an *oversupply* and pressure towards price decreases in free transactions, although marginal. The limited financial resources of the newcomers and the borrowing they had had to resort to, meant that they could not afford any delay in the setting up of production plants or the adaption of the production flow to the specific dynamics of the demand. Instead, these firms were interested in sending as much of their crude onto the market as rapidly as possible, to generate sufficient funds to cover the debt service and recover the huge costs sustained.

Obviously, as a free crude oil market slowly developed, the efficiency of the control of the oil supply by the majors was reduced. In other words, these developments made available excessive quantities of additional crude with respect to the requirements downstream of the large companies, leading to pressure to reduce prices.

The competitive entry of new firms also had an effect on the *fiscal costs* of the producing countries. Their absolute need to guarantee themselves research and development areas which were as far as possible differentiated (to reduce the overall risk on their investments), forced the newcomers to offer producing countries both economic and regulatory conditions (area extension, contract period, control of investments and profits, and involvement of local resources, for example) which were much more advantageous than those so far agreed by the majors. This form of competition between the firms, which sprung from legitimate expectations of the producing countries, was in time to prove even more relevant than price competition.

Helping to accelerate the evolutionary process in the relationship between States and firms were the State-owned companies which were being formed in both producing and consuming countries. Indeed, the greatest innovations came from the direct agreements between these firms – who had far from opposing objectives – which had been taking place since the late Fifties. Historically, the most relevant was the agreement of co-participation signed in 1957 between the National Iranian Oil Company and Enrico Mattei's Agip (ENI group). For the first time a joint company – the Societé Irano-Italienne de Pétroles – had been formed along political, economic and juridical lines which were radically different from those which had regulated traditional concessions. That an insignificant firm like Agip was able to guarantee far better conditions than those so far guaranteed by the majors – in terms of profit-sharing, participation in investments, control of the destination of the profits and so on – could not fail to influence all other contractual relations of the time. This particular joint agreement did not give its shareholders great economic satisfaction, but certainly changed the course of history in the relations between host States and foreign firms (Colitti, 1979).

In the absence of competition between firms and since the States could not themselves conduct oil production activities, contractual relations between States and majors were similar to those occurring in conditions of *bilateral monopoly*. As is known, in such conditions there is no predetermined solution of equilibrium, in the sense that this model is able to indicate only the upper and lower limits within which the essential aspect of the negotiation may be fixed between the parties. In this case, this aspect was profit sharing. The actual solution depended on the contractual force of the parties, on their behavioural strategies, and on other exogenous variables of a political and institutional nature.

Since it was impossible for the States to do any competitive bidding in the allocation or renegotiation of the concessions, because the identity and position of the majors was known and there was an absence of alternatives, their only deterrent in conflicts with the firms was to threaten nationalisation or withdrawal from production. Both these solutions put the governments in a weak contractual position. Withdrawal from production meant in fact abandoning oil profits, their only source of support. A public threat of nationalisation instead ran the risk of making the governments prisoners of their own deeds, putting them in a delicate position as regards public opinion (which tended in general to be hostile to oil companies) or the more radical political groups. Withdrawing the threats without having obtained what they wanted could thus prove very risky for the stability of the governments. All parties were aware of the probable damage to governments forced into positions difficult to extricate themselves from, and this led states to accept requests that they had previously been unwilling to consider, in order to avoid conflicts and political instabilities that would have proved even more costly.

The entry of new firms – even only the threat of their arrival – radically changed the terms of the bargaining process. This change was also brought about by the producing States themselves entering the oil industry. It was clear in fact that property rights would only be granted to third parties to exploit domestic resources when this was shown to be more convenient than direct handling by the States. Competition between the firms was, in turn, to push profit sharing to the level where the concession holder would retain only a "normal" profit for an activity such as the oil industry (Penrose, 1959).

These institutional modifications did not have immediate effect on the whole structure of the existing contracts, but they certainly started off a process of revision. Even the slightest improvement that the governments were able to make in issuing new concessions (within the same country but also in other countries) led to an immediate criticism of the old contracts and a growing political pressure inside the countries for their revision.

The competitive pressures in the upstream also combined with those no less considerable in the downstream, especially in refining, and in international trade, due to the reappearance of consistent export flows from the communist bloc.

In line with the growth of consumer demand, *refining* capacity underwent an approximately fourfold expansion between the Fifties and Seventies (Tab. 1.3.). This expansion was more innovative than in the past as regards dimension, siting and ownership control. With the growth in demand and in export flows, the siting of the refineries shifted away from the producing to consuming areas. There were several reasons for this. One was the lower cost of the transport of crude compared to refined products, another was the consuming countries' intention of limiting the political role of the producing countries, as well as their need to save hard currency and to encourage the process of post-war industrial reconstruction.

The geography of the refining industry saw a radical revolution. Whereas in 1940 70% of the capacity (outside the United States and the Soviet empire) was located near the oil fields, in 1960 this was reduced to 25% and in 1970 to little more than 10%. At the same time the capacity located close the areas of consumption increased over the three years from 30% to 65% and 80%. The remaining share, around 7%, was located in *intermediate* areas between centres of production and consumption.

Growth and relocation of the refining capacity were accompanied by a reduction in the degree of control of this sector by the majors. New independent operators entered the market, attracted by expectations of good profits and by the increased availability of crude oil on the free markets. Forced by the low level of integration to bring in supplies, but also to sell, with short-term contracts, they helped to fuel competitive dynamics where the marginal profit tended to prevail over every other consideration. In a situation of oversupply in the entire oil chain this could not fail to encourage prices to drop.

This result was also fuelled by the *reappearance of Soviet oil* on the international markets. Before the Second World War, oil was already a fundamental part of Soviet exports as a whole, but its share on the international market had been extremely low. Despite the huge commitment towards industrialisation, the Soviet oil industry had developed with extreme slowness, so that the growing energy requirement of the USSR was largely satisfied with the increasing exploitation of their abundant coal resources. It was special war requirements that really boosted the development of the oil industry, whose relative importance became fully evident both from a strategic/military profile and an economic one (Durand, 1962). The rate of export expansion (necessary to increase revenues in hard currency) soon turned out to be rapid, multiplying sixfold during the

Fifties. The main point however was not so much the increase in exports, but the price policy practised. The average cost of Soviet oil was low, and in any case could not be compared with that of foreign oil due to the different economic logic behind it. Also, since most of the production was directed towards the domestic market and that of the satellite countries, the cost of the additional production for export was even lower. The Soviets were thus in a good position for starting off aggressive price competition.[2] This aroused harsh reactions, especially from America: it was the middle of the Cold War, and the Soviet policy was interpreted in a solely political, rather than commercial, manner.

In an alarmed editorial, "The Economist" (19th August 1961) stated that the "Soviets consider and use oil as an instrument of their policy" in order to make the West depend increasingly on their supplies; to appear as the champion of less developed nations, selling them oil at lower prices; and to damage relations between western companies. The Soviets were then accused of resorting to the most abominable capitalist practices: dumping in the West in order to conquer the market, eliminate the competitors and then raise the prices.

There was much propaganda in these arguments, but they were the prevailing sentiments of the time. They thus help to explain the reaction in western circles on the decision of Enrico Mattei to undersign a massive oil purchasing contract with the USSR in 1958. This was a contract of historic breadth since it was the first long-term contract between a producing State and a consuming State, through the respective owned firms; because it constituted a large percentage of Italian imports as a whole (about 15%); and because it provided that the payment on the Italian side should be done by means of barter deals with Italian goods, mainly produced by the companies of the ENI group itself.

Mattei's decision was however dictated by the need to acquire oil supplies at good price, given the majors' refusal to grant him any space of action in the areas under their control. His decision certainly increased the irritation and aversion towards him on the part of the majors (Frankel, 1966).

4.3. THE CONQUEST OF THE EUROPEAN MARKET

Until 1950, oil consumption continued to be mainly (65%) limited to the American area. In Europe it was still nationally produced coal which dominated the energy balance. Oil usage was around 60 mill. tons. This low level was conditioned by the limits of economic activity, by plants obliged to use coal, and by a still marginal penetration of consumer goods with high energy intensity. In this context the price variable was not of crucial importance. Even a large reduction in price would have had little effect on the dynamics of consumption. This was also hindered by competition from other energy sources.

This scenario changed in the Fifties, when the potential of the increases in energy consumption were clearly manifested; the (cyclical) preoccupations about the shortage of oil disappeared; and the difficulties that the coal industry was to encounter were highlighted. The oil companies therefore imposed a strategy of attack on the European market aimed at *maximising* the penetration of oil in the shortest time possible.

There were several reasons for this: (a) the European market was the only one able to absorb the increasing production of the Middle East and of the new producing areas, given the basic closure of the American market; (b) a greater demand would allow full exploitation of economies of scale in the stages of transport and refining, with beneficial effects on costs. An increasing supply and decreasing costs would support a strategy of attack in which price was an essential part.

The substitution of coal with oil was not, however, the simple result of a mechanical and transparent process of competition (Clô and Prodi, 1975). Indeed a determining role was played by the different *rules* within which the oil industry and the coal industry, as well as public policies, found themselves operating. While it is true that in some European government circles there was a preoccupation with possible international energy shortages, and while it is true that in some places the conviction was spreading that it was necessary to maintain a high domestic energy production (as was happening in the United States), there was no relaunch in the production of domestic coal anywhere in Europe.

At the end of the Fifties, through the spread of the new philosophy of abundant and cheap energy, the coal industry began its irreversible decline with a huge destruction of resources.

The collapse of investments, the halt of technological progress in the methods of extraction and usage, the continuing low productivity and high costs are some of the factors behind this decline. Production fell by a factor of 2 between 1950 and 1970, employment by about three times (to little more than 600,000 workers, today there are just 100,000), and mines gradually began to shut down everywhere, despite the huge volume of resources still untapped. And in the coal industry every closure is irreversible. What for two centuries had been the glory of the European economy became merely a social preoccupation.

The European Coal and Steel Community (ECSC), fearing a return of the coal industry to the monopolistic practices adopted before the war (remember the Ruhr Cartel) imposed (under strong pressure from the Americans) an extremely rigid market regulation through a *downward price levelling* in every country. They did this without taking into account the differentiation of the production costs and individual competitive urges towards alternative fuels. What is more, the ECSC forced firms to respect the greatest market transparency, through the dual imposition of non-discrimination to the consumer and the publication of posted prices, prohibiting any form of agreement or merger between the firms. This was basically a whole series of obligations which *denied the mines the benefits of a market economy while imposing on them only the costs.*

On the other hand, there was no regulation of the entry and operation of the oil companies. Their strategy of penetration was based on two fundamental lines: the *refining policy* and the *price policy*. The type of investments and the planning of the refining processes were established in order to *maximise the output in heavy products*, directly competing with coal, and *limit the output in light products* destined to the new areas of consumption (transport and petrochemical industry), which could not be replaced by any other energy source.[3] The price policy was instrumental (and complementary) to the quantity policy, with the common objective of maximising the penetration of oil. This policy was primarily focused on price competition with coal. Despite the similarity of the costs of oil in European countries, the prices of the refined products were *differentiated*, in relation to the prices of the competing products, to the actual possibilities of replacement, and to the demand/price elasticity. This policy of differentiation was sustainable because of two conditions. Firstly, the existence of barriers (particularly logistic) to intra-European trade. Secondly, this price strategy had to be adopted by all the majors. Indeed, as regards the final consumer market of each country, there was no absolute identity in the prices practised, as was the case in the

Figure 4.1. *Price structure of oil products*[1]

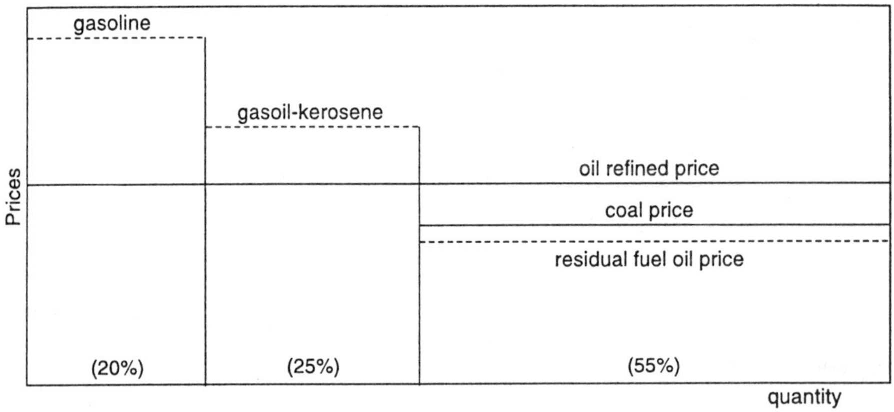

1. For the years 1960-70.

crude markets. Marketing competition was admitted, although homogeneity in costs did not actually allow much liberty in price competition. Especially at a publicity level, less risky and costly non-price competition was preferred. The prices of fuel oil – a product with the best possibilities of quantitative penetration, but subject to competition from coal – were thus quoted in the various countries at *lower* levels than those of coal paid by the various users. On the other hand, in the case of gasoline, a product not subject to competition, prices were defined in a state of quasi-monopoly. Where fuel oil prices were lower, gasoline prices were higher (and vice versa), so that aggregate profitability did not suffer as a result. In the mid-Fifties, fuel oil prices were quoted in a range between a minimum threshold of 22-25 doll./ton in Italy and Germany and a maximum of 30-35 in France and Great Britain. These values corresponded to relatively higher gasoline quotations in Italy and Germany (90-95 doll./ton) and lower in France and Great Britain (70-75 doll./ton).

From the mid-Fifties, under the pressure of international competition, fuel oil prices began to fall almost everywhere. This decline was due to several reasons: the desire to speed up the dethroning of coal; the appearance of crude oil at market prices lower than posted prices; and the competition between the final sellers. The fall in fuel oil prices, given the difficulty of raising gasoline prices, could not fail to affect the oil prices, a determining item in the structure of final costs. This led to a gradual deterioration in the competitive capacity of coal, and a collapse in the financial situation of the coal industry.[4] From a price advantage which on some markets reached 40%, in just a few years coal saw its competitiveness upturned with prices 20%-100% higher than those of fuel oil in 1960 (Thur, 1960).

The only possibility of regaining a balance in the price ratio could have lain in a possible fiscal policy from the European governments, or in a protectionist intervention along the lines of that proposed by the United States with oil import quotas. In fact, the fiscal policy in all European countries adapted perfectly to the price strategy of the oil companies: hitting light products hard (gasoline) and having much less effect on the heavy products (fuel oils).

At the end of the Sixties, the conquest of the European market by oil and the majors could be considered to be complete. The share of coal in the energy balance fell from 80% to 25%, while that of oil rose from 10% to 60%. At the same time, foreign dependence went from 10% to 65%. All this was perhaps inevitable. The economic advantages of moving to oil were undeniable. The political implications were no less relevant but neither the national governments nor the European organisations seemed to worry too much about these. It goes without saying that for the Europe of the Sixties the energy question took on a radically different connotation from the past. While coal use in Europe was an essentially domestic matter, oil use had dimensions and relevance at an international level (at least until North Sea Oil came onto the market in the Eighties). The co-operation that the coal industry required – reaching a common policy within the ECSC – was followed by the diversity of the national interests in the oil industry. Instead of an aggregating industry, energy thus became an industry of dialectics. No common policy was ever to be set up within Europe, remaining a dominion of national States.

Of no less interest is the difference in policies followed by Europe and the United States, as is the close interrelation between them. The severely *protectionist* policy of the United States contrasted with the *laissez-faire* attitude of Europe towards oil imports and to the declining domestic coal resources. The differential costs that the American economy found itself supporting – which may be summarised in the value of the oil import ticket of 1.50 doll./bbl – contrasted with the advantages the European economy could enjoy in drawing from the cheaper Middle Eastern productions. Basically, U.S. policy thus came to "subsidise" the European markets for two reasons. The first was the refusal to import, thus allowing an oil oversupply on the international markets and lower prices. The second was the "protective umbrella" that the U.S. spare capacity allowed against situations of tension in the international supply, as seen in the Suez crisis in 1956-57.

4.4. STATE-CONTROLLED PRICES AND MARKET PRICES

The main effect of the combination of the evolutionary processes we have looked at so far was to *reduce the degree of oligopolistic co-ordination and trade control* by the oil majors. As is always the case on the oil market, the outcomes of this change were to properly make themselves felt only much later in time, but to understand them fully we should examine them from their origins. Three of these outcomes were particularly important: the birth of a *free crude oil market* outside the *integrated circuits* of the majors; the different pricing criteria and the different dynamics of the industry downstream; and the growing role of the producing States.

Even in the periods of the maximum market control by the majors, crude oil trading took place between them or with non-integrated operators. The balance between supply and demand could not always be perfect, so at any time and for each firm there was a surplus or shortage of crude oil. Then there were structurally "short" firms, such as Shell, or structurally "long" ones, such as Gulf and BP. The positions were rebalanced through long-term contracts with quantities which were generally fixed and always, in any case, closely linked to the posted prices. So these transactions too were not ruled by the free market forces. These transactions concerned only a very limited number of marginal lots, which because of their very exiguity could not be called significant – in terms of price – in the oil market as a whole.

It was in fact onto this free market – also known as the spot market because it consists of single short-term transactions – that most of the increasing crude supply outside the integrated circuits of the majors tended to flow. A proper oil market was created, with total quantities that were for many years to remain clearly minor in relation to transactions as a whole, but whose economic significance was certainly not secondary in the actual balance between demand and supply. There is no doubt that the fact that it was a *residual* market affected the trading prices, but these however were the only pointers towards the market conditions as a whole. This market was mainly operated, at the upstream stages, by independent or partially integrated firms, traders and majors for the quantities they required for balance. At the downstream stages were non-integrated refiners, particularly those who owned intermediate and physiologically export-oriented refineries, traders and non-integrated distributors.

The competitive pressures that accompanied the expansion of the free market and the situations of a surplus supply led to prices structurally *lower* than the posted prices controlled by the majors. They were fixed by taking as reference the structure (for quantity and siting) of the supply surplus, on which a certain percentage discount was granted. Table 4.2. shows the relationship through time between the posted prices and market prices of the crude oil Arabian Light, which has always been considered a reference point at international level. As can be seen, until 1956-57 the two price series run almost parallel. The pressure to lower the prices was limited at the beginning by the Korean War (1950-53), by the disappearance of Iranian exports following the nationalisation of Mossadeq (1951-54), and by the Suez War (1956-57). With the end of these exceptional factors, the gap between posted prices and market prices tended to increase after 1958, with discounts exceeding 25% after the mid-Sixties. Faced with competition on the free market, the majors first responded by cutting their posted prices. Later, because of the fierce reactions of the producing countries, they were to keep the posted prices constant, though applying much lower discounts than those recognised on the spot market.

From the late Fifties, the world oil market may be represented as *three sub-markets*:

(a) the *US market*: basically closed within itself and with a price structure higher than the international one;
(b) the *Soviet market* (USSR and satellite countries): also isolated from the rest of the world, apart from the net export flows;
(c) the *international market*, which may in turn be divided into the *market controlled* by the majors, which represented 85%-90% of the overall trade, and the *free* or *spot market* which handled the other 10%-15%.

Phenomena not unlike those examined for the crude oil market could be seen for the market of oil products. The rise of free transactions, the much greater number of operators (independent, refiners, large consumers, traders) led also in this case to the formation of a free market fed daily by spot transactions. Since the end of the Fifties, these were priced every day in the newsletter "Platt's Oilgram Price Service" published by McGraw Hill of New York. We will be saying more about the significance of these prices later. Here it is important to highlight how they managed to represent another type of information about the market trend that could not fail to have an impact on crude oil trading prices. Although it was true that the quantities appearing on the product market were irrelevant relative to final consumption as a whole (no more than 5%-10%), the quotations applied were the only information available on the consumers' willingness to pay, and on the quantities involved. A market certainly not perfect, but the only one which could call itself such.

Table 4.2. *Posted prices, market prices, and Arabian Light fiscal costs (doll./bbl)*

Years	Posted Price	Market Price[1]	Per Cent Discount	Fiscal Costs[2]
1950	1.75	1.71	–2.3	n.d.
1951	1.75	1.71	–2.3	n.d.
1952	1.75	1.71	–2.3	0.52
1953	1.93	1.93	=	0.75
1954	1.93	1.93	=	0.85
1955	1.93	1.93	=	0.82
1956	1.93	1.93	=	0.83
1957	2.01	1.90	–5.5	0.88
1958	2.08	1.85	–11.0	0.88
1959	1.93	1.70	–12.0	0.76
1960	1.80	1.53	–17.7	0.75
1961	1.80	1.45	–19.5	0.75
1962	1.80	1.42	–21.1	0.76
1963	1.80	1.40	–22.2	0.79
1964	1.80	1.33	–26.1	0.82
1965	1.80	1.33	–26.1	0.83
1966	1.80	1.33	–26.1	0.83
1967	1.80	1.33	–26.1	0.85
1968	1.80	1.27/1.40	–22/–29	0.88
1969	1.80	1.25/1.38	–23/–30	0.87
1970	1.80	1.35	–25	0.88

Source: 1. 1950-67: Jenkins (1989); 1968-70: Clô and Peredo (1972). 2. Rïfai (1974).

4.5. THE FOUNDATION OF OPEC

It was the producing countries which reacted to the competition of the newcomers more than the majors. The fall in the posted prices in fact meant a proportional drop in the oil revenue on which their economies were almost entirely based. This happened without the host States being at all involved in the decisional processes of the firms, while the lack of homogeneity in their fiscal policies played no small part in favouring the competitive forces on the market. While in fact in most of the Middle East the taxes received by the States were calculated on the posted prices, in the new producing States (such as Libya or Nigeria) they were calculated on the actual market prices. This disparity had been a great boon to the newcomers on their entry and perhaps it had been engineered by them to a large extent. By letting market prices slide to compete with the majors, they saw the fiscal costs reduce in proportion, while the same could not immediately be said for the incumbents.

Although the reasons for the producing States' dissatisfaction with the majors were long-standing and went far beyond the price question, the fall in prices urged the States to seek common policies and a revision of the contractual relations with the firms. The foundation of OPEC in September 1960 was a *collective political act* agreed by its founders in order to bring an immediate end to the absolute freedom of the firms in the fixing of prices. From its first Resolutions onwards, OPEC stressed that the companies should not be permitted to make any reduction in price without first consulting the States involved. Though timid, this was the first attempt in the history of international economic relations to assert the inalienable right of sovereignty of the states over their own natural resources(Al Chalabi, 1980a).

OPEC had two fundamental lines of action regarding prices. Firstly, blocking their fall through a revision of the fiscal policy. Secondly, standardising the fiscal rules in the various countries, eliminating the harmful differences of the past. This dual action managed to achieve the desired results, or at least prevented the market dynamics from having a negative effect on the oil revenues of producing countries. In fact, the freezing of the posted prices changed the income taxes into indirect taxes, in cents per barrel. Like any other tax of this type they were treated as a cost and became, as we will see in the next section, a minimum base for price. No firm would be able to go below the sum of the taxes plus the costs of production per barrel: the so-called "tax-paid cost". The unification of tax regimes and the fact that the tax-paid cost became more or less uniform and fixed in almost all the countries (bearing in mind the different quality and siting) helped to halt the entry of potential newcomers, limit the current supply of the existing producers, and stabilise market prices. The fiscal policy in the Middle East basically acted in support of the non-competitive organisation of the oil market.

4.6. CRUDE OIL PRICES AND TAXATION SYSTEM

The basic principle of the taxation of profits in the Middle East could be summed up as the "fifty-fifty formula". According to this, the profit made on each barrel produced was (in theory) shared equally between host government and company. Generally speaking, a *fixed share* of the profits made was taxed. In practice, this did not happen for two reasons:

(a) the appearance, first of all, of a *dual price system*: the *posted prices* and the discounted *market prices.* As the volume sold at market prices increased in time, the posted prices – left untouched through the opposition of the producing States – gained a purely nominal significance, becoming to all effects just a base of reference for calculating the taxes, from which the term *tax reference prices* thus derived. Another important consequence was that, by basing the taxes on the *nominal* profits, the taxation system was transformed from a *fixed share* of the profits gained into a *fixed payment* per barrel produced. In relation to the production costs and to the actual market prices, participation in the profits with respect to the fifty-fifty principle varied as a consequence;

(b) the second complication came from the transformation of the royalties (12.5% of the posted prices) from *advances* on income taxes to a deductible *cost* when calculating the taxable income. This transformation involved a *net* increase of the fiscal drag of the producing countries from 50% to 56.25% of the posted price (*net of costs*). Prior to this, with the posted price at (*P*) and production cost (*C*), the total government duties (*T*)

Figure 4.2. *Taxation effects on oil production profits in the Middle East*

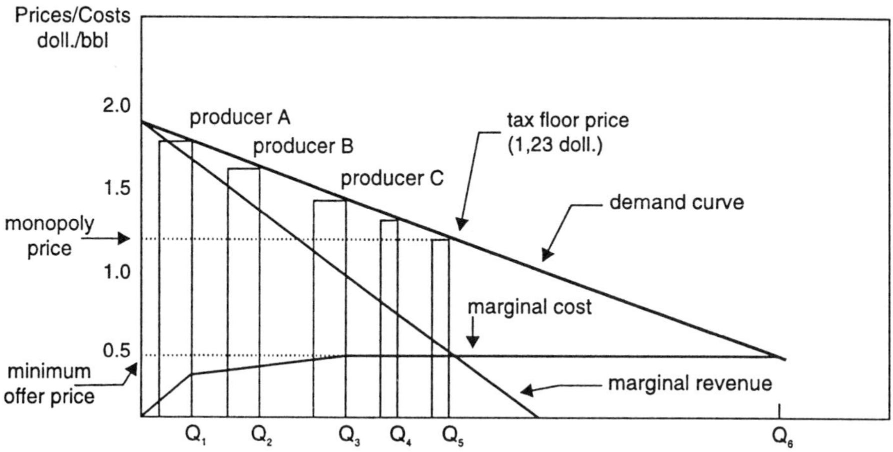

were equal to:

$$T = 0.50\,(P\text{-}C)$$

With the royalties (R) becoming cost and thus deductible, the taxes could be calculated as:

$$T = R + 0.50\,(P\text{-}C\text{-}R)$$

and with $R = 12.5\%$, P was:

$$T = 0.5625\,P\text{-}\;0.50\,C$$

Following the agreements of 1964, which had taken as long as two years to negotiate, the governments were paid 12.5% *ad valorem* plus 50% of nominal profits. To soften the impact of this tax reform on the firms' accounts, the producing countries granted them, *pro tempore*, discount allowances on the posted prices that brought their level close to the market prices. While formally the firms saw their fiscal burdens reduced to a few cents per barrel with this compensation, the negotiation basically marked an important turning point in the role played by the producing countries.

The fiscal reform also had effects of considerable importance on the production policies, on the competitive relations between the companies, and on the allocation of resources (Polanyi, 1966). Let us look at these briefly, outlining three possible situations in Table 4.3.: *producer* A who sells at posted prices of 1.80 doll./bbl, with a relatively low production cost of 0.10 doll.; *producer* B who sells at market prices of 1.55 doll./bbl, 14% less than posted prices, with costs of 0.22 doll.; *marginal producer* C with average costs of 0.33 doll./bbl and prices even lower than the previous at 1.30 doll.

The table shows how the nature of the overall fiscal drag of the government – consisting of a fixed royalty, independent of the market prices and the costs, and of an income tax based on nominal profits which vary with the costs, but not with market prices – leads to *considerable variations in the percentages of taxation of the real profits*: from a minimum of 52.5% for producer A to a maximum of 80.5% for the marginal producer C.

Table 4.3. *Government fiscal drag and companies' profits (doll./bbl)*

	Producer A	Producer B	Producer C
Market price (1)	1.80	1.55	1.30
Production cost (2)	0.10	0.22	0.33
Real profits (3) = (1) - (2)	1.70	1.33	0.97
CALCULATION OF TAXES			
Posted price (4)	1.80	1.80	1.80
– discount allowance (8,5%) (5)	0.15	0.15	0.15
– production costs (2)	0.10	0.22	0.33
– royalties (12,5% posted) (6)	0.23	0.23	0.23
Nominal profit (7) = (4) - (5) - (2) - (6)	1.32	1.20	1.09
Income Tax (8) = 50% (7)	0.66	0.60	0.55
Total govern. duties (9) = (8) + (6)	0.89	0.83	0.78
Companies' residual profit (10) = (3) - (9)	0.81	0.50	0.19
% Governments: real profits (9) : (3)	52.5	62.5	80.5
% Companies: real profits (10) : (3)	47.5	37.5	19.5
CALCULATION OF PRODUCER'S REVENUE			
Capital cost per bbl. per annual production (11)	0.50	0.83	1.15
Minimum profit (12) = 10% (11)	0.05	0.08	0.12
Minimum profit + production costs (13) = (12) + (2)	0.15	0.30	0.45
Market price (1)	1.80	1.55	1.30
Producer surplus (14) = (1) - (13)	1.65	1.25	0.85
– government surplus (9)	0.89	0.83	0.78
– company surplus (15) = (14) - (9)	0.76	0.42	0.07

Source: Polanyi (1966).

In absolute terms, the profit that remained to the firms varied from a maximum of 81 cents/bbl for producer A (with lower costs), 50 cents for producer B (with discounted prices and average costs) to just 19 cents for marginal producer C. It is interesting to think about this last case which to a certain extent reflected the actual situation in which the newcomers found themselves: forced (in general) to operate in new areas, more costly than those seized by the majors, and with the need to make market outlets for themselves by decreasing prices lower than the posted prices. Each reduction of the market prices – with no variation in the technical costs of production and also of the fiscal costs, since the taxable base for their calculation remained unchanged – meant a reduction in the profits gained and the profit rate. In Table 4.3. we have calculated as 10% the *minimum* profit rate which would attract operators to enter Middle Eastern production.

The sum of the production costs, the government duties and the minimum profit (a profit guaranteeing a profit rate considered minimum by the companies) identifies a price limit level (pm) below which no new firm would enter. A price lower (pc) than (pm) thus represents the *exclusion price*. In our example this would be 1.04 doll./bbl for producer A, 1.13 for producer B and 1.23 for the marginal producer C. The taxation system did not only influence the entry of new producers, but also the current production. This could not in fact increase beyond the level on the demand curve corresponding to the limit price of the marginal producer (Q_4 in Fig. 4.2.). The taxation system could not therefore be called *ideal* from the point of view of the optimal allocation of resources, since it had the effect of preventing the drop in prices towards the long-term marginal costs (with production Q_6). It was also iniquitous for the various producers, since it had more effect, in terms of profit-taking, on those with higher costs.

What was ideal for consumers and for the potential new producers was therefore not so for the host States and for existing producers. They were solely interested in maximising their income from the taxation of profits and not the joint value of the surpluses of the consumer and the producer. The ideal condition for them would have been that the price-quantity combination was fixed at the point in which the revenue of the producer was maximised: thus at a level of monopolistic equilibrium, where marginal revenue was equal to marginal cost. With prices much higher than those of competition (in the example 1.17 doll./bbl compared to 0.45 doll./bbl, minimum supply price of the marginal producer) and considerably smaller quantities (Q_5 compared with Q_6 in a ratio of about 1 to 2).

While it affected all firms, this system was thus not at all against the interests of the majors: halting the price fall (and therefore the fall in profits and revenues); restoring order and stability to the markets; eliminating damaging differentiations in the taxation systems, from which the competitive manoeuvres were partly derived. The similarity of interests between majors and producing States was immediately shown by the satisfaction that both expressed regarding the effects of stabilisation of the prices achieved by the fiscal reform, particularly in the areas where their slide had been most pronounced[5] (OPEC, 1967).

4.7. TOWARDS NEW EQUILIBRIUM

The period from the Fifties to the Seventies was when the dominance of the majors was at its height but, at the same time, the processes of evolution which were to lead to their overthrowing could be seen to be developing. The whole world benefited from the stability of the oil markets, in that the supply matched the growing demand, the trade flows were flexible in the face of continuous political micro-tensions, and there were reductions in real oil prices. However the institutional conditions that guaranteed all this were gradually disappearing, for a range of economic and political reasons.

These reasons collectively led to a reduction in the profitability of the firms and a major cutback in investments throughout the oil cycle, but particularly upstream. The excess of capacity which favoured competition between firms and the fall in prices was gradually reduced, aiding the passage from a *buyer's market* to a *seller's market*.

Higher taxes paid to the producing countries and lower prices were certainly not a way to make firms more profitable, during a time of growing inflation. Until the mid-Sixties, technological innovations, and economies of scale, all pushed by the increasingly acute competition, allowed a level of profitability to be maintained which was considered acceptable with respect to the risks of the sector, though lower than that of a decade earlier. But towards the end of the decade, with the reduction of the possibilities of exploiting economies of scale or other innovations to bring down the costs, the firms' capacity to generate sufficient resources to finance their future went into drastic decline. From levels close to 19% in 1956-57, the profitability rate of the majors fell to 13%-14% in the years 1958-63 and to values around 11% in the late Sixties (Pirinc, 1971).

The effect of this reduction was considerable and accentuated by a continued excess in production capacity. No less important was the fact that investments began to be aimed increasingly towards new and more costly oil areas, for reasons of political risk alone. It was in the mid-Sixties that the once unthinkable Alaskan and North Sea "adventures" began. Between 1958-70, the exploration costs amounted to 15 bill. dollars. Of these, only 3.5% was located in the Middle East, 55% in the United States and 7% in Europe, whereas the finding cost was 20 cents/bbl in Europe, 17 in the United States and 0.25 in the Middle East (Roushdi, 1972). If the new production capacity was to be put into operation, an improvement in the margins of the firms and therefore in the crude oil market prices was a *necessity*. The benefits that the competition had managed to gain were, in time, to give the opposite results.

Notes

1. In 1953 two groups of independent newcomers, American Independent-Aminoil and Getty Oil, obtained concessions in the Neutral Zone (controlled by Kuwait and Saudi Arabia) where they found large quantities of oil, while in 1954 nine minor American companies had access, under pressure from the U.S. American government, to the newly formed and very important *Iranian Consortium* which was to control the majority of the Iranian production until the Seventies.
2. The crude oils exported onto western markets had reductions on average between 10% and 20% on the posted prices.

3. Mean prices of petrol and fuel oil (atz) net of taxes (doll./tonn.)

	Gasoline			Residual fuel oil			Gasoline/Residual fuel oil		
	1956	1960	1970	1956	1960	1970	1956	1960	1970
France	74	67	69	35	30	25	2.1	2.2	1.8
Italy	80	57	108	22	16	16	3.6	3.6	6.8
Germany	96	81	51	26	12	16	3.7	6.8	3.2
Great Britain	68	70	50	29	30	28	2.3	2.3	1.8
USA	78	77	89	14	14	25	5.6	5.5	3.6

4. In Germany for example, the price ratio between coal and fuel oil in industrial usage (with the same thermal power) rose between 1957 and 1960 from 0.55 to 1.35 and in the domestic sector, from 0.67 to 1.21.

5. The director of Exxon stated in 1966 that European prices, which had fallen rapidly in 1964 and 1965, had gradually been stabilised at the end of 1965, showing slight signs of reprisal in 1966. Even more explicit was the director of BP who stated that "most of the merit, to the halting of price erosion, goes to the recent changes in the fiscal terms brought by the oil agreements in the OPEC area" (OPEC, 1967, p. 12).

5 ENERGY CRISES: PERCEPTION AND REALITY

5.1. THE TURNING POINT OF THE SEVENTIES

The modern history of the world energy industry, which dates from the second half of nineteenth century, encountered a definite "turning point" in the 1970s. Events of great relevance had certainly happened before this time – indeed the period between 1870 and 1900 featured great technological innovations linked to the discovery, processing and use of modern energy sources[1] – but their combined outcome was always to further the harmonious growth of the energy system, which was to come to a temporary halt only with the Second World War. After this was over, the industry once again began to expand at an even higher rate.

In the Fifties and Sixties the world energy industry gave a formidable demonstration of technological and entrepreneurial capacity in anticipating and supporting extraordinary economic growth, such as had not been observed in the two previous centuries, and in satisfying the new needs expressed by the final consumers (Tab. 5.1.). In no other period apart from those twenty years were the close reciprocal ties running between economy and energy able to express themselves so completely: a virtuous circle providing the former with energy in the necessary quantities and qualities and, no less important, at price conditions which improved as the importance of oil was consolidated.

That circle was broken and upturned with the sudden – *though certainly not unexpected* – outbreak of two "oil crises". The *first* was in 1973-74, when the nominal oil prices (average values) rose from levels close to 2.5 doll./bbl to about 11.5; the *second* was in 1978-80 when there was another jump to 35 doll./bbl with peaks close to 45. During the Seventies, oil prices underwent the largest increase in their century-old history: 20 times in nominal terms and about 9 times in real terms (Tab. 5.2.). Swings in prices had of course been seen before this, but at that time their importance was greatly mitigated by the limited importance of energy to the general public, by the limited role of oil in the energy balance, and by the high degree of self-sufficiency that the majority of western countries enjoyed. These were all aspects that economic development and the gradual rise in importance of oil were to modify radically in the years following the second world war, coupled with the favourable dynamics of the market prices.

When these dynamics suddenly and unexpectantly became less favourable, the economy of every country was seriously affected, as well as the international political equilibrium. This highlighted the *centrality* that energy had taken on in the development of modern societies as well as their extreme vulnerability due to the huge dependence on oil imported from areas with strong political instability and in growing opposition to the western world. For the first time, *energy, and oil in particular, became a critical and priority matter* for both the public and governments' agendas.

Table 5.1. *GNP, energy and population in the world: 1900-1990*

Years	GNP ($'000 bill. 1987)	Energy [1] consumption (bill. toe)	Population (bill.)	Per capita income (doll. '000 1987)	Per capita energy consumption (toe)
1900	0.7	0.550	1.600	0.44	0.344
1950	3.8	1.696	2.554	1.48	0.664
1960	6.1	2.798	3.036	2.01	0.922
1970	10.1	4.527	3.702	2.73	1.223
1980	14.1	6.250	4.454	3.17	1.403
1990	18.8	7.762	5.291	3.55	1.467
% Annual average variations					
1900-'50	3.4	2.3	0.9	1.3	2.5
1950-'60	4.8	5.1	1.8	3.3	3.1
1960-'70	5.2	4.9	2.0	2.9	3.1
1970-'80	3.4	3.3	1.9	1.4	1.5
1980-'90	2.9	2.2	1.7	0.5	1.1

1. Commercial energy (see note 1 at end of chapter)

Source: 1900 figures: our evaluations from various sources; 1950-90: population and income: Brown et al. (1993); energy: United Nations.

5.2. HISTORICAL DEVELOPMENTS

Though the crises of the Seventies broke out following sudden and momentary political tensions (the so-called "Kippur war" between Egypt and Israel at the beginning of October 1973 and the coming to power of Iman Khomeini in Iran at the start of 1979), they were caused by the gradual interaction of three *historical developments* of a different nature but equal relevance. Each of these, though originating from different circumstances and in distant periods, reached its most critical point at the same time in the early Seventies. The nearer this date became, the more their *individual* relevance and their mutual *interaction* grew, in a complex sequence of events whose relationships are still not easy to isolate today.

The *first historical development*, within the energy systems, had several elements: the expanding role of oil in the economies of industrialised countries,[2] the growing technological limits in its use and therefore in the processes of substitution/competition with other sources, the need to increase oil supplies from the more abundant and cheaper Middle East reservoirs (and from Arab countries in particular),[3] and finally the shrinking of the margins of spare capacity on a global scale.

Table 5.2. *Crude oil prices in the international market: 1955-1995 (doll./bbl 1992)*

Year	Current price[1]	Inflation-adjusted price	
		On consumption USA	On consumption: Industrialised countries
1955	1.93	9.6	12.1
1956	1.93	9.4	11.8
1957	2.01	9.5	12.0
1958	2.08	9.5	12.0
1959	1.92	8.8	11.1
1960	1.86	8.3	10.5
1961	1.80	8.0	10.0
1962	1.80	7.9	9.7
1963	1.80	7.8	9.5
1964	1.80	7.7	9.2
1965	1.80	7.6	9.0
1966	1.80	7.4	8.7
1967	1.80	7.2	8.4
1968	1.80	0.8	8.0
1969	1.80	6.5	7.8
1970	1.80	6.3	7.3
1971	2.13	6.9	8.2
1972	2.48	7.9	9.1
1973	2.81	8.4	9.7
1974	10.98	29.5	33.3
1975	10.71	26.4	29.2
1976	11.63	27.1	29.3
1977	12.57	27.5	29.1
1978	12.91	26.2	27.9
1979	29.82	54.4	58.9
1980	35.85	57.7	63.2
1981	34.29	50.0	54.9
1982	31.76	43.6	47.2
1983	28.72	38.2	40.6
1984	28.08	35.8	37.9
1985	27.52	33.9	35.7
1986	13.84	16.7	17.5
1987	17.29	20.2	21.3
1988	13.49	15.4	16.0
1989	16.34	17.9	18.6
1990	21.11	22.4	22.9
1991	17.38	17.9	18.1
1992	17.60	17.6	17.6
1993	15.20	14.8	14.8
1994	14.90	14.1	14.2
1995	16.32	15.0	15.1

1. As a point of reference we have taken crude Arabian Light (34° API). The prices are annual averages of the posted prices up to 1972 and official from 1973 to 1984 FOB Ras Tanura. From 1985 Ras Tanura spot prices.

Source: 1955-59: OPEC, "Annual Statistical Bulletin" 1989 (Tab. 71); 1960-87: US Department of Energy, "Annual Energy Review" 1988 (Tab. 117); 1987-95: "Petroleum Intelligence Weekly" (various issues).

The *second development* was the process of political emancipation of the Middle East and North Africa from western domination and from the factor which had been a main feature of that domination: the old concessions "extorted rather than negotiated" (El Sayed, 1967, p. 48) through which the Anglo-American companies exploited their oil resources. This emancipation had three main aims. Firstly, it was to reinforce the political role of the host States in the decision-making processes. Secondly, it would make use of oil exports for the take-off and diversification of the domestic economies. Thirdly, it would increase the producing States' participation in the allocation of oil surplus, mainly seized by the consuming States and by the oil companies.

The *third development* was the gradual reinforcement of the State of Israel in the territories of Palestine, against the wish of all the Arab countries and aided by support from the United States. Throughout the lengthy period of military and diplomatic clashes between the Arabs and Israel, which had taken place since this State had first been constituted in 1948, a radical position had gradually been adopted between exporting countries, in favour of using *oil as a weapon of political and economic pressure* against the West and in support of the Arab cause. Actions to this end were recorded during the 1956 Suez war and in the Six Days War of 1967, though the effects were of little significance because of the great *flexibility* of the international oil trade on both sides of the Atlantic. The diminishing of this flexibility, with a huge reduction in the surplus of production capacity and the concentration of the marginal capacity in Arab countries, was to give enormous potential to that strategy.

Together, these developments helped to foster a climate in which the rise of oil prices would be both possible and long-lasting. The more demand for oil increased, the more the prices would be forced to rise; the more demand looked to the Middle East, the more their trading role and their monetary demands would increase, while their actual internal sovereignty and their international political weight evolved slowly. Capital was also accumulated in many exporting countries given their reduced expenditure capacity: due to the limited size of their economies, small population and the marked inequality in the distribution of wealth. The reduction – or even the temporary interruption – of oil exports was to have no influence on their economies, giving those countries (where producing capacities and reserves were mainly concentrated) further motives for contractual power.

Therefore a dual conflicting division emerged from the historical developments that we have outlined. The first was between the *actual* role and *essential* role played by the Middle East on the international scene. The second was between expanding *external* needs for that oil and *internal* needs for its revenues, which tended to diminish as the prices increased and as it became more convenient to postpone the exploitation of the resources. The reappearance of this dual divergence eventually radicalised the conflict of interests between producing and consuming with final outcomes that require one more detail for full comprehension: *how western countries reacted to the oil crises.* Their absolute failure to comprehend, in spite of the warnings coming from various sides,[4] the significance of the historical events which were developing, not only meant that a coherent and timely response was not given but also conditioned the *dominant perception* of the two crises, completely focused on two interrelated aspects:

(a) the reduced *political security of oil supplies*, and therefore the serious threat deriving from this to the stability of the international economic and political framework;

(b) the progressive *physical shortage* of oil.

The sharp rise in prices was interpreted, from this second viewpoint, not as a temporary insufficiency of production capacity (given the levels of demand), but as the manifestation of an *absolute* shortage in oil resources. This unexpected event appeared initially to give a gloomy confirmation of the new cycle of apocalyptic predictions generated by the publication of the first studies of global modelling promoted by the "Club di Roma".[5] Both the first study by Joy Forrester, *World Dynamics*, in 1971 and the second, celebrated *The Limits to Growth* by Dennis Meadows in 1972 predicted a forthcoming *catastrophic decline* for the whole of humanity due to the suffocating limits deriving from the imminent "exhaustion of resources", from the "explosion of wars, epidemics, social tensions", due to the effect of changes brought by pollution, and the subsequent "uncontrollable decline of the level of population and the industrial system".[6] The fact that things did not go as feared – indeed, over the following twenty years, the world population increased by 42%, the GDP by 78%, and oil reserves by more than 100% – is of secondary importance here, although it was this type of prospect and *collective perception* of the crises (more than their reality) which conditioned the response of governments, companies and consumers.

5.3. THE WARNING SIGNS, EVENTS, AND OUTCOMES OF THE CRISES

From the end of the Sixties – since oil began to dominate other energy sources – a radical process began of the *transformation of the structures of control of the oil industry* and of *the methods of operation of the markets*. This process would require over a decade to reach completion, and its final outcome was to be the (almost) complete expulsion of western interests from the Middle East; full control of the exploitation of oil by the local States; the vertical disintegration of the international industry, and thus the disappearance of the mechanisms of stabilisation of the market which had been skilfully set up by the powerful organisational structure of the major companies. The Seventies thus witnessed a formidable *redistribution of power* between the North and South of the world and the end of the "international petroleum order" that had been seen over the previous three decades.

This process of transformation had distant roots and drew its origins from the unresolved, but increasingly important question of the right of developing countries to participate in the exploitation of their natural resources. This principle had been fully and officially recognised in Resolution 2,158 of 25th November 1966 passed by the General Assembly of the United Nations and entitled "Permanent sovereignty over natural resources". The Resolution declared (in item 1) "the inalienable right of all countries to exercise permanent sovereignty over the exploitation of their natural resources in the interest of national development" and (in item 5) "to gain a share in the companies under foreign control and increase their quota of the profits gained, on an equal basis". These statements echoed the guidelines that the Organisation of Petroleum Exporting Countries had laid down for itself since its foundation in 1960. It had a condition for the profound *revision* of the system of contracts and concessions which had regulated the relations between States and foreign companies since the Twenties. In the "Declaratory Statement of Petroleum Policy in Member Countries" of 1968, OPEC, with the juridical premise of "changed circumstances", established the fundamental principles of this revision, advising the member countries to follow them as closely as possible in order to preserve a unity of action that more than anything else would maximise their individual and collective interests. Unfortunately, the concrete dynamics of the processes of revision had to take into account the different political ideologies of the individual producing countries, their collocation in the system of international relations with the two major powers (United States and Soviet Union), the form in which the presence of foreign capital in the national industry was rooted, and last but not least, the profoundly different situations and economic interests of the individual countries. All this explains the huge difficulty that OPEC was to encounter – even when it *seemed* to be successful – in combining the various national situations into a single line of action: preventing the different situations from obstructing a common path, preventing short term successes which would be counterproductive in the long term, and thus protecting the Organisation's very *raison d'être*.

The multinational oil companies were, on their part, perfectly aware of the *irreversibility* of the political transformations that were being consolidated in the producing States and the huge repercussions which would come to them from these, since most of their stock of reserves was located in these countries and since the majority of their integrated profits derived from those productions. Since the early Sixties the great multinationals *had* activated a *strategy of re-location of their activities*: almost cutting off their mining investments in the Middle East and directing them towards new areas which were politically more reliable, though much more expensive (as in the case of Alaska and the North Sea). The companies were worried not so much about how to resist what they pragmatically considered as the inevitable course of history, but rather *how to manage the transition* towards new industrial structures, so as to prevent stable market conditions being followed by a chaos which would do no good to anyone. The increasing pace of events was however to take precedence over the intentions of the single players, finding most of them ill-prepared to understand and face what was happening.

The historical development of the oil crisis may be divided into three stages: the first saw the onset of the "clash" between States and companies (1970-October 1973); the second, with the acceleration of the economic and institutional changes accompanying and following the "Kippur war" (October 1973-end of 1978); and thirdly the "Iranian crisis" (end of 1978-80) and the end of the crisis.

5.4. THE ONSET OF THE "CLASH": 1970-73

Although certain signs had already appeared in 1968 and 1969, the onset of the "clash" between States and oil companies may be dated from January 1970 when Colonel Muammar Gadhafi, having risen to power a few months before in Libya, denounced every existing agreement with foreign companies. The Libyan initiative, which mainly affected independent American companies, who were less able than the majors to resist cuts in activity and threats of nationalisation, soon triggered off an *imitative, cascading process* in all the other producing countries: where each tried first to align itself with the conditions gained by the other countries and then to improve them by exploiting their own strong points (Schuler, 1974).

The pursuit of the fiscal costs-prices of crude oil was accompanied by a new development in the traditional system of oil concessions, between two extremes: nationalisation, supported by the radicalism of the Arab League, and participation, supported by the new technocracy represented by the Saudi oil minister Yamani.

The first *nationalisation* of significance (after the minor one in Algeria in 1967 and much later than the unsuccessful one in Iran in 1951) took place in Libya in December 1971 when, without any prior warning, Gadhafi promulgated a decree of nationalisation of all the acquired rights and all the assets of British Petroleum. The only motivation was retaliation against the Iranian occupation of three small islands in the Persian Gulf which had been under British protection until the day before. Proclaiming himself champion of Arab causes, Gadhafi denounced the Anglo-Iranian collusion by hitting BP and threatening the United States with similar actions if they persisted in their pro-Israeli policy. This nationalisation was followed by others in Algeria, Venezuela and Iraq.

Completely different was the strategy of *participation* followed by another group of countries led by Saudi Arabia and including Abu Dhabi, Kuwait, Qatar, Iran and in part Iraq. According to the pragmatic philosophy of Yamani, their aim was to take control of the oil industry *gradually* without threatening the stability of the markets and taking into account the time needed to acquire full awareness of its operating mechanisms. This strategy coincided precisely with the wishes of the western companies. In all cases in history in which it had been used, nationalisation had proved to be, at the crucial time and given the structure of the industry, completely incapable of reaching the desired goals. As Yamani stated succinctly, "The lesson is that you can nationalise oil but not its market" (Penrose, 1959, p. 238). In his opinion the states could not do without the function of equilibrium performed by the majors because they formed "the only means of market stability available in this moment" ("Business Week", 25th October 1969). At the 18th OPEC Conference in Vienna in 1969, Yamani was even more specific: "we are anxious to bring about participation with the majors because we want to help preserve their role as guardians of the prices and clearing house between us and the consumers, and also to learn, from their exceptional experience, how to act on the international markets". Instead of conflict and breakage with the fragile market equilibrium, Yamani and the moderate wing of OPEC were suggesting negotiation and stability. But the changes elsewhere were inevitably to sweep away both.

In New York, on 5th October 1972, Sheikh Yamani and the representatives of three majors (Exxon, Texaco and Royal Dutch-Shell) signed the "General Participation Agreement" which established guide lines for the participation of the producing countries in the activities of the companies, leaving individual countries to adapt them to their specific national situations. In general, the companies agreed to hand over a growing share of their business, with due compensation: starting with 25% in 1973 to reach 51% in 1981. No less important was the extremely slow rate at which it had been agreed to relinquish control of the crude oil supply. Although the governments were entitled to a share of the crude oil equal to their participation, in practice it was agreed to let the companies manage the majority of the crude oil equity of the governments. The aim in doing this was to allow them to respect the supply commitments already made with third parties and to exploit production capacities elsewhere. Besides this, there was also a shared preoccupation with avoiding the consequences for the market of the entry of inexpert and non-integrated sellers. The "Exxon-Yamani agreement", as the radical critics of the Arab world referred to it with scorn (Sarkis, 1972), basically aimed to preserve for the companies the long-term – the agreement with Aramco was scheduled to last until 1999, the one with Kuwait Oil until 2026 – control of the critical *stage of crude oil sale*: in a position of decision-making autonomy that could not however be compared with that of the past.

Political-institutional changes and price increases occured right from the start of the

Seventies in a relationship which should not just be considered in terms of cause and effect, since it was primarily the evolution of the real market conditions which made them possible. *So it was not politics which ruled the economy, but the economy which allowed politics to expand to its full extent.* Prices were pushed higher mainly by the huge increase in the demand for oil, but even more by the fact that this demand could only be satisfied in the Middle East, as the unused capacity in the other areas was gradually exhausted.[7] "The end of the excess of supply in the United States" wrote Daniel Yergin in his monumental work on the history of the oil industry, "meant that the margin of security on which the Western world had always relied was eliminated" (Yergin, 1993, p. 461). On the other hand, the strong increase in American domestic consumption led, in just 4 years, to oil imports *doubling* to 6 mill. bbl/d in 1973. Exports from the Middle East increased by about 18% in 1971, 12% in 1972, 17% in 1973. This growth of 54% over the three-year period was the equivalent of that of the previous six years. A series of unusual and *specific circumstances* was triggered off against these basic trends which combined to help to increase micro-tensions on the market and to reinforce the action of the producing countries.[8]

Faced with the disjointed pursuits of single countries, at the end of 1970 OPEC decided, at the 20th Conference in Caracas, to take a direct role in negotiations with the oil companies: establishing its own package of requests and offering to act as a contractual body to represent the member countries. The majority of companies were happy with this decision, seeing the formal recognition of OPEC as the only way to limit the individual countries' pursuit of increases in price and unilateral revisions of the existing contracts. On 10th January 1971 in New York, the majors and some of the main independents sent a joint message to OPEC in which they declared their unwillingness to negotiate separate contracts with the individual member countries. No less important was their decision to adopt a plan of aid against possible production cuts, as a means of resistance to pressure, attacks and retaliation against single companies. This enabled the *Teheran Agreement* to be reached on 14th February 1971 with the Persian Gulf States, and shortly afterwards other similar agreements were made with the other producing countries.[9] These had dual significance. On the one hand, the economic relations between companies and States were reconsidered in full, to the advantage of the latter and to the total detriment of the consumers (to whom the companies entirely shifted their increased fiscal costs).[10] On the other hand, OPEC's status as an *international political entity* was ratified for the first time in relation to the oil companies as a whole and implicitly in relation to their states of origin. The management of oil prices, from then on, would no longer be handled unilaterally by the companies, but *jointly* by the companies and OPEC together: the latter thus asserting itself as a new centre of power in international economic relations.

The *Teheran Agreement* created a *new pricing system*, which was to be applied for 5 years and then reviewed. Its basic contents were: a generalised and uniform increase of 35 cents per barrel for all the crude oil posted prices in the Gulf area; an annual increase of 2.5% in these prices to compensate for erosion in the buying power of oil revenues; an increase from 5% to 55% of the income tax rate: resulting in an increase in government duties per barrel from 56.25% to 60.625% of the posted prices; abolition of all the discounts that the companies had enjoyed; and revision of the system of differential prices between the various crude oils according to quality and location.

Table 5.3. *Crude oil market prices (doll./bbl)*

Crude oil	°API	1967	1968	1969	1970	1971	1972	1973 1st quarter
Iran	34°	1.45	1.27-1.42	1.25-1.38	1.35	1.78	1.97	2.15
Kuwait	31°	1.40	1.22-1.39	1.20-1.39	1.27	1.66	1.80	1.97
Saudi Arabia	34°	1.53	1.27-1.40	1.25-1.38	1.35	1.78	1.95	2.10
Qatar	41°	–	–	–	1.52	1.94	2.06	2.25
Libya	40°	–	1.80-1.90	1.65-1.70	2.15	2.75	2.85	3.15
Algeria	40°	–	1.90-1.92	1.90-1.98	2.40	2.64	2.80	3.20

N.B. Where two quotations appear, one is the maximum and the other the minimum.

From the start of 1970 to the start of 1973 the posted prices increased by 40-50%, rising from 1.80 to 2.59 doll./bbl in the case of Arabian Light. *Fiscal costs* increased even more, by 70-80%, eventually representing almost 2/3 of the overall revenue. *Market prices*, applied to (marginal) free transactions,[11] increased by up to two times, to values exceeding the posted prices, where before they had been 20-30% lower (Tab. 5.3.).

5.5. THE OIL CRISIS OF 1973-74

The policy of *joint* price administration between OPEC and companies ratified in the "Teheran Agreements" soon revealed intrinsic weaknesses, showing itself unable to cope with the market dynamics and the development of the power equilibrium. The increasing tensions registered in the demand-supply ratio in the early part of 1973 pushed the market prices of many crude oils well above the posted prices on which the tax revenues were calculated. An atmosphere of growing unrest was spreading through the markets, among the operators and in the mass media. "The oil crisis: psychosis or reality?" was the headline in "Le Monde" of 24th August 1973. Since the beginning of the year, the entire structure of market prices had been rising, beginning with the prices of oil products negotiated at Rotterdam and in the Mediterranean. Fuel oil prices rose from 17 doll./ton (the 1972 average) to about 30 dollars in June 1973, diesel prices went from 27 to 66, gasoline prices from 39 to 112 doll./ton. These were increases of between 80% and 180% which inevitably reverberated on the market prices of crude oils but only marginally on the posted prices. In other words, it was not the increases in posted prices – though dominant in the transactions – which led to increases in the prices of oil products. In fact if anything it was the opposite. Another fact of great importance, which more than anything gave rise to *expectations* of crisis from the operators, was the explosion of the spot freight rates. From average values in 1972 of around 5-6 doll./ton (from the Persian Gulf to Northern Europe), they leapt to 12 dollars in January 1973, to 24 in June and to 31 in September. This was a race to seize the available shipping, to anticipate product purchases and thus face the crisis that the operators, though not the governments, expected. The exceptional profits that the companies were gaining persuaded the governments of the producing countries to denounce the agreements of two years before and demand that new negotiations begin. The date of the meeting was the 8th October 1973 in Vienna.

While the delegations of the companies and the States were on the way to that important gathering, something happened which was to influence its outcome once and for all. On the evening of the 5th October 1973 – the day of the most important Jewish festivity, Yom Kippur – Egyptian and Syrian troops suddenly invaded Israel. From that moment politics was to have clear predominance over economics. The oil companies' proposal to increase the official posted price by 15% was rejected as "ridiculous" by the producing countries, whose counterproposal was a decisive *doubling* to 5-6 dollars per barrel. Refusal by the western governments marked the failure of the Vienna negotiations. Thus the path was cleared for *unilateral price fixing by OPEC*, which took place in Kuwait on 16th October 1973 at the meeting of the producing countries of the Persian Gulf. In "compliance with the right of sovereignty of each State over its natural resources", oil prices were drastically increased: that of Arabian Light – the benchmark for the other crude oils – by 70%, from 3.01 to 5.12 dollars/barrel.

On the following day, the 17th October 1973, the Arab members of OPEC united in the *Organisation of Arab Petroleum Exporting Countries* – OAPEC (Saudi Arabia, Kuwait, Iraq, Qatar, Libya and Algeria) decided (a) to block exports of oil to the United States and Holland, "guilty" of having provided economic, political and military support to Israel; (b) to reduce the oil production in October by 5% with respect to September, and by a further 5% in the following months "until Israel had totally withdrawn from all the Arab territories occupied in June 1967 and until the legitimate rights of the Palestinian people had been restored". The non-Arab OPEC countries – Iran, Venezuela, Indonesia – declared, for their part, that they would not increase their output so as not to interfere with the decisions of OAPEC.

Although there was no formal connection (due to the diversity of the protagonists and the motivations) between the decision to reduce production and the decision to increase prices, the relation of close *functionality* running between them was clear. *Generally speaking, the oil embargo had led to the increase in prices.* In the same way, while the United States was the main target of the attack (despite being their main political ally) in the intentions publicly manifested by the Arab States (by Saudi Arabia and Kuwait in particular), it was clear that Europe and Japan were the hardest hit. While an embargo towards the United States could, indeed, mean a possible energy deficit of just 1.4% for that country, the impact on Europe was much more serious with a reduction of 40%, and 28% for Japan. What is more, in the United States the negative impact would mainly be felt in the transport sector, but in Europe and Japan it was mainly supplies to industry which were threatened, with not much chance of reducing demand without affecting levels of activity. A further element of weakness for Europe came from the discrimination shown towards individual countries: *total* embargo towards Holland; *freezing*, at September 1973 levels, exports towards "friendly" nations (France, Spain, Great Britain); and proportional *redistribution* of the remaining quantities among the other countries. The embargo against Holland had particular significance, not for the extent of its consumption, which was in truth very small, but due to the strategic importance that Rotterdam played in the European oil system. Indeed, 2 mill. bbl of oil and oil products were sent out from there every day towards the neighbouring states (Clô and Prodi, 1975).

The situation that Europe found itself in was certainly quite delicate but not *dramatic*. The degree of autonomy it enjoyed amounted on average in the various countries to 90 days of operating stocks and another 30 days considering the shipping already at sea. These were joined by 80 days' worth of coal stocks. What is more, Europe and the other consumer countries gained advantage from the enormous confusion that reigned at the time about how to interpret the decisions made by the Arab nations and, above all, from the fact that in the majority of cases they were absolutely incapable of managing the reduction of the production as they indicated, or even of checking the actual destination of their exports, so the companies could clearly send them where they pleased. Significant, and not lacking in humour, was the decision of the Algerian government to send a government official with the tankers setting sail to check the final destination of the load. Holland's request for solidarity from her European partners, and on the other hand, France and Great Britain's anxiousness to avoid any confrontation with the Arab nations, made it impossible to define any common action at a European level. Thus a vulnerable state of indecision was created, with a tacit tendency to delegate to the oil companies the responsibility of allocating the "shortage" of oil to the various countries. In spite of the changes which had occurred, the seven majors in fact still controlled 65% of European supplies. The failure of both producing and consuming countries to face the development of the crisis meant that it was more or less managed by the international oil companies.

It is not surprising, therefore, that the re-allocation of the available supply on a world scale responded, first of all, to their specific commercial interests and that these did not necessarily coincide with an equal distribution of the reduced supplies available among the various countries (Penrose, 1975, p. 42). The majors tended on the one hand to discriminate in favour of their own subsidiaries, to the detriment of the independent companies, and on the other hand, to favour those countries with the highest domestic price for oil products. Unlike in previous crises, that of 1973-74 saw no serious form of co-operation between the large companies, between the companies and the governments, nor between the governments themselves (again revealing the changes that had been observed on the oil market).

Relations between States and companies were also worsened by the belief (not always supported by facts) that the companies were in some way conditioned by the interests and politics of the US government, though operating autonomously. There was also widespread dissatisfaction about the ambiguous role of the companies in managing the crisis itself. There was a widespread impression (later confirmed by the enquiries of the American Congress) that the companies had been preparing for a long time to face a reduction in oil exports. It was difficult to verify the accuracy of this suspicion but it only increased the governments' sense of frustration and the aversion of public opinion towards them.

However the fact remains that European governments, and those of the consuming countries in general, never really knew what was happening, what the real needs were, or who and what had to be dealt with. Many specialist periodicals and journals even printed on the blackest days of the crisis news about the trouble the European ports were having in receiving all the incoming tankers or about the inadequacy of the storage facilities for excess oil. And this was while all the governments were announcing the need to adopt measures of rationing and austerity. The question that "The Economist" asked itself in the midst of the embargo (in the issue dated 15th December 1973) was significant: "How scarce is oil?"

The production of crude oil was cut between the third and the fourth quarter of 1973 throughout the Arab nations. The reductions ranged between 20% in Saudi Arabia, 16% in Kuwait, 13% in Algeria, and 11% in Libya. The fears that these cuts would lead to an oil shortage (as hardly ever happened) created a situation of absolute panic, especially in the United States, and the market prices shot up to over three times the posted prices. Afraid of being strangled by the embargo, the offers of independent refiners, State companies and major consumers pushed the auction price of crude oil to incredible heights.[12] In this climate of agitation and great confusion an OPEC conference was held in December 1973 in Teheran at which a further increase of 130%-140% of the oil price (according to the qualities) was decided. Arabian Light jumped from 5.04 to 11.65 dollars per barrel: 6.2 times the level of 3 years before.

5.6. INTERPRETATIONS

As the days passed it became increasingly clear that, if real, the shortage was only limited and short-lived. Only very rare cases of shortage occurred and these were due to sluggishness in the distribution chain rather than to an actual shortage of crude oil on the markets. In the first quarter of 1974, oil was easily available throughout the world, albeit at the new price levels. The overall production of the OPEC countries in 1973 was on average completely in line with the expected and actual increases in demand. The reductions in supply by the Arab nations in the last quarter of the year compensated for the huge and highly irregular increases recorded in the first three quarters,[13] while the production of non-Arab nations remained unaltered. So there was no actual oil shortage, or at least not such as to justify the explosion of prices to more than sixfold. However this only became clear after the event.

It was this very observation which made some authors claim a close *convergence of interests between producing countries and oil companies*. The increase in the production rate in early 1973, especially in Saudi Arabia, was apparently planned specifically by the majors in view of the expected breakout of hostilities; the producing countries would then reduce the output to compensate for the previous increase. "The commonly advertised and accepted explanation for the events of 1973-74 is profoundly steeped in folklore. Through the insistence of every communication, the responsibility was effectively shifted onto the governors of distant and undeveloped lands, with the result that the high prices are invariably put down to crude oil shortages" (Blair, 1977). Blair and others were convinced that, though they had different aims, both producing countries and majors fully agreed with the increase in prices. The majors aimed to increase the profitability of the investments that they were making (and particularly the investments of the future) in marginal areas to create a new base for themselves outside the Middle East. The *producing* countries wanted to maximise their revenues, bringing oil prices in line with the user-cost of the alternative sources, in order to seize the whole consumer surplus for themselves.

Observation of the positive effect that the price increase had on the profits of the majors has led some authors to suggest that the crisis was in some way favoured or at least not prevented by them and by the United States government. This theory attributes the leap in prices to a strategic manoeuvre by the US government with the aim of improving the competitive position of the American oil industry. This theory is certainly difficult to verify since we must be very cautious in interpreting any apparent causality between the numerous events which characterised this period; however there is no doubt that a sharp and prolonged rise in oil prices was *instrumental* to the energy policy that the American administration had started at the beginning of the Seventies.[14]

On the 2nd of February 1970 the "Cabinet Task Force on Oil Import Control", chaired by George Schultz, sent the US President Richard Nixon the final report entitled *The Oil Import Question: A Report on the Relationship of Oil Imports to the National Security.* In 400 printed pages he outlined to the President and the country the situation in which the United States would find itself, just a few years onwards, if the energy policy of the country did not undergo drastic modification. Domestic oil production would be unable to cope with the expected growth in consumption and thus the administration would inevitably be obliged to lift the ban on imports in force since 1959. In little more than a decade – stated the Report – imports would increase 4 fold (to 10 mill. bbl/d) and foreign dependence would increase 2.5 times (to more than 50%). The United States ran the risk of losing their "energy independence" just as the Soviet Union was consolidating its position as net exporter. "The risk of disruption in oil supplies is a severe threat to the functioning of the American economy with possible serious effects on the nation as a whole". This severe admonishment concluded the Schultz Report. It caused a huge stir among Congress as well as in the mass media and among the American public.

Action had to be taken to avoid a situation which would place the country in an unbearable condition of external vulnerability. This required a drastic intervention: increasing the domestic supply, both of oil and of other sources; limiting and reconverting the demand towards non-petroleum sources; and resorting to oil imports. However for this strategy to work there had to be a sharp rise in domestic oil prices to allow new oil fields with higher marginal costs to be developed, and to reach the threshold of profitability for other energy sources. These circumstances could not come about, however, given the negative effects that the liberalisation – though necessary – of *oil imports at much lower prices than domestic oil* would have on the structure of domestic prices. A study by two economists, Burrows and Domencich, of Cambridge (Mass.), arrived at this conclusion, and formed the main economic basis of reference for the political conclusions of the Shultz Report. In their opinion the increase of the US demand for oil towards OPEC countries would make a reduction in their prices convenient, since the elasticity of the new aggregated demand curve was greater than the previous one (Burrows and Domencich, 1970) and since American foreign demand was more elastic than the European demand. A liberalisation of imports would thus reduce the international prices (so the American administration argued), making it more difficult to bring about a structural adjustment of the American energy system. The United States thus had an objective interest in an increase in the international oil price. This does not mean, *in itself*, that they were in some way to blame for this increase, but rather that they did not do anything to stop it.

In a similarly well-known report by the National Petroleum Council of December 1972, the need for an increase in oil prices of between 60% and 125% according to the different scenarios of the growth of the economy was explicitly stated (NPC, 1972, p. 7). On 18th April 1973 President Richard Nixon sent the country a dramatic "Energy Message" in which a black future was painted for the American nation. On the thirteenth line he declared that "in the next few years we will have to face the possibility of an energy shortage and energy price increases". Few paid attention to those prophetic words.

However, it is a fact that the international prices of crude oil – under the pressure of growing fiscal costs and of increasing market tensions – began to rise steadily in early 1973, gradually reducing and then overturning the differential they had previously had on the more expensive American crude oils. The spread, for example, of Arabian Light crude oil on the east coast of the United States with respect to a homogenous East Texas crude went from -0.60 dollars per barrel in 1968-70 to close to zero in 1972 up to

+1.40 doll./bbl in August 1973. This upheaval of the relative prices enabled (exactly in August of that year) the American government to *eliminate the oil import quota* . Domestic prices and profitability immediately began to rise, inducing American companies once more to invest huge amounts in their country of origin. Exactly as the Washington Administration wished.

Another way to interpret the oil crisis of 1973-74, as well as the no less disruptive crisis due to occur at the end of the decade, is given by the theoretical analysis extrapolated by the numerous microeconomists who concerned themselves with the world of oil at that time. The most widely shared theory suggested that prices rose due to the cartellisation of the market by OPEC, which had used its market power to reduce production and force prices over the levels of competition. Robert Pindyck (1978) made the most complete extrapolation of this interpretation, using an applied model of wealth maximisation, in reference to the 1931 contribution by Hotelling. According to this theory, the owner of exhaustible resources fixes the prices or the quantities in time, in order to maximise the present net value of the future flows of profit. In a competitive regime, the quantities will be fixed in such a way that the prices (less the marginal production costs) increase in time at the rate of interest; while in a monopoly regime it will be the marginal revenues (again, net of the marginal extraction costs) which will increase at the interest rates.

This approach was for a long time considered the best way to explain what happened in 1973-74 and to predict the future. The subsequent delusion was equal to the initial approval. The difficulty in adopting simplifications necessary to the mathematical formalisation of the model; the fact that it did not take into any account the uncertainty with which the OPEC countries had to deal (as regards demand, elasticity, income, etc.); the difficulty of these models in explaining the initial price level; the limits to upholding the hypothesis of wealth maximisation: these are just some of the reasons behind the rapid decline of this analytical approach. "From a theoretical point of view," stated Pindyck, disappointedly, "models of OPEC oil pricing have reached a practical limit as tools of analysis. ... Economic rationality probably applies even less to OPEC producers than to many other real world economic agents" (1982, p. 179).

5.7. THE OIL CRISIS OF 1978-79

In the mid-Seventies the oil market, after the deflagrations of the first crisis and the heavy economic recession which followed it, seemed to regain a relative equilibrium with prices which were steady in nominal terms but strongly declining in real terms because of the high rates of inflation in those years. Using the export price index of industrialised countries as a measure of inflation imported from OPEC countries, the price of 12.70 doll./bbl of the marker crude in force in June 1978 was worth 7.70 when expressed in 1973 dollars, with a drop in real terms close to 40%. The erosion of the purchasing power of oil revenues (which dominated their total available funds) pushed producing countries once more to raise the question of the *international terms of oil trade*, advancing the old idea (applied for a brief period) of linking its price to that of a basket of imported goods. This idea clashed, however, with the depression of the international market and with the considerable erosion of OPEC in the international market share after 1973: falling from a peak of 54% of world oil production to 48% at the start of 1978.

Another, and only apparently contrasting trend consolidated in those years in the culture of the Arab countries, was that towards a more careful *conservation* of their oil resources. The concept of the optimal exploitation of oil, which had always been debated by theoretical economists and oil engineers, began to spread, encouraging several Arab countries to introduce *ceilings* on their annual production. This decision was more nominal than real, given the drop in demand, but for the first time showed the intention to set up a restrictive policy regarding oil supply not only aimed at supporting the prices.

The relative stability of the nominal oil prices in the mid-Seventies was due to the large companies' ability to control structural transformations, regaining at the crude oil trading stage the central role they had lost for the production stage. While it was certainly true that juridically, the producing States had gradually taken over full control of the exploitation of their resources, it is also true that they had remained substantially extraneous to the stage which had always been the most critical of the whole oil industry: the trading and management of the world crude oil supply. This stage had tended to remain in the hands of the largest multinational operators. The risk of actually having to reckon with the markets had in fact pushed almost all the producing countries, following the nationalisations or the participation agreements, to support what had hitherto been considered their historical and natural counterpart. In other words they preferred to resell almost all the crude oil they owned to the companies themselves rather than market it directly. What Yamani in 1973 called the "catholic marriage" between producing countries and major oil companies actually *cancelled every area of free market* and therefore the very significance of the prices practised therein.

While on a *formal* level the power of oil price fixing was officially and unilaterally attributed to the producing countries, *in actual fact* it was the major companies who guaranteed the respect and steadiness of these at world level. The detailed logistic system that they could rely on was an essential instrument of the global and centralised management of the supply flows, and this was the only way to anticipate or prevent falls in prices caused by oil surplus. The integrated and co-ordinated structure of the oil companies was a formidable protective mesh within which the market forces could easily be restrained, as was widely experimented in the years 1974-78 featuring low demand, abundant supply and substantially firm prices. That it was the companies who supported the prices was confirmed by the fact that the few discounts on the official quotations were conceded in the marginal sales carried out directly by the producing countries or by their national companies.

The joint administration of the market by exporting States and companies was once more not to last. As in 1973-74, it was a politics that precipitated events: the destitution in Iran of the Shah Reza Pahlevi – a traditional ally of the western world – and the ascent to power of the Ayatollah Ruhollah Khomeini. Iran was then the fourth largest oil producing country in the world with 6 million barrels per day (Tab. 5.4.), and the second largest exporting country after Saudi Arabia with 4.5 mill. bbl/d. At the beginning of November 1978, this flow was reduced to just 1 mill. bbl/d due to the serious strikes and rioting that broke out throughout the country. At the end of December it was almost zero. On 16th January 1979 the Shah abandoned the country. On the 1st February the elderly Khomeini returned and gained control of the country, immediately triggering a harsh political conflict with the western world. Panic-stricken, the markets reacted in an unreasonable way.

The events of those days were (once more) perceived and interpreted as a classic problem of *shortage of supply* which inevitably pushed prices higher until the excess in demand was cancelled. Only a few thought differently and one of those was this author

Table 5.4. *World oil production in the second oil crisis (mill. bbl/d)*

	1978									1979	Producing
	I	II	III	IV	Sept.	Oct.	Nov.	Dec.	Year	Jan.	capacity
Saudi Arabia	7.9	7.6	7.5	10.0	8.4	9.3	10.3	10.4	8.3	9.8	11.8
Iran	5.5	5.6	5.9	3.8	6.0	5.5	3.5	2.4	5.2	0.4	6.9
Iraq	2.3	2.4	2.7	3.1	2.9	3.0	3.1	3.1	2.7	3.1	4.0
Kuwait	1.8	1.9	2.3	2.3	2.6	2.1	2.6	2.2	2.1	2.6	3.3
Venezuela	1.8	2.2	2.2	2.3	2.3	2.3	2.2	2.3	2.2	2.2	2.5
Nigeria	1.5	1.7	2.0	2.3	2.1	2.1	2.3	2.4	1.9	2.4	2.4
Libya	1.8	1.9	2.0	2.1	2.1	2.0	2.2	2.2	2.0	2.1	2.5
Algeria	1.2	1.2	1.2	1.2	1.2	1.2	1.2	1.2	1.2	1.2	1.2
Others	4.6	4.6	4.5	4.4	4.4	4.4	4.3	4.5	4.3	4.6	5.2
OPEC	*28.4*	*29.1*	*30.4*	*31.4*	*32.0*	*31.9*	*31.7*	*30.7*	*29.9*	*28.4*	*40.2*
Var. %[1]	–9.5	–7.0	0	0	–3.2	+5.2	+0.6	–5.5	–4.1	+3.0	–
USA[2]	10.1	10.5	10.5	10.4	10.4	10.4	10.4	10.4	10.3	10.4	10.4
North Sea[3]	1.3	1.4	1.5	1.6	1.4	1.5	1.6	1.6	1.4	1.7	2.3
USSR	11.3	11.4	11.5	11.8	11.4	11.7	11.8	11.8	11.5	11.7	n.a.
Others	9.0	9.3	9.3	9.1	9.4	9.2	9.0	9.1	9.8	9.6	n.a.
Non-OPEC	*31.7*	*32.6*	*32.8*	*32.0*	*32.6*	*32.8*	*32.9*	*32.9*	*32.5*	*33.4*	*n.a.*
Var. %[1]	+5.3	+7.2	+6.7	+4.1	+5.5	+4.8	+4.1	+3.0	+5.2	+6.7	
Total	*60.1*	*61.7*	*63.2*	*64.5*	*64.6*	*64.6*	*64.5*	*63.6*	*62.4*	*61.8*	
Var. %[1]	–2.4	0	+2.9	+2.1	+3.7	+4.4	+2.1	0	+2.2	+4.9	

1. % variation over the corresponding period. 2. Including gas natural liquids. 3. Great Britain and Norway.
Source: OPEC: "Petroleum Intelligence Weekly" 29 January 1979; USA: "Petroleum Situation" - CMB; other countries "Petroleum Economist".

(Clô, 1979). Let us see why. The intention of the OPEC countries to recuperate the loss in real terms of the value of their crude oil exports had clashed, as was said above, with a situation of *relative abundance of supply* due to the stagnating international demand and the increase of non-OPEC production.

Back in mid-1978, the oversupply could be evaluated as 2-3 mill. bbl/d, more than 10% of the overall OPEC production. In such a situation, OPEC could not hope for large increases in prices, although a rise in prices was now expected and accepted by all. The huge increase in the buying of crude oil by the operators in the third quarter created conditions favourable for prices to rise on the market. The OPEC Conference at the end of 1978 voted for a 10% rise in official selling prices throughout 1979 by means of quarterly increases starting from the beginning of the year.

However, this realistic, not to say moderate, behaviour from the OPEC countries encountered unscrupulous reactions from other producing countries (particularly Great Britain) and from some international operators. OPEC's decision to stagger the price increases, although lessening the impact of a single and immediate rise, in fact created conditions favouring continuous speculative manoeuvres (purchasing advances, delays in deliveries). It is even more important to bear in mind what was happening on the spot

Figure 5.1. *World oil production and oil stocks*[1]

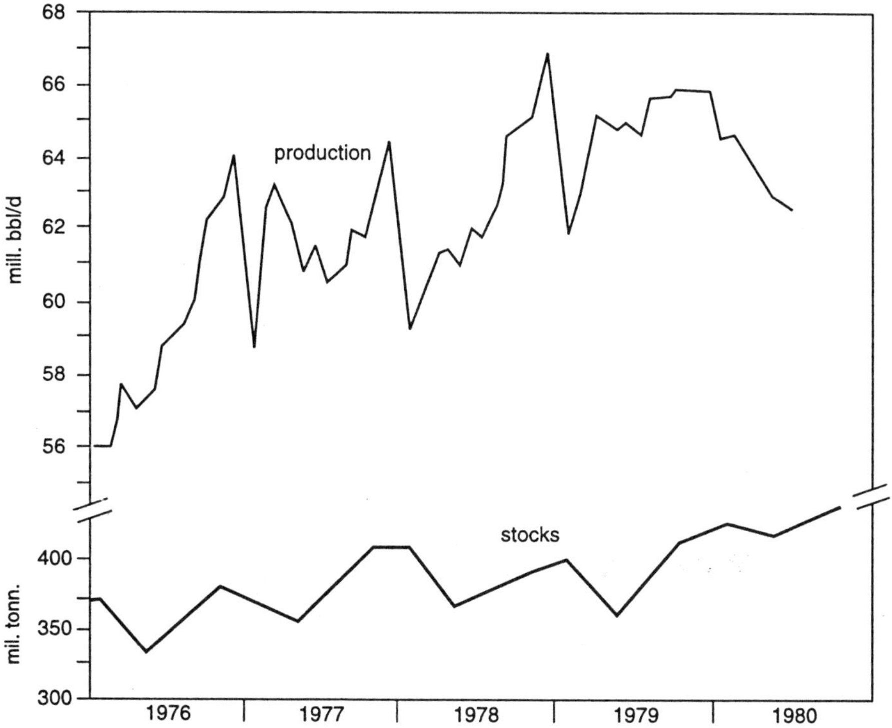

1. In the member countries of IAE, Paris.

market of oil products which, moreover, moved a completely insignificant quota of the overall final consumption. It is not wrong to state that it was the high quotations reached on this market that first anticipated and then accelerated the upstream rise in crude oil prices. Already in November 1978 a refined barrel of Arabian Light could be sold in Rotterdam at a price of 17.1 doll./bbl, against a purchase cost of 12.7 with a resulting profit of 4.4 doll./bbl. In February this margin had risen to 11 doll./bbl (Tab. 5.5.).

The manifested willingness of the consumers to acquire refined products at much higher prices than those implicit in the cost of the crude oil purchased from OPEC countries could not fail to encourage the producing organisation to increase their quotations, bringing forward the increases initially decided for the end of the year to the 1st April 1979, while leaving the individual member states free to practice any surplus charge that they were able to gain.

The race between spot prices and official selling prices (Fig. 5.1.) seemed to be unstoppable and completely unrelated from the real market dynamics. The average OPEC prices (Tab. 5.6.) jumped from 12.9 doll./bbl in December 1978 to 18.7 in June 1979, to 26 in December 1979, peaking at about 35 in December 1980, with an increase of 270% in 24 months. The lighter crude oils were well over 40 doll./bbl.

Table 5.5. *Oil price and net-back value in the Rotterdam market (doll./bbl)*

	1978						1979		
	I quarter	II quarter	III quarter	Oct.	Nov.	Dec.	Jan.	Feb.	Mar.
Arabian Light price	12.70	12.70	12.70	12.70	12.70	12.70	13.34	13.34	13.34
Net-back value[1]	12.70	12.86	12.91	13.55	17.10	14.94	17.47	24.69	22.44
Profit[2]	0	0.16	0.21	0.85	4.40	2.24	4.13	11.35	9.10

1. Gross product worth (weighted with refinery yield pattern) minus incremental refining cost and freight cost.
2. Difference between net-back value and official selling price.
Source: "Petroleum Intelligence Weekly".

Not without irony, OPEC rebuked the consuming countries for not managing to intervene on the international market to prevent the speculative activity of the operators, thus confirming that the astronomical prices were certainly not due to a situation of shortage of supply or of decisions made by OPEC itself. To placate the market it would have been sufficient for the central governments to set up a strict monitoring of the oil companies, refusing to grant them the highest quotations, or that even small amounts could have been drawn from the huge compulsory stocks that the countries of the International Energy Agency said they possessed (Verleger, 1982).

Figure 5.2. *Oil prices in the second oil crisis (doll./bbl)*

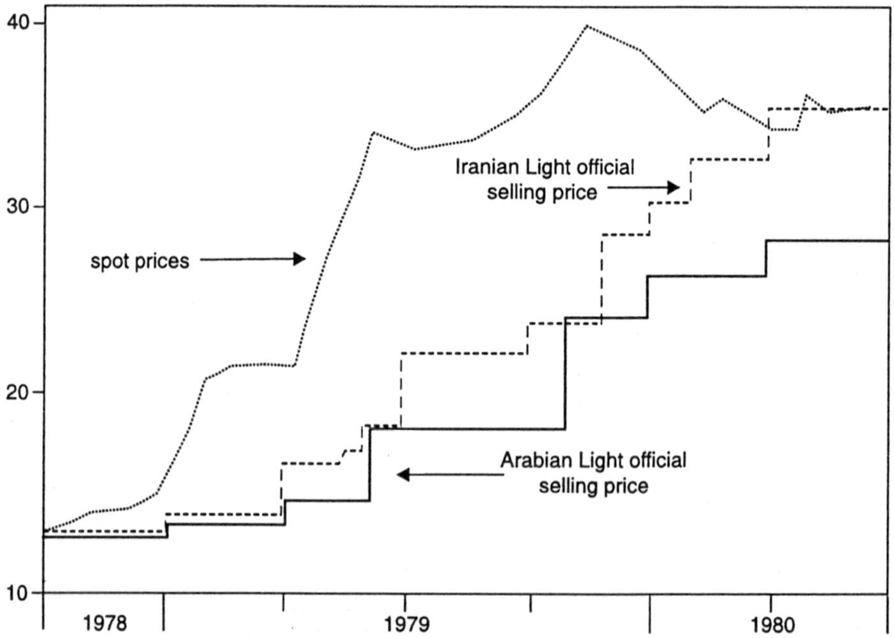

Table 5.6. *OPEC official prices in the second oil crisis*[1]

	doll./bbl	Index
December 1978	12.9	100
March 1979	14.0	109
June 1979	18.7	145
September 1979	20.3	158
December 1979	26.2	203
March 1980	29.1	226
June 1980	31.0	240
December 1980	34.7	270
June 1981	34.1	265

1. Average values weighted on the volumes of production of the individual countries.

The truth is that the Paris Agency was miserably unsuccessful in the tasks for which it had been set up, failing to give the consuming countries even the slightest correct information about what was happening as regards the supply and about the actual (and non-existent) situations of shortage.

The race between spot prices and official selling prices continued throughout 1979, completely *disconnected* from the market fundamentals. What elsewhere I have indicated as the *1979 paradox* was indeed to go down in the annals of oil industry as a year of *exceptional production* (+4.6% with regard to 1978), despite the Iranian crisis; of *low demand* (+0.2%), despite the sharp rise in GDP (+3.3%); of *high stocks* (+7%), despite the Iranian production decline; and *strong price increases* (+103% in the average OPEC values), despite all of this.

The profile of world production shown in Figure 5.2. efficiently summarises the dynamics of events. Production registered an *all time high* close to 67 mill. bbl/d, *coinciding exactly with the Iranian crisis*. The supplementary action of the other OPEC countries fully compensated the reduction in Iranian production by 6 mill. bbl/d between September 1978 and January 1979. The drop in global production of 5.2 mill. bbl/d at the start of 1979 was, indeed, completely in line with the seasonal oil production trend, even resulting slightly lower than that recorded in the previous two-year period. As in the last part of 1978, throughout 1979 production remained at high values, with a seasonal decline at the start of the Eighties which was actually reduced to less than a third of normal levels, while stocks, *not consumption*, saw a continuous and consistent increase. This series of events shows us how the price increases could *not* be due to cartel manoeuvres, nor even situations of aggregated shortage of supply, but does not tell us the *real reasons* behind them: there were several, in particular:

(a) *panic buying by consumers*: due to a mistaken perception of what was actually happening, as was candidly admitted later by the IEA of Paris itself;
(b) *reinforcement of the speculative component* of the market following the weakening of the dollar, the adoption of flexible exchange rates and expectations of continuous and further increases in prices (Sylos Labini, 1981). Speculative purchasing of oil or title-deeds was always good business, thus explaining why stocks continued to rise in the first quarter of 1980 despite the huge drop in consumption (Clô, 1980a);
(c) *changes in the structure of oil supply control*, which, as we have seen, were begun

in the early Seventies and completed at the very same time as the Iranian events took place. These changes as a whole were to lead to a *different configuration of the oil market structure* on which the pricing mechanisms most depend, and together they were to accentuate the reasons for the crisis.

5.8. SUPPLY AND DISTRIBUTION CONTROL

After the peak reached in the early Seventies, the overall availability of the oil of the majors underwent a massive slide, both in absolute terms and even more so regarding international transactions as a whole. From about 40 mill. bbl/d in 1970 (about 70% of the world production, excluding the countries with State-planned economies), it dropped to 30 in 1973, 23 in 1978 and about 20 in 1980. That was a drop of about 50% in one decade, with a loss of 27 points of market share at 43%. (Tab. 5.7.).

No less important is the change observed in the *structure of the availability* of the oil of the majors. At the end of the period, only half of this consisted of their own crude oil (known as equity), while the other half derived from agreements made with the producing countries at official selling prices. These agreements gave the majors a formidable *competitive advantage* over the other companies (forced to purchase at the higher spot market prices), but nevertheless they restricted their degree of decisional autonomy with respect to what they could do with their equity crude.

During the second crisis, the majors found themselves facing an unexpected series of negative events. First of all, the Iman Khomeini decided to break all contracts with the western companies of the Iranian Consortium, which since 1954 had controlled almost the total national production.[15] Secondly, Nigeria nationalised the assets of British Petroleum, which had already been hit heavily by events in Iran.[16] Thirdly, Saudi Arabia, Venezuela, Abu Dhabi and other countries decided to reduce the quantities of oil they had previously granted to the majors, preferring to market it directly through agreements with the governments or companies of the consuming countries. The sum of these events meant for the majors a further, unexpected and large reduction in the availability of crude oil, and the dual necessity to (a) cut, to just a fifth (from 3.7 to 0.8 mill. bbl/d) the supplies existing with third-party contractors (independents, refineries, traders, etc.); and (b) to draw directly from the more onerous spot market, in order to compensate situations of unbalance that had been created in their integrated circuits.

For the first time the majors began to operate largely on the free market, from which they had hitherto kept away, considering it not unreliable and a cause of instability. The implications of this went far beyond simple fact – it indicated an overall change in the strategy of the majors in the international market, given their altered and reduced role. Their main objective, in the conditions, was to defend the positions they had obtained in the final markets rather than encourage (as in the past) the stability of crude prices. This was a responsibility by now outside their scope, and to which they would have been able to contribute only by abstaining from purchasing on the spot markets at marginal prices (to cover themselves for possible shortages of product) and sacrificing shares of final market. No longer able to act as price makers – a role which was covered fully by the market, rather than by the OPEC countries – they positioned themselves so as to exploit any profit opportunity. Despite the considerable drop in the quantities of crude oil traded, the total profits of the seven majors increased 2.8 times between 1973 and 1980 to 23 bill. doll. and 3 times in unitary terms to 3 doll./bbl (Tab. 5.7.).

Table 5.7. *Majors' oil and derivatives supply control (mill. bbl/d)*

	1973	1978	1980	1985	1990	1995	1998
Oil supply:	*30.0*	*23.3*	*198.6*	*9.3*	*7.5*	*7.9*	*8.9*
– Equity	25.5	13.4	9.7	7.3	7.5	7.9	8.9
– Preferential agreements[1]	–	7.7	7.5	2.0	n.a	n.a.	
– Other agreements	4.5	2.2	2.4	–	–	–	
Oil destination:	*30.0*	*23.3*	*19.6*	*9.3*	*7.5*	*7.9*	*8.9*
– to own refineries	23.3	19.5	17.9	9.3	7.5	7.9	8.9
– to a third party	6.7	3.7	0.8	–	–	–	
– to others	–	0.1	0.9	–	–	–	
Product sales:	*24.3*	*22.4*	*20.4*	*18.1*	*19.5*	*22.3*	*25.3*
– from own refineries	22.9	19.1	17.6	9.3	7.5	7.9	8.9
– net purchases	1.4	3.3	2.8	8.8	12.0	14.4	16.4
Integrated profits:							
– total (bill. doll.)	9.1	21.7	23.0	16.4	20.1	22.6	21.0
– unit (doll./bbl)[2]	1.0	2.6	3.1	2.7	2.8	2.8	2.3
Market share[3]*:*							
– oil supply	65%	49%	41%	22%	15%	14%	14%
– product sales	55%	44%	43%	42%	36%	37%	40%

1. In 1978: 3.6 mill. bbl/d for Aramco and 4.1 for Iran Consortium. In 1980: 7.5 mill. bbl/d for Aramco and zero for Iran Consortium. 2. Of crude oil bound to refineries or of products. 3. Ratio between integrated profits and sales. 4. Excluding countries with state-controlled economy.

During the second crisis, the normal circuits of distribution of crude oil underwent profound laceration. The entire structure of the existing contracts – between producing States and majors, between these and the independent companies, and even between branches of the individual majors – was affected. This triggered off a *series of tensions* which also reached the final consumption markets. Besides the true ratio between aggregated demand and supply – which had been seen in full equilibrium – the commercial circuits, the individual refineries and the individual operators found themselves having to cope with a *shortage of product*, due not so much to the Iranian crisis as to the breaking of the normal long-term supply contracts and to the timing lag that occurred in the conclusion of new contracts on the spot market or in direct purchases from the producing States. The additional demand, unexpected and copious, that flooded the spot markets inevitably caused pressure towards a price rise, which was to stop only when new circuits of distribution had been completely redefined.

Another important effect of this redefinition of distribution was the *differentiation of the costs of supply* forming between the various companies, depending on whether they could purchase at official selling prices, at spot prices or at preferential prices. While the majors could still widely draw on equity crude oil or count on preferential agreements with the producing States (at official prices or lower), the other companies depended to a greater extent or completely on the spot market.

The advantages from which the majors benefited allowed them to accept a "security premium" on marginal lots which could reach 6-8 doll./bbl (on a base price of 34 doll./bbl). Given the minimal quantities concerned, their *average cost* was in any case lower than that of the other companies. This diversity of situations and each company's haste to cover themselves was of enormous importance. As regards the price dynamics, *the particular equilibrium of the individual operators was more important than the general equilibrium of the market.* Those who found themselves in situations of actual or anticipated shortage tended to purchase at increasingly high prices, in order to reconstruct or preserve their supplies, while those who had excess supplies preferred to increase stocks by postponing sales.

The differences in costs then led to differences in the *profits* of the companies. Profits for independents or integrated minors were always *lower* than those of the majors or even *negative* when the price rise of refined products was delayed with respect to that of the costs of the raw material. Indeed, the final prices (net of taxes), though at higher levels than the official oil quotations, did not manage to fully repay the maximums paid on the spot markets. This occurred particularly in those countries, including Italy and France, where there was a system of administrative control of the prices. On the other hand, the alternative was whether to fix the prices to the levels of the *average* costs remunerated by all the *system* (thus penalising the marginal operators) or define them on the *marginal costs* of the non integrated operators (thus creating enormous revenues for the majors and fuelling inflation, which was already high). Price control seemed for many governments to be the only tool which could halt the spiralling prices. Controlling the final prices meant rejecting that oil policy favoured by Great Britain *at all costs* (since it was also a seller) and not unfavourable to Germany, though there would be inevitable repercussions (or even just threats of repercussion) as regards supplies.[17]

To conclude, the differentiated structure of the costs allowed the majors high profit margins while penalising the independent companies until some were even forced out of the market. The redefinition of the circuits of oil distribution, associated with the disappearance of the single price system, thus paradoxically reinforced the relative position of the majors which the action of the producer countries wanted instead to penalise.

5.9. DISTRIBUTION OF THE OIL SURPLUS

It is generally thought that the central element of the oil crises of the Seventies was the increase of the oil prices *at origin*, entirely under the responsibility of the producing countries. Little or no attention has instead been given to the dynamics of prices *at consumption* of oil products, especially as regards the quota owing to the fiscal authorities of the consuming countries, particularly the European ones. The same error has also been made by a number of scholars. Most of the mathematical or descriptive models of oil demand have, for example, taken the price of crude oil as its principle determinant, completely disregarding the price of the final products from which it derives and to which it responds (Fischer et al., 1975). This is one of the reasons why so many forecasts of demand trends have been unsuccessful.

In reality, most of the refined barrel is sold to the final consumers at prices which have little connection with the prices it fetches on the international markets, with the specific industrial costs, and even more with those of the incorporated raw material. This is due to the high percentage of taxes on consumption of the oil products: in other words the part of the oil surplus – in this case the consumers' surplus – taken by the consumer States. The most striking case is Italy. At the end of 1995 the average weighted price of oil products was about 121 dollars per barrel in Italy, compared with 35 of the average price net of taxes and just 18 of cost of raw material. 86 doll./bbl or 71% of the final price was made up of taxes. From an annual petroleum consumption of around 2 million barrels per day, the Italian State collects taxes equal to what Saudi Arabia takes altogether with a volume of exports of 8 million barrels per day. The figures speak for themselves.

Oil product prices in almost every European country are linked more to taxes than to costs: especially in the case of products used in the household and in transport sectors. This was true in the Sixties and is still so in the Nineties, despite the two oil crises that we have analysed. In the competition between producing States, consuming States and companies to increase oil revenues, the former certainly cannot be said to have won, although they have been on the receiving end of most of the criticism and discontent (Angelier et al., 1983).

In reality, if we consider the whole period from 1970 to 1995, we may see that the blame that can be attributed to the consuming States (especially European) is much greater than that attributable to the producing States, despite more attention having been given to this latter.

Table 5.8. supports this statement. In 1970, with values close to those of a decade earlier, the unit revenues of the producing countries amounted to just 0.86 dollars per barrel, compared with about 1 of company profits and 8.3 of fiscal revenues of the European States (while the overall industrial costs amounted to 4.3). In ten years the value and allocation of the petroleum revenue has changed radically. The increase of about 4 times the final average price – from about 14.50 to 63 dollars per barrel – was to the advantage above all of the producing States, who saw their participation leap from 6% to 52%. The reduction of the quota received by consuming States and companies was still accompanied by a considerable increase in their revenues in absolute value. The taxes on consumption increased, in fact, about 2.5 times, from 8.3 to doll./bbl. while

Table 5.8. *Price structure of the oil consumed in the European countries (doll./bbl)*

	Unit revenues of the producing	%	Fiscal revenues of European countries	%	Industry costs countries and margins	%	Final price	%
1961	0.76	6	7.10	52	5.70	42	13.60	100
1970	0.86	6	8.30	57	5.30	37	14.46	100
1975	10.10	30	14.90	45	8.20	25	33.20	100
1980	33.00	52	22.00	32	10.00	16	63.00	100
1985	33.00	46	28.00	39	10.50	15	71.50	100
1990	25.00	26	55.50	58	15.00	16	95.50	100
1995	20.00	18	78.00	69	15.00	13	113.00	100

Source: 1961-75: "Petroleum Intelligence Weekly", 16 May 1977; 1980-95: our calculations.

the integrated profits of the companies tripled to 3.0 doll./bbl (Tab. 5.7.).

During the period 1980-95 there was a new rearrangement of positions. The final average price – despite the fall (which we will be examining later) in the oil prices – increased by 50 dollars per barrel (much more than in the two previous crises), owing to the approximately fourfold multiplication of the taxes on consumption to 78 dollars per barrel, while the producing countries revenues fell sharply to 20 doll./bbl. In the redistribution of the oil surplus today, the consuming States are in a completely dominating position, with a share close to 76% (if we subtract about 10 dollars per barrel industrial cost from the final price), while the producing States hold just a 20% share, although oil for the majority of them is the only and often insufficient source of income. But as we said at the beginning of this section, little or no attention has been given by observers and scholars to this aspect. It is almost as if the tax burden was light on the economies and the consumers, and as if the struggle for the redistribution for the oil surplus could be considered irreversibly concluded.

5.10. CONCLUSION

While it cannot be denied that the events of the Seventies in the oil market were most immediately motivated by the exhaustion of the margins of *flexibility* permitted to the international system by an *excessive* production capacity, it is also true that the *outcome* of those events (particularly on the prices) could be attributed to the modifications which had been developed in the *structure* of the oil industry. The old order that the majors had been able to establish was followed by a "situation of confusion which leads rapidly towards a state of general chaos" as Paul Frankel commented bitterly (1982). A *convergence* of interests was replaced, among all the participants in the "new oil game" (as Chevalier defined it in 1973), by *conflicting* behaviour incapable of generating stability. While the substantial stability of the prices in the Fifties and Sixties had been a sign of the global order guaranteed by the majors, "the current chaotic, irrational and capricious behaviour of the price is the consequence of a disordered system in which no force in the field – and not even OPEC – is able to impose itself on the others so as to recreate a new situation of equilibrium and stability". This was the inevitable situation which was to come about with the *birth of a true international oil market*.

Notes

1. "Modern" sources mean commercial sources, traded on the international or domestic markets. Biomasses are also excluded (especially wood) which still represented the majority of consumption in the first two decades of the 20th century.

2. On a *world scale* the structure of energy consumption changed between 1950 and 1972 as follows (in %):

	1950	1960	1970	1972
Coal	55.7	44.2	31.2	28.7
Methane	8.9	13.5	17.8	18.4
Primary electricity	6.5	6.4	6.5	6.9
Oil	28.9	35.8	44.5	46.0
Total	100.0	100.0	100.0	100.0

	Growth rates (% a.m.)	
	1950-60	1960-72
Coal	2.5	7.8
Oil	7.1	7.8
Methane	9.4	8.3
Primary electricity	4.8	6.2
Total	4.9	5.5

3. For the three main consuming areas the proportion of oil imports took the following trend in the decade preceding the first crisis:

	% oil import on energy consumption		% oil import on oil consumption		% Middle East oil import on oil consumption	
	1962	1972	1962	1972	1962	1972
United States	9.1	13.5	20.7	29.7	16.0	15.0
Japan	43.6	72.6	100.0	99.6	73.5	78.9
Europe	37.2	59.0	99.0	99.0	73.2	80.4

4. For a historical reconstruction of these events, including the many elements which helped to "forecast them", see first of all D. Yergin, *The Prize*, Simon & Schuster Inc., New York 1991, and the famous article printed a few months before the crisis by J.E. Akins, *The Oil Crisis, This Time the Wolf is Here*, in "Foreign Affairs", vol. 51, April 1973, pp. 462-490 and by M. Adelman, *Is the Oil Shortage Real? Oil Companies as OPEC Tax Collectors* in "Foreign Policy", Winter 1972, pp. 69-108.

5. A private cultural organisation set up in 1968 to call the attention of governments and the world public opinion to the complex and increasingly critical interactions between population dynamics, economic development, increase of pollution, availability and consumption of natural resources.

6. For a review of the global models drawn up in the Seventies, including in particular those of Forrester (1971) and Meadows (1972), see M. Fortis, *Modelli globali e scenari di sviluppo mondiale*, in "Energia", 1981, no. 2 pp. 32-57.

7. In the United States in particular, it fell from 4 million barrels per day in 1965 to close to zero in 1970.

8. In particular a relative shortage of oil refining capacity in Europe and in Japan; a shortage of tanker capacity for transport, aggravated in 1970 by Syria's blocking of the Trans-Arabian Pipeline

for political reasons; a relative shortage of fuel oil for electricity generation in the United States (due to the growing environmentalist opposition to the use of coal and to its stricter quality standards); a concomitant crisis in the supplies of gas on the east coast of the United States; and also in the US, delays in starting up nuclear plants and the blocking of the exploitation of the oil fields in Alaska due to environmentalist opposition. This opposition was to cost the whole world a figure of between 1,500 and 2,000 billion dollars (of that time) and which, ironically, was to contribute to an increase of 13 times the world's electronuclear power over the following twenty years.

9. On the basis of the complicated mechanism used, the governments would first *manage and collocate* just 10% of the crude oil they were entitled to (in theory equal to 25%) with a 30% expected growth in 1976-77, gradually increasing until it reached 100% of the equity crude only in 1992 (when it would represent 51% of the total).

10. On 2nd April 1971 the "Tripoli Agreement" was signed in Tripoli, and later the "East Mediterranean Agreement" which regulated the new contractual relations for the North African producer countries and for those who exported via the Mediterranean.

11. As a whole, these provisions led in 1971 to an increase of the tax paid costs (production costs plus fiscal charges) of 40-50 cents per barrel in the Persian Gulf. This increase was immediately transferred by the companies onto the market prices which recorded an increase of 30%-35% in the mean of 1971, while the posted prices rose by 27%-32%, greatly exceeding 2 dollars per barrel.

12. This upward spiral culminated in the record sum offered in December 1973 to the NIOC, who had auctioned 500,000 bbl/d for 6 months: the independent buyers even offered as much as 17.34 doll./bbl for Iranian crude compared with the official selling price of 5.25. Fired with enthusiasm, the Shah actually stated that this was the true value of the oil market.

13. Variations in the crude oil output of the countries (selected intervals, 1972-73)

	IV quart. 1974 III quart. 1973	first 9 months 1973 first 9 months 1972
Saudi Arabia	−19.5%	36.4%
Kuwait	−16.1%	−5.9%
Abu Dhabi	−12.9%	30.4%
Qatar	−19.2%	31.1%
Libya	−11.0%	1.3%
Algeria	−13.4%	5.8%

14. "If this hypothesis of interpretation turns out to be true", I wrote *before* the 1973 crisis broke out, "the result will be that until the American domestic supply can guarantee the country an ample margin of autonomy, the *prices of the international market may tend to be drawn upwards*" (Clô, 1973). This is what in fact happened.

15. The nationalisation law adopted by Mossadeq in 1951 was actually never abrogated. The Iranian Consortium however benefited from an exclusive advantage in the country.

16. In 1978 the share composition of the Iranian Consortium was as follows: BP: 40%; Royal Dutch-Shell: 14%; Gulf, Mobil, Exxon, Socal and Texaco: 7% each; CFP: 6%; other minors: 5%. Nationalisation in Nigeria took another 400 thousand bbl/d from BP, thus lowering its total supply from 3.5 mill. bbl/d before 1978 to just 0.7.

17. In Italy, for example, fears grew about a supply deficit calculated for the whole of 1980 as 20 million tons by some Ministries up to 30-35 by others (with a budgeted consumption of 100-102 mill. tons). All the worry disappeared as if by magic, however, when in a dramatic meeting of the Council of Ministers on 29th December 1979, increases in the already remunerative prices were agreed (Clô 1980b).

6 THE EFFECTS OF THE CRISES AND THE PROCESSES OF ADAPTATION

6.1. THE ECONOMIC EFFECTS OF THE OIL PRICE SHOCKS*

Although many years have passed, a reading of the effect of the increase in real oil prices on industrial economies is still highly complex and vigorously debated by economists. Difficulties arise on several counts: (a) *the impossibility of completely isolating the role of the "energy factor"* from the set of variables that was influencing economies at the time, considering there were some not immediately quantifiable aspects in which the relation expressed itself; (b) the difficulty in keeping these effects separate from those that could be imputed to *monetary and fiscal stabilisation policies* adopted by governments to cope with the price shock; (c) the steady decline in the driving force, already felt by the end of the Sixties, on the economies, generated by the post-war technological paradigm. We are far from making a comprehensive interpretation of these phenomena. In the paragraphs to follow, we shall restrict ourselves to a description of the main causal relationships between the energy shock and the economy, in what today appears to many to be a "consensus" opinion unsupported by adequate data, but which in those years undoubtedly shaped the decisions made by governments, companies and consumers in response to the new situation.

6.1.1. The data of the crisis

At first sight, statistical evidence seems to confirm the hypothesis of a close cause-effect relation between the increase of oil prices and the recession phases immediately following this. These phases featured four main factors of imbalance: considerable reductions in economy growth rates; exceptionally high inflational dynamics; a great increase in unemployment; and sizeable deficits in the balance of payments. Despite the fact that there were noticeable differences in the way the economies of the six most important industrialised countries related to energy, they saw a drastic cut in their *rates of income growth*, down to almost zero or even below zero. This was something that had not happened since the war (Tabs. 6.1. and 6.2.). The drop was especially sharp after the first shock, when average levels of GDP growth for OECD countries went from 5%-6% over the two-year period 1972-1973 to a modest +0.5% in 1974 and a contraction of 0.3% in 1975. The drop was immediate in the United States, Japan and the United Kingdom, with figures already below zero in 1974 (around – 1%), and in other countries coming later but hitting even harder (with a peak of -2.7% in Italy in 1975).

From that point the recovery started, though growth rates were consistently lower

Table 6.1. *Energy prices and economic indicators in the OECD countries during the two oil crisis (% var.)*

Year	Real oil price[1]	Final real prices[2]	Consumer prices	Real GDP	Unemployment (mill.)	Balance (bill. doll.)[3]
1960-73	1.0	0.6	3.9	5.0	–	–
1973	3.2	–0.6	7.8	6.0	11.3	13
1974	166.1	25.9	13.5	0.5	12.2	–25
1975	–4.2	6.9	11.3	–0.3	17.4	6
1976	–1.0	5.4	8.7	4.71	8.5	–16
1977	–1.0	5.5	9.3	3.6	18.3	–21
1978	–6.3	–2.0	8.1	4.11	8.7	12
1979	27.6	11.3	10.4	3.2	18.9	–27
1980	55.3	15.9	13.5	1.2	21.4	–68
1981	2.4	5.2	0.8	1.77	5.3	–24
1982	–12.0	–4.5	8.0	–0.3	30.1	–23

1. Crude oil import prices in the seven leading OECD countries deflated with OECD deflator of GDP. 2. Weighted mean of the prices of oil products for the seven leading OECD countries deflated with consumer price index for the residential sector and transport sector and with wholesale price index for the industrial sector. 3. Current part.

Source: first two columns: OECD-IAE (1982); other columns: OECD "Economic Outlook - Historical Statistics (1960-90)".

than before the crisis. William Nordhaus has calculated the cost of the first oil crisis between 1973 and 1979 for the real incomes of OECD countries at $713 billion (at constant 1975 prices).[1] The unexpected and violent outbreak of the second shock brought economies back to a phase of stagnation which appeared however less generalised (Japan remained relatively unaffected) and less intense than the first, though lasting longer in time. GDP growth slowed to around 1% in the period 1980-1982, three points lower than average figures for 1976-1979. This drop could again be explained in terms of the combined effect of the deterioration of the terms of trade with OPEC countries (2% of the GDP), of the reduction of aggregate demand (3%-4%), and of restrictive economic policies. The accumulated loss of real OECD income was calculated for 1981, compared to the levels that would otherwise have been observed, at *1,000 billion current dollars*.

Compared to those of the past, the peculiarity of the recessions in the Seventies lay in their intensity, and in their combination with an inflation growth that had not been

Table 6.2. *Principal economic indicators of the most industrialised countries (1960-90)*

Year	USA	Japan	Germany	France	UK	Italy	OECD
Real GDP[1]							
1960/66	4.6	10.2	3.9	5.5	2.9	5.6	5.1
1967/73	3.4	9.7	4.5	5.5	3.2	5.3	4.7
1974/75	–0.9	0.7	–0.7	1.4	–0.9	1.3	0.1
1976/79	4.1	5.0	3.8	3.6	3.1	4.9	3.9
1980/82	–0.2	3.5	0.1	1.6	–0.6	1.7	0.9
1983/85	5.0	4.0	2.1	1.4	3.2	2.1	3.6
1986/90	3.0	4.6	3.0	2.9	3.2	3.0	3.3
Inflation[2]							
1960/66	1.7	5.7	2.7	3.5	3.4	4.4	2.7
1967/73	4.6	6.3	3.9	5.4	6.4	4.8	5.0
1974/75	10.1	18.0	6.5	12.7	20.0	18.0	12.3
1976/79	7.8	6.3	3.7	9.7	13.5	15.2	9.1
1980/82	9.9	5.1	5.7	12.9	12.8	18.5	10.7
1983/85	3.7	2.0	2.6	7.6	5.2	11.5	5.4
1986/90	4.0	1.4	1.4	3.1	5.9	5.7	4.4
Unemployment[3]							
1960/66	5.1	1.3	0.7	1.2	1.4	4.9	3.1
1967/73	4.4	1.2	1.0	2.4	2.4	5.7	3.4
1974/75	6.8	1.7	3.2	3.5	2.8	5.6	4.7
1976/79	6.6	2.1	3.8	5.1	5.0	7.1	5.5
1980/82	8.0	2.2	4.7	7.3	8.3	7.9	6.9
1983/85	8.0	2.6	7.9	9.4	11.3	9.8	8.2
1986/90	5.8	2.5	7.2	9.9	8.4	11.4	6.9

1. Average annual change. 2. Annual average change of the consumer price index. 3. Unemployment as a percentage of the total labour force.

Source: our calculations of OECD data "Economic Outlook - Historical Statistics (1960-90)".

seen since the war. Already high, due to the combined expansion of the economies and the tensions of a speculative nature over raw materials, *inflation* during the first crisis went from two to five times higher. From 1972 to 1974-75, it went from 3% in the U.S.A. and 7% in Great Britain, to 10% in the U.S.A., and 18-20% in Japan, Italy and Great Britain. Germany was the only country relatively unaffected, maintaining its price dynamic around 6%. The severe response of the economic policies of western governments allowed price rises to slow down, as well as the mechanisms of endogenous activation that fed them, and in some cases they were brought down to a lower point than before, or very nearly, and in other cases to levels a great deal lower than the highest levels reached, if still higher than before the crisis. By 1978 the growth of the consumer price index of OECD countries had fallen to around 8%, five percentage points less than the 1974 maximum, but three points higher than trend levels seen previously. However, these results, achieved after a great struggle, were soon to be swept away by the second shock, with inflation rising in the majority of countries (United States, France, Great Britain and Italy) at rates near to or higher than the double figures that had been reached in 1974-75. The extremely rigid monetary policies adopted by the Bundesbank in Germany, and a rigid wage freeze in Japan in 1979, meant that these two countries remained relatively unaffected.

The third macroeconomic variable most useful in expressing the sense of the two recessions of the Seventies was *unemployment*. Insignificant in countries such as Japan, France, Germany and Great Britain, and at rates not far from normal in the United States and Italy, in the course of the decade from 1973 to 1983 it rose uninterrupted (except in Japan) to between 8% and 11%, in terms of the labour force, the highest since the Great Depression. In absolute terms, the number of people unemployed increased by 65% between 1973 and 1976, to 18.5 million in the OECD as a whole (Tab. 6.1.), while at the same time the labour productivity dynamic dropped by half almost everywhere. By the end of the Seventies and the beginning of the Eighties, this led to a huge reorganisation of the production processes of modern industry, with the further shedding of labour and a reduced capacity to absorb new labour as it reached the labour market. In 1982 OECD countries had 30 million people unemployed: three times the amount of ten years previously.

The last, but perhaps the most widely felt factor of disturbed equilibrium was in the *current balance* of goods and services of the balance of payments, which in OECD countries went from a surplus of 13 billion dollars in 1973 to a deficit, never before experienced, of 68 billion dollars in 1980. All in all, this was a fairly modest deterioration, considering that the (net) oil bill had increased in that time by eight-fold, to 290 billion dollars. However, it raised everywhere the spectre of pathological crises in foreign accounts and their financing, re-opening the debate between expansion and re-equilibrium of international payments.[2]

6.1.2. Effects or coincidences?

Besides the *similar direction* of macro-economic indicators in the various countries, important distinctions in particular situations and for specific indicators support the opinion[3] that the energy variable was *neither the only nor the main one* behind the severe recessions of the Seventies.[4] Indeed, Germany managed to avoid high inflation after both shocks, with Japan avoiding high unemployment rates, while Great Britain failed on both counts. These differences cannot be explained simply by the degree of exposure of individual countries to the events going on in the oil markets. If this had been the case, Japan and Great Britain would have performed in exactly the opposite way to what actually happened. Besides these factors, others seem to have played a decisive role:

a) the *economic policies* pursued by the various countries before and in response to the oil crises;
b) their *institutional specificity*, above all in terms of industrial relations and mechanisms of index-relating prices and wages;
c) the *extent of the increases of energy prices* borne by each country, in relation to the mixture of energy sources utilised, the structure of the uses, the criteria of price formation, and the taxation on final consumption;

d) the *response capacity* permitted by the structure of the energy system, by the availability of national energy resources, and by industrial and governmental capacity in this field;
e) finally, the *macro-economic trends* already present when the first oil crisis struck.

Economists are generally persuaded[5] that the energy crisis was at any rate the first explanation of what O. Eckstein defined in his book "The Great Recession" of 1974-75. On retracing its various phases, he wrote: "The energy crisis was the single largest cause of the Great Recession. Without it, the economy would have suffered no worse than a year of small GNP decline in 1974, and would have seen 1975 as the first year of recovery. But when the energy crisis was superimposed upon an already highly vulnerable economic situation, it was sufficient to turn the beginnings of recovery into the sharpest decline of the post-war period" (Eckstein, 1978, p. 124).[6]

The crisis occurred at an extremely critical moment in the international economic cycle. Overall, 1973 had been a year of great expansion, with a higher growth rate in the seven leading OECD countries than had been seen for two decades. In many countries production levels were near to available capacity and full employment, thus fostering, on the supply side an inflational dynamic already growing through excess demand[7] from the middle of the Sixties. This growth was essentially due to the disorder in international monetary conditions and excessive creation of currency (especially on Euro-currency markets), that had been seen with the steady breakdown of the 1945 Bretton Woods agreements, based on fixed parities adjustable only for the correction of situations of fundamental imbalance. Monetary expansion was so pronounced at the end of 1973 that it determined a general coincidence of maximum levels of the economic cycle of OECD countries,[8] something that had not happened since 1963.

It was inevitable that these countries, almost at full utilisation of their capacity, were unable to transform the additional monetary base into expansion, thus channelling it towards speculation in the field of raw materials, especially metals and other primary products, leading to exceptional increases in their prices and thus pushing up inflation.[9] Since the dollar no longer had a role as reserve currency and value accumulator, with the US President Richard Nixon's decision on August 5th 1971 to suspend the dollar's convertibility into gold, some raw materials were made to perform this role instead. The price index of the 35 principal commodities traded on international markets increased from 100 in 1967-69 to 108 in 1972 and 160 in 1973. At the start of 1974 inflation everywhere reached all-time highs, just when the decisions to increase oil prices were emerging in the costs paid for imports, while there was no knowing their possible *amount* or *duration*. The potential risks were exaggerated by monetary authorities, to whom the contradiction between a demand believed to be excessive and a supply also restricted exogenously seemed increasingly obvious. Right from the start the reaction of the policy makers was to consider the increase in oil prices as a *normal* external inflationary impulse, justifying the adoption of restrictive, fiscal and above all monetary policies, or the tightening of those already adopted, in order to limit internal demand at levels more consistent with supply potential, thence reducing the price dynamic and the current account deficit of the balance of payments. This had a depressive effect on economic activity, hitting family consumption and expenditure on investment in the first place, and thence the labour demand. This process then continued because of the expectations generated and because the productive base was not broadened. The severity of these restrictive policies would indeed turn out to be greater than desired or required, because of the *specific features* of the oil price increases, which in themselves would lead to a substantial reduction in aggregate demand.

6.1.3. Energy and economy at a glance

Increases in the *relative prices* of imported energy determine a complex chain reaction affecting the costs of production, aggregate demand, the combination of product on factors, and the structure of production. An initial effect may be seen in a *reduction of aggregate demand accompanied by cost-push inflation*. Since oil, or more generally energy, is the input for productive processes, an increase in its price leads to an increase in the production cost at every level of wages and output. In other words, the curve of aggregate demand shifts upwards and left, expressing the availability to produce for every level of cost (and hence of price in a competitive regime). The greater the intensity with which energy is used, compared to the capital and labour factors, the higher the increase will be, and the greater the importance of oil in the number of sources employed, the less chance there is of substituting imported oil with domestic oil or other, less expensive, sources of domestic production. The effects on aggregate demand are less certain, although there are very good reasons to believe that its curve will shift downwards (and to the left). The final result of the increase in oil prices will hence depend on the relative shifts in the curves of demand and supply, which in turn are a function of the amount of the increase and of the relative shape of the two curves. There are two reasons to motivate the reduction of aggregate demand: low energy demand price elasticity in oil-importing countries, associated with a low import-income elasticity of oil-exporting countries (in the conditions of the early Seventies).

A low demand price elasticity determines an increase of total expenditure on this commodity, causing a reduction of available income for the purchase of other commodities, mainly to the disadvantage of low income categories. In this respect energy (and particularly oil) is a very special case compared to other commodities. The response of its demand to price variations is (as we have seen in chapter 1) strongly limited by the modest range of the set of available techniques (to which energy is associated in production, transformation and consumption); by the rigid energy/output ratios defined by the design of capital stock; and by the high degree of necessity associated with its use. Two marked consequences result from these conditions: a very low *short-term* elasticity of final demand to prices, and a much greater *long-term* elasticity, though referred to an interval of time from at least 5 to 25 years and beyond. Both in the short and long term, the negative values of the elasticity of primary demand (of crude oil) turned out in addition to be higher, in a ratio of roughly 1:3, than those referred to final demand, in relation to the significant influence that the items of cost other than raw materials (particularly taxes) had on the final price. The conclusion is that over a period of time up to 20 – 30 years, the constraints of plant equipment would be the greatest determinant of the interaction between energy and the economy. Rapid and marked price increases could only mean a discharge of capital in advance or in a contraction of the aggregate demand.

The recession effect would have been greater the more oil-producing countries showed a propensity to increase savings with the growth of their revenues, causing a contraction of world demand in goods produced by oil-importing countries. It has been estimated that in the 3 years after the first crisis of 1973, little more than 50% of the additional revenue of the OPEC countries as a whole became a demand for imports of goods from industrial countries.[10] This sharp subtraction of wealth, comparable in its effects to *indirect taxation* on the part of the producing countries, would have been all the more evident among consuming countries: (a) the greater the share of oil imports with respect to domestic production, and the less chance there was of increasing the latter; (b) the greater the difficulties in limiting the use of oil (and of energy) and, conversely, in increasing the domestic supply of alternative sources at less cost; (c) the greater the commercial exposure towards oil exporting countries at low import-income elasticity.

A second range in the effects of a change in the relative prices of energy can be seen in the *optimal combination of resources employed in the production* of commodities and services. "Substitute" factors of energy have to find greater employment, unlike the productive factors which are "complementary" to them, while the absolute level of their utilisation will depend on the scale of economic activity. These effects have to be added to those examined beforehand, helping to change both the *form* and *position* of demand and supply curves, in relation to the characteristics of productive processes and to the adjustments made by companies. Thus, for example, if capital can substitute energy, then the investment demand should rise in reply to increases in energy prices, while the opposite would be true if energy and capital were complementary. This is a point over which much controversy has developed. Econometric analyses supporting the thesis of complementarity were opposed to engineering analyses arguing for replaceability.[11] To resolve the matter, Dohner[12] advanced the plausible hypothesis that, regardless of the energy sectors (obvious candidates for additional investments), energy may be complementary to capital in the *existing* plant equipment, but may be substituted with it in *additional* investments. Before these new combinations can show their effects, in the transition phase there are cost increases (in the approximate proportion of energy against total costs) associated with a redistribution of income between the other factors of production. Where there is rigidity in the prices of factors (especially in wages, because of the presence of index-linked mechanisms) and in their mobility, additional negative consequences are generated for levels of output, for employment, and for the productivity of factors. This was indeed just what was seen in the majority of industrial countries in the Seventies, and particularly in Italy.

Lastly, increases in the relative prices of energy determine effects on the *composition of final demand and production* in relation to the specific content of energy for commodities and services; of the degree of complementarity with the energy for their use; and of the degree of transfer of the greater costs to the final prices, and of the demand elasticity of individual goods. The complex game of sector interdependence would have given rise in each country to a compression of aggregate demand and of internal output which was all the greater, the greater the specific importance of *energy intensive* sectors on both sides, and the less chance or time there was to shift resources towards sectors that could profit from the new relative advantages (starting with the energy sector itself).

6.1.4. Economic effects and political motivations

Theoretical analysis and empirical evidence seem to confirm, in conclusion, that the *direct* effect of the increases in energy costs on industrial economies in the Seventies might not have been either significant or decisive for the heavy recessions that followed.

The *indirect* effects of those increases could have had greater significance. These effects include (a) the effect of reducing employment, because of structural rigidities particularly in wages; (b) the effects on flows of capital services, because of the more rapid obsolescence of capital stock; (c) the spreading out in the market of the factors and in production, due to the changes occurring in the composition of demand; and (d) the reduction in output and in employment due to the contraction of aggregate demand. A further (but perhaps not final) explanation of those crises lies in the *policies of macroeconomic stabilisation* of the western countries, applied even before the crises exploded, which helped to make them still more severe, while economies were already moving or were about to move into a depressive phase of the economic cycle.

To deduce from all this that the energy crises were irrelevant to the economic events of those years, or to go even further and say that there was a weak relationship between economies and energy would, however, be wide of the mark. Most governments, international organisations or economic operators certainly did not believe this. At the June 1980 summit meeting in Venice of the seven most important industrial countries, it was stated that "the key for successfully overcoming the main economic challenges to be faced by the whole world lies in a greater commitment towards reaching and maintaining a balance between energy supply and demand, at reasonable levels and at tolerable prices". What is more, the general perception of most observers was that the energy crisis had lost its merely economic dimension and had taken on the aura of a terrible "natural calamity" and the definitive "end of an epoch", necessarily giving rise to new models of development, different from those of the consumer society.

It was generally believed as well that the rise in oil prices was likely to continue in the long term (and indeed illusions to the contrary nourished after 1973 were dispelled with much bitterness in 1979-1980), with levels forecast at $100 a barrel and beyond.[13] Consumers and companies changed their behaviour, with a strong commitment to investment, certain that the consumer models of the past would not be relevant to the future. Likewise governments adopted both sectional and general economic policies "out of all proportion" to the external events from which they were derived. If such collective reactions could be included among the "effects of the energy crisis" – in addition to those attributed only to economic mechanisms – it might be seen that their final impact was noticeably different to and greater than what has emerged from the most recent econometric reviews of that period.

As well as economic mechanisms, a no less important role was played by the logic of politics. Right from the time the oil crisis first burst upon the scene, it was interpreted and experienced as a threat to both the *security* and the economy of the western world. In other words, to use the expression of US President Jimmy Carter during the second crisis, as "the moral equivalent of a war"! There were four aspects of central concern to the leading powers. *Firstly*, the conviction that the increase in oil prices was essentially politically motivated and an adequate reply should be found at this level. *Secondly*, that it was OPEC that was responsible for the price increases and not market forces; and OPEC was seen both as a political coalition with strongly anti-western characteristics and as an economic cartel. *Thirdly*, that the exclusion of the majors from oil producing areas led to the loss of the main instrument of mediation between the interests of producing and consuming countries. *Fourthly*, that the radical change in the international "balance of power" could strengthen the strategic position of the Soviet Union. This latter was for two reasons: (a) the completely self-sufficient energy situation from which ther Soviet

Union was able to benefit; and (b) the strong economic and strategic relationships it had established with the countries and political movements of the Middle East (Levy, 1974).

These convictions, rooted above all in the American administration, were behind western policies of response to the oil crisis. On a *domestic level*, they were aimed at the reduction of dependence on OPEC-imported oil, with policies of energy conservation, incentives for domestic production, and public support for alternative sources to oil (particularly nuclear). On an *international level*, the aim was to create a common front of importing countries to set against the exporting countries. This was to prevent the search for specific advantages in bilateral relations being set up to strengthen the bargaining power of OPEC. Henry Kissinger, US Secretary of State, was the architect of this strategy of building up face-to-face opposition between the two divisions as the foundation on which to construct an eventual dialogue later. Starting from the premise that "our economy and well-being are hostage to decisions made by nations thousands of miles away", that "the energy crisis ... has mortgaged our economy and made our foreign policy vulnerable to unprecedented pressures", that it "has weakened the industrial democracies economically and undermined the political unity that is basic to the security of free nations" (1982, p. 16), Kissinger proposed setting up an organisation to co-ordinate the domestic policies and international action of the industrialised countries. This organisation, the International Energy Agency, of which all OECD countries were members except France, met for the first time in Paris in February 1974. With respect to the initial intentions its role turned out to be extremely marginal, yet it was in this climate of "moral war" that western countries organised their policies.

6.2. PROCESSES OF ADJUSTMENT

Market mechanisms and public policies would display their effects to the full: generating a drastic adjustment of energy systems to the new relative prices of energy (towards other commodities) and oil (towards other sources). This adjustment started immediately after the first oil crisis, but was only completed in full (because of the time lags implicit in investment decisions) after some time had elapsed. Its effects therefore eventually added to those provoked by the second crisis. To fully understand the nature and entity of the phenomena involved it is useful to examine them first as regards demand and supply. It will thus be easier to understand the pressures that were to lead to *new* prices of equilibrium on the oil market, to which the OPEC cartel could only offer vague and costly resistance.

6.2.1. The collapse of demand

The obsessive attention paid in the Seventies to the limits of the *physical availability* of energy resources had resulted, amongst other things, in attracting relatively less attention to the mechanisms that the demand for energy would have shown in reply to a burgeoning growth of its prices and a lasting reduction in the speed of economic growth. Elasticity to the prices of demand appeared to be too low for its very evident contractions to be foreseen, in the absence of an undesirable income dynamic, and before changes in the technology of use and in the preferences of consumers could occur. Of course, there was much serious and well-argued criticism of the econometric evidence for these conclusions, but they were normally restricted to a few methodological aspects of the problem, and never went beyond the narrow limits of professional controversy. The *general conviction* was that the demand for energy would continue on its present course, though at a slower rate compared to the past, even after the second oil crisis. This growth would influence the limits of the physical availability of energy, fearing deficits in supply and thence further inevitable price increases[14] (Hafele, 1981).

The forecasts formulated on the basis of a great number of economic and engineering models at the International Energy Workshop, organised in June 1983 by the prestigious Institute for Applied Systems Analysis of Vienna, projected a growth of world energy consumption until 2000 at around an average 2% per year, with a 3% growth rate in economies (Manne and Schrattenholzer, 1983). This growth would influence the consumption of oil (though at more modest rates) and, given the limits of supply, force its nominal prices to rise to around 50 dollars a barrel by 1990 (from \$29 of 1980), and to \$90 by the year 2000. Another study, developed at Stanford in 1982 by the Energy Modelling Forum, reached similar results, forecasting (in the reference case) a still greater growth of energy consumption and an increase in crude oil prices well beyond \$100 by the end of the century (EMF, 1982). Investors, consumers and governments all put their faith in these forecasts and expectations, trusting in the general consensus they enjoyed, in the prestige of those who formulated them and in the apparent rigour of the models used. The reality, however, was to go far beyond the wildest imagination, even econometrically.

Total consumption of energy, after the uninterrupted threefold growth seen in the post-war period to a relative maximum of around 4.4 billion toe (N.B. tonnes oil equivalent) in 1973, first underwent a slight contraction (mainly due to the income effect), and thenceforth a modest rise to levels in 1979 of 4.7 billion toe. This level then *stabilised* at this point until the mid- Eighties, despite an average annual growth in OECD economies of around 2.5%. The intensity of the use of energy (unit of energy per unit of real output) diminished in OECD countries by 20% between 1973 and 1985 for activities as a whole and by 35% for industrial activity.

Still more overwhelming is the adjustment process to be seen on the demand for oil alone, since it is here that the three types of effect caused by the increase of its real prices are concentrated the most: the *price* effect, the *revenue* effect, and the *substitution* effect. As regards the price effect, it has been said on several occasions, especially in chapter one, that the reactions of the consumers initially could only be marginal, given the existing capital and the low or non-existent flexibility this allowed. Of course the use of cars could be limited, or the temperature of the heating inside buildings reduced, but the practical consequences were negligible. The "austerity" measures opted for by many countries during the first crisis, with the banning of the use of cars on Sundays, often turned out to be counterproductive because they led to a higher consumption of energy inside houses than which was saved outside. The margins of reaction to the increases in prices were basically restricted to the variation of the rate of utilisation of existing capacity, and were indeed limited, unless the standard of living was to fall.

In a graph (Fig. 6.1.) showing oil prices and quantity, the short-term demand could be shown, fairly realistically, as a vertical D_0 at zero elasticity at all points, with an implicit level of energy efficiency of e_0. An increase in prices from P_0 to P_1 keeps demand levels unchanged at Q_0, thus allowing a proportional increase in the total sales

Figure 6.1. *Oil demand reactions to oil price increases*

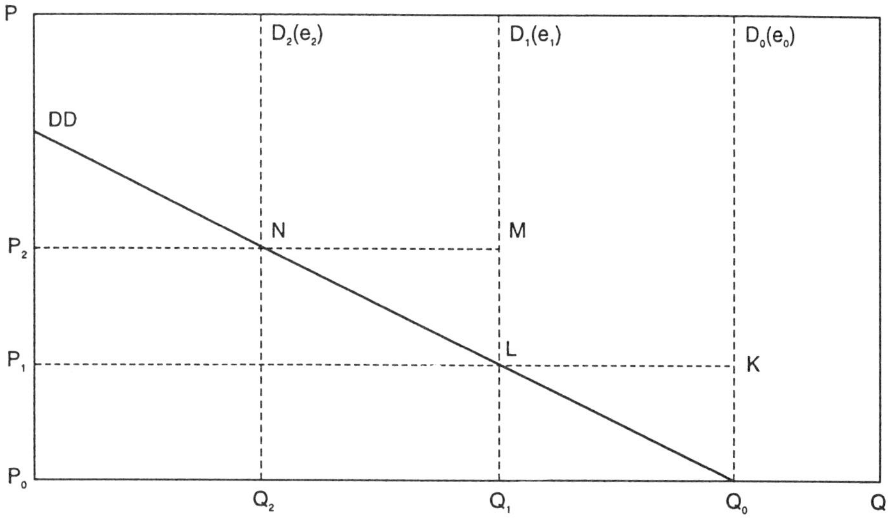

revenues. This is more or less what happened in the early Seventies.

The process of adjustment which began at that time was slow but *structural*. Several phenomena were responsible for it: (a) there was a contraction in the growth levels of the economies; (b) the structure of economic activities changed, so that less oil was used; (c) there were developments in technological innovation, with a consequent reduction in the intensity of the use of oil per output unit; (d) oil was substituted by alternative sources. The interaction of these phenomena in the long term caused the demand curve to shift to the left to D_1, with a higher level of energy efficiency equal to e_1 and a price-quantity combination which went from (K, Q_0) to (L, Q_1). A further increase in prices from P_1 to P_2 likewise led to a new short-term combination equal to (M, Q_1) and in the long term equal to (N, Q_2), with further reduction of the quantity of demand for oil to D_2, due mainly to a further increase of energy efficiency to e_2 and to the substitution of oil with other sources. If we join the three price-quantity combinations obtained with the increase in prices from P_0 to P_1 to P_2, we can roughly represent the demand curve of oil in the long term as DD. This representation simplifies reality, since the elasticity of the demand to the variation of prices tends to grow with their absolute level. Indeed, their increase from P_0 to P_1 allowed greater opportunities for investment than in the passage from P_0 to P_1, and shorter return times. The variable elasticity with the growing level of prices is confirmed by experience, though it is extremely difficult to attribute variations in demand (delayed in time) to the variations in price which actually caused them. What is certain is that the cumulative effect of the variations on the demand side was felt in its entirety in the early Eighties, straight after the second crisis.

After increasing from around 47 million bbl/d in 1970 to 56 in 1975 – despite prices in the meantime increasing over fivefold, from \$1.8 to \$10.8 a barrel – and to 65.2 million bbl/d in 1979, with futures prices in that year at around \$30 a barrel, the world oil demand fell in just three years by 6 million bbl/d, to 59.5, then settling at this level for a few years. The effect was as if an industrialised country the size of Japan had suddenly left the oil market.

The illusion which had guided the action of the OPEC nations (that consumers had no power over the market), was to be bitterly dispelled, after the latter had learned how to gain complete control over their demand.

6.2.2. The increase in supply

No less formidable was the variation occurring in the supply of oil in response to the increase in prices. Two aspects make this phenomenon in some way symmetrical to the demand phenomenon, thus enhancing the effect of their combination. First of all, the *structural* nature of the phenomenon, thus making it not immediately reversible on modification of the conditions of advantage that had given rise to it. Secondly, it took a *long time* to materialise fully, due to the low elasticity of the supply curve in the short term. From the second half of the Sixties, thus even before the crisis started, investments in the oil extraction industry began to grow steadily, after a decade of stagnation. This growth became exponential with the rise in oil prices, and above all, with the rise in profit levels accompanying this. Net profits (after tax) of the sample of 26 companies (including the majors) analysed by the Chase Manhattan Bank went from \$6.6 billion in 1970 to \$16.4 billion in 1974 to \$35.2 billion in 1980. The cash-flow went from 13.8 to 28.8 to 66.8 billion dollars; and profitability on invested capital from 9.9% in 1970 to 17.3% in 1974 to 20.6% in 1980 (CMB, 1982). The drop recorded for the latter from the beginning of the Sixties (cf. 4.7) was recuperated in full and more, so that companies could make greater investment commitments than ever before.[15] In the sixteen years between 1970 and 1986 world investment (excluding former Communist countries) stayed at a constant 1,666 billion dollars, compared with 752 billion dollars in the previous twenty years (Tab. 6.3.).

Table 6.3. *Capital expenditure of the world oil industry by activity: 1950-1986[1] (bill. doll. 1986)*

	Total	Upstream[2]	Transportation	Refining	Others
1950-1959	306	189	39	44	34
1960-1969	446	218	60	91	77
1970-1979	843	416	163	184	80
1980-1986(3)	823	570	56	146	51

1. Excludes countries with centrally planned economy. 2. Includes exploration, development and production. 3. 1985-86 estimates.

Source: CMB (various years).

Of all investment 60%, around \$1,000 billion, went towards the expansion of the production capacity of oil. The distribution of these immense resources was the opposite to Ricard's model of the second phase of exploitation of land, in inverse relation to its level of efficiency. The investments actually did not go where they appeared to have greater chance of success (with lower unitary costs and better quality of the crude oil), but where companies could build up a base of oil resources which enjoyed greater certainty over their property rights, profits, and flow of supply. Where, in short, there was less *political risk* that host States would modify the terms of contracts to their own advantage, or would subordinate the companies' freedom to manoeuvre to political interests. This was to a certain extent a necessary choice (given the exclusion of western capital from OPEC countries) but it nonetheless meant a huge waste of economic resources for the world economy.

About 75% of all mining investment between 1970 and 1986 was directed towards the marginal areas of North America and Northern Europe, where there were only 8% of proved deposits and 15% of probable resources, while just 15% was invested in OPEC countries, despite the fact that they controlled 77% of the reserves and 50% of resources, and despite the costs of discovery of oil being ten to fifteen times lower (Desprairies et al., 1984). Though, as we shall see later, the distribution of reserves and resources did not undergo any particular change, the same could not be said for the production capacity (Tab. 6.4.). While the production capacity of OPEC steadily fell from the second half of the Seventies, passing from maximum levels of around 39 million bbl/d in 1977 to around 28 a decade later, the overall non-OPEC capacity grew steadily in the same period from 24 million bbl/d in 1970 to 28 in 1975 to around 41 in 1985.

In the Seventies, it was not only new big oilfields to which investment had already been directed in the Sixties which went into production (the North Sea, Alaska, Egypt and China), or new fields discovered in older areas (such as Mexico). There was also a large number of marginal finds which became profitable with the new price levels. It was mostly the majors that activated these new sources of production, having had to reconstruct their drilling bases elsewhere after abandoning what they had in OPEC countries. Exxon, Shell, and BP increased their production in Europe from zero in 1970 to around 0.5 million bbl/d (each) in 1985. The overall non-OPEC production of the majors increased from 0.3 million bbl/d in 1970 to around 7.0 million bbl/d in 1985.

The new production progressively undermined the OPEC production. It is interesting to see that this did not necessarily happen through prices competitive with those fixed by OPEC. Other factors entered into it. In particular, these were: (a) the advantage for companies to gain from their investments, pushing the rate of exploitation of the new oilfields to the highest economically possible; (b) the recognition of a bonus at those sources of supply believed to be politically safe, or at any rate not bound by either direct or indirect political conditioning; (c) the competitive advantage which most of the new sources of production could display in terms of location (with resulting savings in the cost of transport), currency (with the product being purchased in local money), and trade.

6.3. OPEC AS RESIDUAL PRODUCER

The OPEC countries had to face the two limitations to monopolist power identified by economic theory: the fall in demand and the arrival of new competitors. The combined effect of the two phenomena was great, all the more so because it occurred over a short period of time. OPEC production collapsed from the all-time high of 31 million bbl/d reached in 1979, to 19 in 1982, and to a relative minimum of 16 in 1985. 60% of this was caused by the fall in demand for oil (including the variation of stocks), and 40% by the increase in non-OPEC production. From a total (Tab. 6.4.) of 23 million bbl/d in 1970 it increased to 26 in 1975, to 33 in 1980, up to an all-time high of 38 in 1985. In this year, non-OPEC production was 2.4 times that of OPEC, while in 1973 it was equal to 0.8%. The figure that sums up all the rest is the divergent rate of utilisation of production capacity. Whereas in the non-OPEC areas it remained constantly higher than 90% (on a steadily increasing basis), in OPEC areas it fell from 90% in 1979 to 55% in 1985. At the same time the market share of OPEC countries in world production (outside the former Communist bloc) fell from 65% in 1973 to 38% in 1985: entirely overturning down the position of domination in which they had begun their challenge to the majors. Challenging the markets turned out to be much more difficult for them.

Table 6.4. *Producing capacity and oil production in OPEC and non-OPEC countries (mill. bbl/d)*

Year	Capacity		Production		Utilisation rate	
	OPEC	non-OPEC	OPEC	non-OPEC	OPEC	non-OPEC
1970	25	24	23	23	93	92
1973	34	27	31	25	92	92
1975	36	28	27	26	75	92
1980	33	36	27	33	82	92
1985	30	41	16	38	55	93
1990	28	40	24	37	85	92
1995	29	37	27	36	92	97

Why did the OPEC countries lose control of the market, despite holding most of the crude oil reserves and almost all the low-cost reserves? There are two parts to the answer. *First of all*, the policy of price increases adopted in 1973, at an excessive rate compared to that which could be justified by the rising costs of utilisation which could be attributed to the future shortage of a non-renewable resource. *Secondly*, the strategy of a rigid defence of fixed prices practised by OPEC straight after the second crisis, when the processes of adjustment were already making their effects felt. In short, in a situation of great confusion over what was actually happening in the market (we may recall the idea of consensus on the increase of demand), OPEC adopted the strategy which caused the most damage to itself. It defended the high prices emerging from the second crisis by adjusting its own production downwards to leave room for new arrivals. The entire OPEC production became virtually *residual*, between the falling total demand and the growing supply of its competitors. High State-controlled prices had a dual effect; on the one hand, they made the arrival of newcomers advantageous with much higher cost functions; on the other, they allowed them (in the position of price-takers) to place on the market any unit of crude oil they were allowed to produce. As always happens, it was those outside the cartel that had the most to gain from its actions (in terms of prices and quantity) without having to pay anything at all for the privilege of belonging to it.

OPEC placed itself in the position of residual or swing producer not just for oil but for the entire world energy system. This position was wholly functional to the energy diversification and conservation strategies that the western countries were applying. "(...) a high degree of government involvement is essential to the success of such efforts – ", wrote Walter J. Levy, a man who had always been close to the US Administration, in "Foreign Affairs", "including the probable necessity of government guarantees putting a floor under the selling prices of alternative energy sources (...) if there is a chance that foreign oil prices will fall" (1974, p. 699). Ironically, OPEC countries themselves were to oblige here.

The strategy of the rigid defence of fixed high prices (to the detriment of market share) was the result of a bitter conflict of interests of both a political and economic nature inside OPEC. On one side, there was the *radical* element (particularly Libya, Algeria and Iran), strongly hostile to the West, (often) with a reduced mining base, high population density and low income per head. Their interest lay in maximising oil revenues *in the short term*, without considering the risks of reducing the centrality of oil in the long term. A completely different perspective was shared by the *moderate* wing (Saudi Arabia, Kuwait, and the Emirates) not just because they were politically closer to the western bloc and sensitive to the fate of its economies (in which moreover, they were in the process of making heavy investments), but because they had long-term interests in extending the exploitation of their oilfields over a long period.

The conflict between these two different components of OPEC exploded violently straight after the second oil crisis. That was really the moment of OPEC's *greatest crisis*, and certainly not its greatest success, as the newspapers generally described it at the time. OPEC certainly benefited from the rocketing oil prices, but it had neither wanted it not been decisive in provoking it. Saudi Arabia provides the clearest example of this, trying between 1980 and 1981 to slow down the price rises by every means possible: pushing up its production to maximum levels[16] and keeping its prices fixed at $30-32 a barrel, compared with $35-40 for the other (homogeneous) OPEC crude oils and with the $36-42 at which Arabian Light was quoted on the spot markets.

During the second crisis OPEC no longer had any conceivable role. Each country was left free to decide what it wanted. All discipline concerning the levels and structure of prices collapsed miserably. It was only at the end of 1981, faced with the collapse in production, that the differences seemed to mend, though by then it was rather late. The spot quotations had already begun to drop as from the end of 1980, to levels lower than OPEC's official ones. The international market had fully recovered both adequacy and flexibility in the flow of supply and no damage nor fears had arisen from the new cycle of political instability in the Middle East (starting from the outbreak of the "Eight Years War" in September 1980 between Iran and Iraq which left a million dead). The OPEC conferences at the end of 1981 marked the return to a realignment of the official quotations on a *marker crude* of $34 a barrel, again Arabian Light. The situation remained the same until the end of 1982. OPEC's strategy of defence of fixed prices started here, under the *regained leadership of Saudi Arabia*.

Market mechanisms were however soon to show that it was no longer possible to limit the damage caused by past decisions; that the role of controller of the market could no longer be taken by OPEC; and that the costs of this strategy were increasingly prohibitive, particularly for those who had most supported it.

6.4. OPEC AS A CARTEL

Commentators and observers began to speak of the "OPEC cartel" right from the start of the first crisis; however this was inappropriate. The objective of cartels is to exercise monopolistic control of production and exports with the aim of raising prices. This may fit the majors in their golden age better than OPEC in the Seventies. Also as regards the determining features of cartels, the idea is inappropriate. Cartels are normally set up as a reaction against devastating forms of competition; due to the appearance of new sources of supply; to reduce the risk of losses due to market volatility or new arrivals on the market; to deal with excesses of production capacity and drops in demand (as took place at the highest ever levels in the Twenties and Thirties when it was calculated that 40%-50% of global commerce was some way or another in the hands of cartels); or to provide governments with the chance to increase their revenues. None of these reasons were to be found in the Seventies, but rather in the depressive phase that followed the second oil crisis.

To make the OPEC defence of price strategy feasible, it had to be associated with a parallel and consistent policy of restrictive control of production and an effective allocation among its various members. An attempt to begin was made with the Vienna Conference of 1982, but the experiment failed after only six weeks because of the impossibility of reconciling the positions of Iran and Iraq, who drew military strength from the level of production. A second more successful attempt was made at the London Conference of March 1983. Faced with the continual drop in spot prices, OPEC on the one hand took the historic decision to lower the price of the marker crude for the first time, from $34 to $29 a barrel (in line with spot values); while on the other hand, fixing a ceiling to its overall production, at 17.5 million bbl/d (an amount which corresponded to the estimated demand at the new price levels). Production quotas were also assigned to each member country, with the exception of Saudi Arabia that took on the role of swing producer with respect to the other members of the cartel. This meant that the State accepted the absorption into its own production of eventual discrepancies between *estimated* demand (*ex-ante*) and *actual* demand (*ex-post*) referred to OPEC countries.

Though Saudi Arabia had tried to attenuate the irresistible rise in prices during the most acute phases of the crisis, it now changed its attitude. It was now worried by the repercussions that the price reduction would have on the economies of Arab countries, and feared the start of a price war, whose only theoretic limit was the covering of marginal costs. In an attempt to protect the nominal price of oil, Saudi Arabia thus permitted its production to be reduced to steadily decreasing values, despite having greater reserves and despite its negligible costs.[17]

Saudi production (previously in a globally dominant position) found itself having to bear the burden of a doubly "residual" situation: that of the entire OPEC (of which much has been said above), and also of Saudi Arabia itself towards the other members of the cartel. OPEC's decisions to modify the price-quantity combinations to bring them more into line with general market conditions[18] had no effect. Only by overcoming the strategy of the defence of fixed prices could the negative effects be arrested, but not through adapting it (*ex-post*) to exogenous dynamics. The most evident and acute sign of the OPEC cartel's defeat lay entirely in the *collapse of total revenues*, which the price strategy had been supposed to defend. After increasing nine-fold between 1970 and 1974 from $14 to $119 current billion, and again twice over between 1974 and 1982 to an all-time high of $262 billion, prices fell to $70 billion between 1982 and

Figure 6.2. *The "boomerang effect" of prices on OPEC oil revenues: 1973-1986*

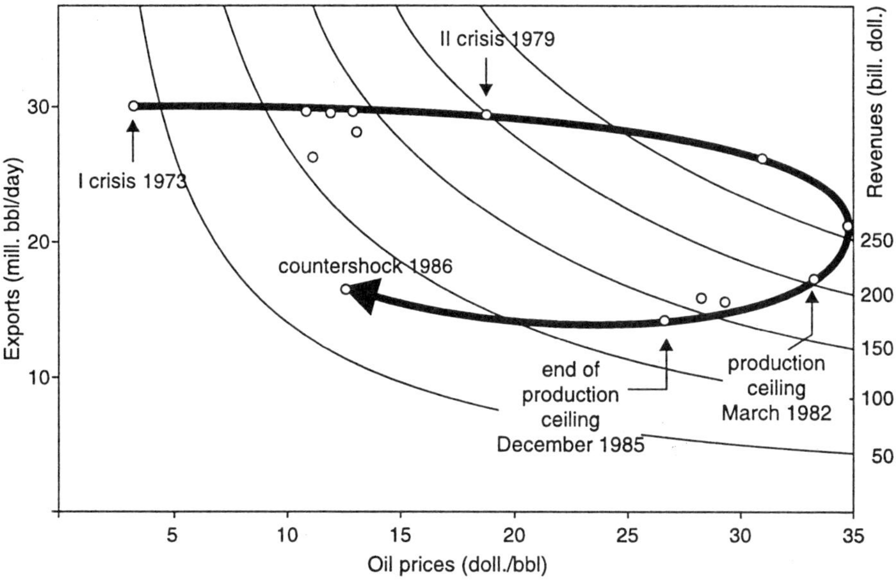

1986. In real terms, this amount was not far off what it had been at the beginning, and much less if calculated at per capita levels. The "boomerang" effect (set out in Fig. 6.2.) of the decisions of the Seventies, whether taken or suffered, continued inexorably. Even the most rigid discipline within the OPEC cartel would not have been able to stop it.

Before analysing what happened afterwards, some brief considerations on the reasons for the failure of the OPEC cartel would seem to be in order. History has taught us that a cartel survives only to the extent that demand reveals itself to be inelastic to prices, so that higher prices increase total profits. This condition certainly existed in the Seventies (although it was inappropriate to speak of cartel practices in the strict sense), but was lacking in the Eighties, when cumulative and delayed effects of the price increases materialised in their entirety on levels of demand and supply. Other more specific reasons combined with this general reason.

First: the fact that external producers both of oil and of other sources of energy were the beneficiaries of the cartel's actions. Such benefits accrued because the prices fixed by OPEC did not constitute a *limit price* with respect to the long-term curve of supply of the group of external competitors (though their costs were far higher than OPEC's). OPEC's fixing of monopoly prices (in relation to its curve of demand, not coinciding with the total) was not a deterrent to the entry of new producers. The more this happened, the more OPEC's residual demand dropped (for every price level), and the more the amount necessary to guarantee the general equilibrium of the market was reduced. It was impossible for them to imagine they could condition it with a market share down to 40%, unless they were willing to reduce total profits and quantities to zero.

Second: the surplus of production capacity afflicted the oil industry not only in its extraction phase (for which the producing countries were responsible), but also in refining and distribution (for which the oil companies were responsible). The surplus of refining capacity drove the companies to process quantities of crude oil in excess of their needs, so that the prices from the spot sales of products ensured coverage of the marginal refining costs (cf. section 1.6.1.). Hence there was a surplus of products, with a consequent fall in their prices which then affected the price of crude oil. The excess of OPEC supply materialised, therefore, in reply to a surplus of primary demand (on the part of the refiners) about which OPEC could do nothing.

Third: the objective difficulty of defining *ex-ante* the ceiling of OPEC production in the absence of informational feed-backs on the dynamic of primary and final demand, as had been allowed to the integrated majors beforehand, therefore became very important. This difficulty was increased by the great changes taking place in the levels of demand growth, in its geographical location, and in its qualitative composition (oil substitution took place for the most part in the heavy fractions used in thermoelectric and industrial systems). The demand forecasts on which OPEC based its supply decisions were almost always too high, and this accentuated the surplus of supply on the markets and lack of discipline within OPEC. Two other no less important aspects complicated the management of supply. In the first place, the inefficiency of the fixing of production ceilings on an annual or even six-monthly basis, given the huge variations in demand in the course of the year. Supply should have been defined on a quarterly or monthly basis, reducing problems of co-ordination of levels and quotas. A further difficulty arose from the need to determine the supply 30-60 days beforehand according to the distances from the areas of consumption, so as to make its control effective in accordance with oil planning times (cf. section 1.7).

The *fourth* and final reason for the failure of the OPEC cartel lay in the natural lack of discipline of its members. The quotas were normally fixed according to the production capacities and budget needs of individual countries. Each time subject to long and laboured political conflict, those countries with high population, low per-capita income and low reserves normally came out best (in terms of the rate of utilisation of capacity). It was Saudi Arabia, as we mentioned, who had to do the impossible. Thus in no way, in order to understand the decisions taken by OPEC, could use be made of traditional economic models, especially the model most commonly used, which assimilated cartels with *many-sided monopolies* with different marginal costs: with production (which maximises global profit) fixed by the intersection between aggregate curves of marginal costs and marginal profits, allocated between the various plants to the point at which each of them has its own marginal cost equal to the aggregate marginal profit (Griffin and Steele, 1980, pp. 121-122).

To conclude, both general and specific reasons explain the failure of the OPEC cartel, as occurred with the previous attempt by the majors in 1928 or in the infinite number of cases history has handed down to us.[19] A few conclusions may be drawn from the unfortunate experience of the OPEC cartel: (a) its action was indeed very slight, because of the technical impossibility of effectively controlling supply in a non-integrated situation and the economic impossibility of bearing the costs, in terms of loss of income; (b) in reality the action of the cartel (control of supply to keep prices high) was more nominal than real and took place over a limited period of time (1983-85); (c) the responsibility for the high level of prices was not only of the OPEC cartel, though it gained from it and in the period of its operation was its main driving force.

6.5. THE TURNING-POINT

Halfway through 1985, when Saudi Arabia's position had become increasingly impossible, the situation came to a head. The lack of discipline of various OPEC members and, particularly, the unwillingness of non-members (especially Mrs. Thatcher's Britain) to listen to Saudi Arabia's warnings, forced Saudi Arabia to produce at levels approaching 2 million bbl/d compared with the 10 million bbl/d of some years before, and 12 million bbl/d and more of production capacity. If they continued on this course, after another two years they would have to start importing oil!

After having pushed its export flows to the maximum to deal first with the Iranian revolution and then the Iran-Iraq war, Saudi Arabia found itself almost unable to finance its own current needs, with export revenues reduced to a third.[20] These difficulties were rendered all the more acute by two elements. *First*: the decision of central bankers to reduce the exchange value of the dollar (the currency of oil trading), whose rise over the previous years had more than compensated for the fall in the *nominal* value of oil, though not the fall in total revenues. *Secondly*: the collapse of Saudi financial reserves held abroad which went from $160 billion in 1981 to around $50 billion in 1985, of which only half could be utilised immediately. The combination of these elements generated much panic and anger in the royal family.

The turning point came in June 1985, when Saudi Arabia announced unilaterally it was abandoning the fixed price system and the role it had played up till then as residual producer. Its exports from that moment onwards would take place at the prices practised on the spot markets. From *price maker* Saudi Arabia became *price taker*, ready to collocate whatever amount the market required. On the technical side, the decision was made by modifying the type of contracts for the sale of crude oil. The formula chosen was net-back pricing (Tab. 6.5.). In other words, the buyers would pay for the crude oil in equal measure to the average price they would have been able to gain from the sale of derived products (after having detracted the full costs of refining and of transport and taking into account the quality of the crude oil). In other words, the price of the crude oil at the time t_0 was calculated backwards (net-back) starting from the weighted mean revenue of the various products sold at time t_1, on the basis of the quotations they registered on the spot markets (published daily in "Platt's Oilgram Price Service" or "Petroleum Argus").

Table 6.5. *Arabian Light net-back value: 1995[1] (doll./bbl)*

Products	Mediterranean spot prices (doll./bbl)	Product yield	Value (doll.)
Virgin naphta	23	7	1.6
Gasoline	26	21	5.5
Gasoi	22	33	7.3
Residual oil fuel	14	34	4.8
Gross Product Value			*19.2*
minus			
• Refining cost			–2.0
• Freight cost			–0.6
Net-back value			*16.6*

1. Product prices (cif Italy) from "Platt's Oilgram Price Service".

The net-back contracts, which spread rapidly in a chain reaction to all countries (OPEC and non-OPEC), generated the following effects:

(a) prices were no longer *controlled* unilaterally. Instead they were connected with the price level expressed by the international markets;

(b) all *market risk was unloaded on to the producer-exporters*, given the advantages for refiners in purchasing and refining crude oil *whatever* the price level, since they were in any case guaranteed a positive refining margin;

(c) in the presence of an inelastic demand, a surplus supply of products was generated, which led in turn to a fall in prices and thus a fall in crude oil prices.

The aim of Yamani, like that of the other OPEC countries (which officially adopted the same position in December 1985), was to pass from a *defence of prices* with restrictive control on production to a *defence of market shares*, with prices left free to find their equilibrium in a supply-demand situation. By attaching prices directly to the market (as well as giving buyers the advantage of fixing them retroactively), OPEC countries dealt a very heavy blow to their external competitors, who were always extremely reluctant to accept any form of limitation of their production. If the latter had not reduced their supply to make room for OPEC's additional supply, prices would inevitably dropped to a *competitive* level, in other words, where they equalled the marginal costs of production. Since OPEC's prices were by far the lowest,[21] this would have meant *eliminating* the less efficient producers from the market. Though the final intention was not quite so drastic, the expected results would only be achieved if OPEC held out when faced with the collapse of prices (Pearce, 1983; Terzian, 1985; Parra, 1986).

After an initial drop in January 1986 from $26-27 to $22 a barrel (over an average of transactions), prices fell even further in February, to 17-18 dollars. A period of stability came in April-May at $13-14, followed by a fall in July to $8-11, around 70% less than their point of departure (Tab. 6.6.). Of the various conjectures possible regarding the behaviour of others, for OPEC the most pessimistic turned out to be true: *competitors were not willing to adjust their levels of supply downwards.*

The dramatic decline of prices, despite OPEC production having risen from 13-14 million bbl/d in June 1985 to 17-18 in April 1986, and to 19 in June, had a very negative effect on currency revenues, halving them throughout 1986 compared to the already low levels of the previous year.

Each month Yamani's strategy burned around $1.5 billion of Saudi Arabian financial assets, and the consequences for other countries were no less devastating. If we consider that only half of Saudi Arabia's assets could be mobilised in the short term, it is clear that not even the strongest of the OPEC countries was capable of withstanding that situation for long. In August 1986 OPEC decided to go back on its decisions in part and bring back a production ceiling of 16.8 million bbl/d, eliminating the most immediate cause of the collapse in prices (the net-back contracts) in December, after the departure of Yamani, held responsible for this collapse, and reinserting a somewhat general reference to government-controlled prices at around $18 a barrel. OPEC's decision, in short, was *to leave market forces to set the price of equilibrium and thus its market share*. OPEC would continue to discuss and decide about ceilings and shares, though with a formal rather than practical application. Market mechanisms would start to work in the opposite direction to that of the past, encouraging a reappraisal in consumption and OPEC production.

Table 6.6. *Crude oil quotations (doll./bbl)*

		Official selling price				
		OPEC	non-OPEC	Spot price	Net-back	Average
1985	December	27.81	26.33	26-27	25	26-27
1986	January	27.70	24	22	22	22
	February	27.20	19.6	15-17	17.5	17-18
	March	–	13-14	12.4	15.2	13-15
	April	–	13	10-13	13.1	12-13
	May	–	13	11-14	13.2	12-13
	June	–	13	10-12	11.2	11-12
	July	–	11	8-10	9.1	8-11
	August	–	12	11-12	11.4	11-12
	September	–	13	13-14	12.9	13-14
	October	–	13	13-14	12.6	13-14
	November	–	13	14-15	12.6	13-15
	December	–	14	14-16	13.4	14-16
1987	January	–	17	17-18	17.1	16-17
	February	17.52	17	17-17	14-8	16-17
	March	17.52	17-18	17-18	16.2	16-18
	April	17.52	17-18	18	16.7	17-18

Notes

*This first part is a reworking of my essay *Crisi energetica: consumi, risparmi e penetrazione elettrica*, in Zanetti G. (ed.) "Storia dell' industria elettrica in Italia. Gli sviluppi dell'ENEL 1963-1990", Rome, Editori Laterza 1994.

1. Cf. W.D. Nordhaus, *Oil and Economic Performance in Industrial Countries*, in "Brookings Papers on Economic Activity", 1980, n. 2, p. 375.

2. The conflict in ideas was particularly strong in Italy, which was more exposed than other Sstates to the vicissitudes of the oil situation. The energy deficit increased three-fold with the first crisis and a further eight-fold with the second one (also through the multiplying effect of devaluation), from 1,800 billion lire current in 1973 to the maximum of 42,000 billion in 1985. Economic policy, in reply to what had been seen at the beginning only as a "problem of the transference" of resources and its financing, and not as a "problem of terms of trade", came with the compression of domestic demand, thus delaying the reallocation of productive resources that alone would have been able to guarantee a new and stable equilibrium in the balance of payments, at an acceptable level of aggregate demand.

3. Cf. W.D. Nordhaus, *Oil and Economic Performance* cit.; D.R. Bohi, *Energy Price Shocks and Macroeconomic Performance*, Resources for the Future, Washington 1989; A. Stagni, P. Bosi, P. Golinelli, *Implicazioni di lungo periodo di shocks petroliferi*, in C.M. Guerci, G. Zanetti (eds.), *Sviluppo economico e vincolo energetico*, Il Mulino, Bologna 1988.

4. W.D. Nordhaus, *Oil and Economic Performance* cit. pp. 374-376, estimates a fairly slight impact of the oil crisis on OECD countries for 1973-1979 for almost all the main macro-economic variables. Without the crisis, GNP would have risen by just 0.15% *per annum* (from 2.68 to 2.83), inflation would have fallen by 0.59 (from 9.3 to 8.71), and productivity risen by 0.14 (from 1.66 to 1.80).

5. Cf. K.A. Mork, R. Hall, *Energy Prices, Inflation and Recession 1974-1975*, MIT Energy Laboratory

Working Paper, 1979; E.A. Hudson, D.W. Jorgenson, *U.S. Energy Policy and Economic Growth 1970-2000*, in "Bell Journal of Economics", Autumn 1974, pp. 461-514; A.M. Okun, *A Post-mortem of the 1974 Recession*, in "Brookings Papers on Economic Activity", 1975, n. 1, pp. 207-221.

6. Of the same opinion are Fried and Shultze (eds.) *Higher Oil Prices and the World Economy*, Brookings Institution, Washington 1975, p. 21, according to whom the reduction of the growth rates of income in the United States and Japan in 1973-1974 compared to previous expectations were *for more than half* to be accounted for by the effect of the oil crisis, respectively 2.5% and 4.2%, relative to a contraction in the growth of the GDP from +2.6% to –2.1% for the United States and from +5.2% to – 201.8% for Japan. For Western Europe, the fall in GDP growth from +4.8% to +2.3% was totally due to the crisis (-2.7%).

7. In OECD countries overall consumer prices increased from 4.7% in 1972 to 7.7% in 1973, against a trend rate in the decade 1961-1971 of 3.7%. In Europe the acceleration had been from 6.7% to 8.7% against a trend of 4.2%.

8. Cf. OECD, *Economic Outlook*, December, 1973, p. 24.

9. Especially accentuated were the price increases of agricultural products, also because of the poor harvests in the Soviet Union. For the correlation between raw material prices and the crisis of the international monetary system, see Sylos Labini, *L'instabilità nei prezzi delle materie prime e il problema dell'oro*, in "Bancaria", 1982, pp. 7-24.

10. Cf. J. Fabritius, et al., *Opec Respending and the Economic Impact of an Increase in the Price of Oil*, in Matthiessen (ed.), *The Impact of Rising Oil Prices on the World Economy*, The Macmillan Press, London 1982.

11. Cf. E.R. Berndt, D.W. Wood, *Engineering and Econometric Interpretations of Energy-Capital*, in "American Economic Review", 1979, n. 69, pp. 342-354.

12. Cf. R.S. Dohner, *Energy Prices, Economic Activity and Inflation: a Survey of Issues*, MIT Energy Laboratory Working Paper, 1979.

13. On the limits and risks connected with price forecasts, see the "entertaining" text by Arthur Andersen & Co. and Cambridge Energy Research Ass., "*The Future of Oil Prices. The Perils of Prophecy*", Cambridge, 1984. In the days of the first crisis, two economics gurus such as Paul Samuelson and Milton Friedman confronted each other on the theme: the former adopting a statement by Nordhaus, according to which "the true value of a barrel of oil should increase by an average of 4%-5% per year over the next 50 years" ("Newsweek International" 26 November 1973) and the latter claiming that "oil prices are weakening. They will soon fall" ("Newsweek International" 4 March 1974, p. 46).

14. Half way through 1979, the influential *Office of Economic Research* of the CIA was still forecasting that over the next two years the demand for energy in industrialised nations could rise from a minimum of three to a maximum of 7 million bbl/d, with possible supply deficits of 2.6 to 5.8 million bbl/d. Rather 20than increase, the demand for energy fell by 2.0 million bbl/d: 5-9 less than had been forecast. (CIA, 1979).

15. *Principal budget data of the CMB company sample (bill .doll.)*

	1970	1974	1980
Revenue	86.2	244.5	602.9
Costs	62.6	178.4	470.1
Write-offs	6.6	10.1	22.7
Taxes	10.4	39.6	74.9
Net profits	6.6	16.4	35.2
Rates of return (%)			
Shareh. equity	10.4	19.2	22.4
Capital empl.	9.9	17.3	20.6

16. From average levels of 8.3 million barrels a day in 1978, it went to 10.5 in 1980-1981.

17. Yamani wrote: "In 1981-1982, when OPEC exports began to decline sharply following the shrinking of demand and the increase of non-OPEC supplies, we understood too late that the oil was over-priced. In other words, the price had reached unjustifiably high levels which had upset the virtual equilibrium between demand and supply. This is why Saudi Arabia (which had increased its production in 1980-81 to slacken the upward pull of prices) decided to change its attitude, reducing production in support of the world price structure. Inside OPEC we realised that if we wanted to avoid a sharp drop in short-term income, a ceiling on production had to be fixed, thus consolidating the price level". (Yamani, 1986).

18. From $29 a barrel – 17.5 million bbl/d in March 1983, by October 1984 it had passed to $29 a barrel – 16 million bbl/d, and in January 1985 to $28 a barrel – 16 million bbl/d.

19. Like the salt cartel of 1301 established by Philip of France and Charles II of Naples, under the auspices of Florentine bankers. Or the no less interesting cartel that the papacy and King Ferdinand of Naples licensed on June 11th 1470 in the production and sale of alum (a raw material essential for dyeing and tanning of cloth and leather), to raise the prices and force out Turkish production to which Italian traders paid very substantial royalties.

20. Yamani, at that time still minister, wrote: "Given such a low level of production, it was clear that the burden of supporting the price structure could no longer be borne by OPEC alone. OPEC simply no longer had the power to do this. Nor could Saudi Arabia continue to support the regime of fixed prices taking upon itself all the effects of the decline in demand: at that point even the possibility of financing its own economic development was seriously at risk." (Yamani, 1986).

21. With levels then valued at between 0.20 and 1.0 dollars a barrel compared with the *average* of 2 for North Sea and around 5 of the United States (Adelman, 1986).

7 THE OIL MARKET CAUGHT BETWEEN ECONOMICS AND POLITICS: THE PERSIAN GULF CRISIS*

7.1. THE TERMS OF THE CRISIS

The "Gulf crisis" began on the night of Wednesday 1st 1990 with the invasion of Kuwait by Iraqi military forces. It ended at dawn on February 28th 1991 with U.S. President George Bush's announcement of the order for a cease-fire to the coalition army, acting under the auspices of the UN, but to which the United States had made by far the biggest contribution and, in the main, directed. The cease-fire occurred after the complete liberation of the tiny Arab Emirate, after a demonstration of overwhelming military superiority, and after Iraq had accepted every one of the resolutions adopted by the United Nations against its action ("Energia", 1990, 1991).

After 211 days of brutal Iraqi occupation, 205 days of total embargo, 42 days of bombing and 100 days of land campaign, Kuwait regained its legitimate sovereignty. 100,000 Iraqi soldiers died on the battlefield, to which perhaps another 100,000 dead should be added, caused by the Baathist repression of popular uprisings (especially that of the Kurds, as in 1975, when driven to anti-Iraq rebellion and then tragically abandoned to their fate), as against 165 allied dead; 3,700 Iraqi tanks were destroyed, as against 2 of the allies, and 141 planes, against 37,800 oil wells which went up in flames in Kuwait (out of the 858 in production) and crude oil was lost to the tune of around 6-7 million barrels a day: around four times its previous production and at a daily cost of $120 million.

Now time has past and, despite the all-pervasive and sensational role of the media which helped to involve the entire world at unusually high levels of participation, there seems to be very little trace left in our collective memory of this train of events. At the most, they seem to appear as a distant nightmare, though to be contemplated with "complacent pride", because of the exemplary ability shown by the entire international community, and particularly the industrialised West, in facing its first serious crisis since the end of the Cold War in 1989, united and unharmed. Despite the high cost in human and economic terms, and despite the fact that the war, as usual, was unable by itself to turn success into positive political outcomes (particularly if we recall that the dictator Saddam Hussein is still firmly in power), there can be no doubt whatsoever that the West, and above all the United States, emerged from the test with their morale high: due to the regained confidence in their own capabilities, and the proven poor qualities of the enemy; the unchallenged domination acquired on a global scale under the renewed leadership of the United States; and the strong internal cohesion demonstrated.

The oil question itself, the main reason for the anti-Iraq reaction, seemed to emerge considerably downsized. The international oil market gave a demonstration of remarkable efficiency and rationality, passing unharmed through a "supply vacuum" such as had never been seen before. At the end of the crisis, prices were at exactly the same level as before, around $18-20 a barrel, thus wholly re-absorbing the peaks of double that figure that had been recorded mid-way through the crisis. The closer collaboration between industrialised countries within the Paris Agency; the breathing space guaranteed by the ample stocks available to the West; the greater opportunities for dealing with risk situations with the new financial and commercial tools available to the operators; the reduced importance of oil in the overall demand for energy and of the latter in the formation of income: these were some of the reasons for the different course this latest crisis took compared to those in the past. Nor in particular, was there any of the *panic buying* that had produced so much harm then. Also, with a view to the future, the oil issue loomed less large, because of the improved chances, it was now argued, of a guarantee of those conditions of political stability and security of access to the supplies of oil in the Middle East, at which western foreign policy had always aimed.

The lesson given to the Iraqi *ras*, the strengthening of the moderate Arab regimes, and the "return" in more or less obvious forms of the American presence in the Middle East, change prospects and make the picture more reassuring, while feelings, as mentioned already, varied from a proud complacency to a new-found confidence. However understandable these feelings are and however valid the reasons for them, they nonetheless risk leading us to mistaken convictions and choices, since they do tend to obscure what actually happened on the oil markets in those 211 days of crisis. They exaggerate the value of the positive short-term outcomes of the war, compared to other outcomes that may be derived from them in the long-term. They also underestimate the significance of issues in the field of energy that the whole world will eventually have face up to in the future. The Gulf Crisis, for the reasons that gave rise to it and for those that have determined its later development, could turn out to be neither fortuitous, nor limited, nor easy to write off by considering only its tragic final outcome. No military victory can provide the complete solution to or government of the problems of this world, certainly not those of the world of oil, and the experiences of the past should provide confirmation of this.

7.2. FACTS AND PRIOR EVENTS

Despite the many differences to be met with during the Gulf Crisis, compared to those of almost cyclical recurrence that have marked the beginning of the two past decades, there are certain similarities. One is the sudden way these events burst upon world public opinion, the business community, political observers and, to judge from appearances, governments in general. The surprise may be due to the brutality of the Iraqi attack, but also the culpable negligence with which the West greeted the numerous signals of tension which, in a relation of reciprocal causality, had been emerging from the political scene and the world of oil. It was perhaps not possible to foresee that their final outcome was to be as initially threatening, then actually to take place, but nothing or very little was done to put a stop to things by that same international community that would afterwards move into action with such incredible efficiency and speed.

Signs that the political climate was at a critical turning point in the Middle East and in the oil market could be observed by the beginning of 1990. A worried article in "Business Week" of April 16th denounced the clear intention of Saddam Hussein "to become the dominant power in the Persian Gulf". Warnings about the risk of a new conflict in the Middle East began to be heard on many sides, showing that this area was still virtually extraneous to the East-West relaxation of tensions, from which, paradoxically, further reasons for destabilisation derived. The gradual superseding of the "automatic alliance" between Arab countries and the Communist bloc resulted in bringing about a fairly significant shift in the region's delicate balance of power, to the exclusive advantage of Israel, which saw its already overwhelming superiority reinforced. "Orphans of perestroika", as "Le Monde" defined them on October 10th 1990, the Arab countries of the PLO contemplated with ever-increasing apprehension the definitive victory the United States and its allies in the Cold War.

Saddam Hussein, strengthened within the region by the increasingly obvious weakness of Syria and by his July 1988 victory over Iran, (after eight years of brutal warfare and thanks to the full support of the West and the moderate Arab regimes – with Kuwait in the forefront), but weakened externally by the collapse of the Soviet system and the supremacy of America, tried to turn the growing resentment of the Arabs towards Tel Aviv to his own advantage. He was aided by the violence shown by the Israeli army towards the Palestinian *intifada*, the new expansionist tendencies feared imminent through the expected immigration to Israel of a million Soviet Jews (with an Israeli population standing at the time at 4.7m.), and the intransigence demonstrated by the Israeli premier Shamir towards any kind of concession on the Palestinian question.[1] After a period of prudence lasting almost two years, because of the need to reconstruct and strengthen his productive and military forces, Saddam Hussein felt himself strong enough to "return to the scene as protagonist". In doing this, he completely overturned the system of foreign alliances he had built up since he came to power in 1979, and which had made Iraq into a valuable and much sought-after partner of the Arab countries in the Gulf, and hence strategic interlocutor of the United States (Rodman, 1991).[2]

On February 24th 1990, – after the foundation of the *Arab Cooperation Council*, a new regional club close to the interests of Jordan but of marginal interest to its other members (Egypt, Iraq, and North Yemen) – starting with the weakening of the Soviet role and influence in the area, Saddam Hussein then launched a first violent attack on the United States for its immense power over the Middle East, with the Gulf still patrolled by its fleet despite the conflict with Iran having ceased eighteen months before, and still more for the position the US had gained from the new international balance of power.[3]

In the crescendo of "signals" following that first attack (the "Economist" afterwards would count at least five of especial significance),[4] one element increasingly emerges: the appeal, in the rhetoric hostile to the West, to utilise oil as an "instrument of political pressure", and if necessary of threat, to force the United States and its allies to review their policies and moderate their ambitions, according to the pattern already seen in the Yom Kippur War of October 1973.

In the depressed conditions of the oil market of 1990, threats of its utilisation for political purposes seemed to be simply rhetorical wishful thinking. To Saddam Hussein's eyes this was not the case, however, if only the contingent causes of the market depression were removed; *i.e.* the "criminal" excess of supply made available by the richer countries in the Gulf, starting with Kuwait. By pumping out more than OPEC had planned and more than the market demanded, they inevitably pushed down the price of oil, to the exclusive advantage of the West, while damaging the Arab economies, already in disarray. If such a policy continued, Saddam Hussein declared, it would be interpreted as a deliberate act of "economic war" against which Iraq would be forced to defend itself.

His political designs on the region, supported by a clear military superiority, risked being crippled by the country's increasingly disastrous economic situation. He was crushed between enormous *foreign* debts, and a *domestic* situation threatened by oil prices not high enough to service these debts as well as to pay for the importation of essential commodities (though it was certainly not these that were the target of all of the available resources). The chance of seeing his current deficit paid off in the future through application to the international financial system seemed to be fading due to the caution and lack of confidence that foreign banks were showing towards Iraq, after years of incredible extravagance. In order to correct the increasingly obvious imbalance between his military force and economic weakness, only one option remained open to Saddam Hussein: raising his voice within OPEC.

Despite the distortions in Iraq's accusations against the rich countries of the Gulf, there was also a good deal of truth. From the beginning of 1989, they had systematically ignored the agreements signed within OPEC,[5] and had resorted to obvious cheating. By doing this, they had more than helped to cancel the advantages concerning prices that could have been derived from the expansionist cycle of oil consumption that started in the late Eighties. After the rock-bottom minimum of below $10 a barrel, reached momentarily in the counter-shock of 1986, oil prices had fluctuated until the middle of 1989 between $13-17 a barrel. From that time onwards, the entire structure of international prices had started to rise, reaching peaks of $22-23 a barrel by the beginning of 1990, something that had not happened for four years.

The peak of the economic cycle and the restriction of margins of spare capacity caused by high consumption were the basic reasons behind this rise in prices, reinforced at its peak by the simultaneous occurrence of several events that, however short-lived,[6] testified that the international oil system had once again begun to display situations of operational and commercial rigidity. What might have seemed the first sign of a new, lasting cycle of expansion in prices – incidentally taken for granted by most current forecasts – dissolved in the twinkling of an eye. From January prices of $22-23 a barrel, prices actually dropped to $13-14 in June 1990. The reason for this collapse was the intensification of competition between OPEC countries, sparked off by those with spare supply capacity who aimed at increasing utilisation by forcing the conditions of sale downwards. In a classic example of reaction-counter-reaction, the other oil producing countries were forced to follow suit, so it was the entire price structure which slid downwards, all to the benefit of the consumers and the oil companies, who found themselves benefiting from increasing unit profits because of the greater strength of final prices compared to the cost of the raw material. As can be seen in table 7.1., in the first quarter of 1990, despite the *programmed* supply within OPEC being 3.5 million barrels a day higher than the year before, the *actual* supply turned out to be 1.7 million barrels a day higher still, because of the production excesses of Kuwait, the Arab Emirates and Saudi Arabia. The damage they provoked could have been even higher if Iran and Iraq had not voluntarily sacrificed around 8% each of the productive quota to which they were entitled.

Table 7.1. *OPEC supply surplus in comparison with total and individual agreed values (mill. bbl/d)*

	OPEC supply		Kuwait	UAE	Saudi Arabia	Iraq	Total surplus (A) - (B)	OPEC prices (doll./bbl)
	Actual (A)	Agreed (B)						
1988								
III	20.3	16.6	0.5	0.7	0.8	1.1	3.7	12.77
IV	22.8	16.6	0.7	1.0	2.3	1.1	6.2	11.95
1989								
I	20.6	18.5	0.3	0.6	0.3	–	2.0	16.2
II	21.7	18.5	0.9	0.7	0.5	–	3.2	17.2
III	22.6	19.5	0.8	0.9	0.4	0.1	3.1	15.8
IV	23.8	20.5	0.8	1.2	0.6	–	3.3	17.5
1990								
I	23.7	22.0	0.6	0.9	0.3	–0.2	1.6	17.5
II	23.9	22.0	0.4	0.9	0.2	–	1.8	14.0

Source: "Petroleum Intelligence Weekly".

Despite the crescendo of accusations and threats by Iraq towards its ex-allies in the Gulf,[7] their behaviour did not change over the second quarter of 1990, with prices sliding down to $13 a barrel by the beginning of June, while the profile on the markets of deferred delivery prices and futures prices remained at the values of the beginning of the year, around $20 a barrel. This proved that in the expectations of the operators, short-term dynamics did not find any correspondence in the medium to long-term fundamentals, even where politics were not taken into account (Verleger, 1990a, p. 19).

In the middle of June 1990, despite the persisting situation of oversupply (which became even more acute in July when Iraq too pushed its productive possibilities to the maximum – perhaps because it had already decided to attack Kuwait), the entire price structure halted its downward course and began to increase again, with renewed vigour. At the end of July 1990, spot transactions were back to $20-21 a barrel with bonuses in longer-term contracts of over $2. This inversion was not brought about by any actual events, but rather by the increasing attention to and belief in Saddam Hussein's threats on the part of the markets (much more so than the objects of these threats), with the expectation that they would soon result in at least a reduction of current production in the OPEC countries.

7.3. ECONOMIC AND POLITICAL DETERMINANTS

The series of events preceding the invasion of Kuwait, and those to follow, show a mixture of political and economic determinants of the Gulf Crisis whose causal interconnections and final outcomes may not be appreciated fully unless one basic factor is considered: the critical conditions – besides the short-term conditions described above – that had been reappearing on the oil market. These conditions in themselves gave an explosive force to every crisis situation in the Middle East. This may be illustrated by a few facts. From 1983 to 1990 demand for oil in the areas of market economy increased daily by around 7 million barrels (equal to twice the entire production of the North Sea). 91% of this increase was satisfied by OPEC countries, and of this, 80% by Persian Gulf countries. At the same time, there was an increased dependence on foreign supplies for the West, particularly in America, which in 1990 reached a new all-time high of nearly 50% of domestic consumption, twice the level of 1983.

Although the aggregate data of the international oil market recovered, and in many cases went beyond, the "critical level" that had accompanied the second oil crisis, the West was reluctant to pay the slightest attention to it. This indifference was in no way affected by the growing tension between Iraq and Kuwait, which indeed led to hopes of the possible positive effects on the prices of oil. Until August 1st 1990, the "energy question" was largely seen as having no significance. Every aspect concerning the political and economic vulnerability of foreign supplies of oil had been cancelled from the collective memory, and removed from the agendas of western governments. It was almost as if the favourable trends in the markets were resistant to every profile of growth in the economies; as if this did not depend on the difficult process of adaptation the industrial countries had managed to put into practice after past crises; and as if the serious social and economic problems which many producing countries were having to deal with after the counter-shock of 1986 had nothing to do with market stability.

Warnings to the contrary went unheard, such as the various reports published in the United States on the subject of energy security (DOE, 1987; NPC, 1986, 1987; Toman, 1990). These all agreed that the combination of low energy prices with the expansion of the economy and with the widespread environmentalist opposition to every kind of investment, would finally push up oil consumption and reduce domestic supply, thus re-creating situations of dependence on foreign supply "incompatible with every need of national security".

However, after August 2nd the crudest realism replaced the illusions prevalent up to that time on both sides of the Atlantic. Though it cannot be denied that the determination to restore full sovereignty to Kuwait, and enforce full respect for international law was the reason behind military intervention in the Gulf, it would be naive to believe that oil played a secondary or an entirely coincidental role. And yet in the abstract debate between "interventionists" and "pacifists" throughout the long months of the crisis, as well as in the various motivations given in support of the intervention by the US Administration, – in the words of President Bush: "for our jobs, our way of life, our own freedom and the freedom of friendly countries around the world" ("The Economist", November 17th 1990), the existence of economic causes linked to oil was either ignored entirely or in the minority opinion (Ginzburg, 1991).[8]

7.4. THE DAYS OF CRISIS

As we look over the days of the crisis once again, we must first of all keep the *facts* separate from our *perception* of them, a perception created under the influence of propaganda as inevitable as it was imponderable.[9] Secondly, to try to understand what happened and why, we need to separate the final outcomes, which today are known but then were impossible to imagine, when nothing was known about how the crisis would develop or how long it would take. We should start with the sudden *supply vacuum* on the international oil markets following the UN Security Council resolution on August 6th 1990 – 13 votes in favour out of 15, with two abstentions, Yemen and Cuba – when all the countries of the world, even non-UN members, were asked to blockade all economic transactions to and from Iraq and occupied Kuwait, thus placing a total embargo on these two countries. The day afterwards, Saudi Arabia authorised the use of its own bases and territory by the US army (and its allies) as part of the "Desert Shield" operation, aimed at preventing any further advance by Saddam Hussein's troops, already massed on Saudi Arabia's borders, and at enforcing the embargo decreed by the UN. The rigid blockade imposed on land and sea traffic led after very few days to the total *cancellation* of oil exports from Iraq and Kuwait.

The international markets were flung into a turmoil. Cash prices rose $8 in five days to around $27.5, then dropped by $2 on the announcement of the American military intervention. Compared to the current flows of July, the supply vacuum was 4.9 million barrels a day, equal to 9.2% of the entire world supply (outside the ex-communist bloc) and to 16.2% of the entire international trade in crude oil and refined products (tab. 7.2.). The Gulf Crisis instantly appeared as one of the most serious to take place since the Second World War (tab. 7.3.), particularly in consideration of the possible developments (involving Saudi Arabia) and the reduced spare capacity.[10]

Table 7.2. *Supply deficit during the Gulf crisis*[1] *(mill. bbl/d)*

Countries	Oil production (A)	Domestic consumption (B)	Exports (A) - (B) Crude	Products	Total
Iraq	3.40	0.5	2.80	0.10	2.90
Kuwait	1.95[2]	0.1	1.15	0.70	1.85
Saudi Arabia[3]				0.15	0.15
Total	5.35	0.6	3.95	0.95	4.90

1. Calculated on levels of July 1990. 2. Including the Neutral Zone. 3. Export reduction due to larger domestic consumption required by the army.

Source: "Petroleum Intelligence Weekly".

Further critical elements compared to the past arose from (a) the great importance, mainly for environmental reasons, given to the quality of the crude oils, resulting in a much greater rigidity in international trade; (b) the changing composition of the flow of exports from the Gulf, to a significant extent now refined products of high quality. The dynamic of the crisis developed therefore from the complicated interrelation between two shortages: concerning the quantity and quality of both crude oils and products. Both were made more critical by the simultaneous saturation of production capacity and the refining of oil outside the scene of the crisis.[11]

Table 7.3. *Comparison between crises: oil supply*

Date	Cause	Supply deficit mill. bbl/d	Supply deficit % consumption[1]
1951-1954	Iranian nationalisation	0.7	5.3
1956-1957	Suez Crisis	2.0	11.4
1967	Six Days War	2.0	5.0
1973-1974	Yom Kippur War	1.6	2.8
1978-1979	Iranian Revolution	3.7	6.9
1980-1981	Beginning Iraq-Iran War	3.0	5.6
1990	Gulf Crisis	4.9	9.2

1. Excluding countries with centrally planned economy.

Source: our processing of various sources.

A third dimension to the crisis was added by the geographical location of the demand–supply imbalance: this gave rise to a long-term dynamic of tensions over prices in the various regional markets. A significant number of individual purchasers found themselves short of products and therefore had to cover themselves elsewhere. In a market where everyone paid marginal costs, the sudden appearance on the scene of some deficit operators forced everyone to pay higher prices, although, hypothetically, the system had recovered its arithmetical equilibrium.

The transmission of the effects of the crisis was matched the intensity of the impact it had on the energy systems of the more industrialised areas. As regards total energy consumption (tab. 7.4.), the hardest hit was Japan, then Europe, and only marginally the United States, while for overall oil imports the situations were relatively more homogeneous. This data shows an evident contradiction between the degree of *vulnerability* revealed by the three greatest industrial powers, and the actual *involvement* that each of them would accept in its subsequent development. Indeed, although it was the US who were the first to react to Saddam Hussein's attack, and who were the ones actually to carry out the liberation of Kuwait, there is no doubt that it was Europe and Japan above all (not to mention Italy) who had the most to fear from the consequences. This concern did not appear, however, to affect them very much (nor did this argument ever enter into the debates on the position to take), with the exclusion, paradoxically, of England, the only country self-sufficient in oil, which provided the firmest and most prompt support to the United States.

A positive annexation of Kuwait would have allowed Iraq to almost double the amount of oil reserves it directly controlled, to around 30% of the Middle East total and 19.5% of the world's, while its overwhelming military superiority in the region would have easily allowed it (in the absence of any outside intervention) to exercise a leading role with regard even to the still greater Saudi Arabian reserves, extending its control to 45% of the world reserves (tab. 7.5.). This was a risk that did not cause the slightest worry to Europe or Japan, whereas it did get an immediate reaction from the country more closely threatened, Saudi Arabia. The Saudi government agreed to bear almost the entire financial burden of the Desert Shield operation, estimated at 5 billion dollars a month before the war,[12] thus making the American soldiers (in the words of a journalist in the Gulf, reporting the opinion of some Saudi Arabians) a "kind of mercenary army", or even "a new kind of foreign worker (just like) the Pakistanis who drive the taxis" (Schlesinger, 1991).

Table 7.4. *Oil crisis ratios in the main areas (%)*

Ratios	USA	Japan	Europe	Italy
Oil/energy consumption	42	58	45	58
Oil import/oil consumption	45	100	81	95
Oil import/energy consumption	19	58	36	55
Arab oil import/oil import	30	61	58	69
Iraq-Kuwait import/oil import	9	13	11	14
Iraq-Kuwait import/energy consumption	1.7	6.4	4.7	7.7
Energy consumption/GDP	4.8	1.8	2.6	2.3
Oil consumption/GDP	1.9	1.0	1.2	1.4
Net oil import/GDP	0.8	1.1	0.8	1.2

Source: BP, "Statistical Review of World Energy", 1991.

Table 7.5. *Effects of Kuwait annexation on the control of oil resources*

	Oil reserves		Production[1]	
	mill. bbl	%	mill. bbl	%
Iraq	100	10.0	3.4	5.2
Kuwait	97	9.5	1.9	2.8
Total Iraq-Kuwait	197	19.5	5.3	8.0
Saudi Arabia	258	25.5	5.3	8.0
Iran	93	9.2	3.0	4.5
Middle East	663	65.6	16.2	24.6
World	1,009	100.0	65.6	100.0

1. Data of July 1990.

Source: BP, "Statistical Review of World Energy", 1991.

The fact that this crisis was the most critical yet experienced in the Middle East was objectively illustrated by the geopolitical dimension it took on, with a coalition of a number of countries and military forces that had not been seen since the Second World War (before the final military showdown there were around 1,310,000 soldiers from 29 different countries); the "supply vacuum" that had been created in the markets with the embargo, and the vacuum feared even more from the conflict, if the Saudi oilfields were involved; and finally, the absolute uncertainty about the possible duration of the crisis. All this was confirmed by the markets' reaction, with price increases in the first four weeks twelve times higher than those recorded in the 1978-79 crisis, and ten percentage points higher than those of 1973-74 (tab. 7.6.). Despite these facts, the policy makers' reaction to the crisis, as well as that of most companies and commentators, was right from the start imbued with a great sense of optimism and confidence. This was understandable on a political and military level, but much less so as regards the oil situation (at least on the basis of the information available at the time). Whether it was right or wrong, this perception of the situation was in itself hugely important in influencing the expectations of the operators, playing down the "buyers' hysteria" bitterly experienced in the past, though the later positive course of events naturally cannot be traced back to it.

Table 7.6. *Comparison between crises: prices*[1]

	Yom Kippur 1973		Iran 1978		Gulf 1990	
	doll./bbl	Var.%	doll./bbl	Var.%	doll./bbl	Var.%
At the beginning[2]	5.65		13.81		17.39	
30 days	8.48	+50	14.46	+5	27.85	+60
60 days	14.28	+152	16.61	+20	41.00	+135
90 days	18.92	+234	15.98	+16	34.35	+98

1. Arabian Light net-back in Rotterdam market due to Kippur and Iran crisis; Brent dated for Gulf crisis. 2. Average value for September 1973 for Kippur crisis; September 1978 for Iran crisis; July 1990 for Gulf crisis.

Source: our processing of data from various sources.

There were four reasons for optimism:

(a) the wide availability of commercial stocks held by the international oil system, and the strategic ones accumulated by certain governments (United States, Germany and Japan);
(b) the oil producing capacity remaining in some of the producing countries of the Persian Gulf, in the hope of being able to count on them as the need arose;
(c) the fact that economies were less sensitive to oil price increases;[13]
(d) the perfecting of the mechanisms of international co-operation, provided for in IAE for emergency situations, after the positively non-brilliant experiences of the 1979-1980 crisis.

These reasons also include the absolute confidence (especially on the part of the US Administration) in the virtues and functionality of the "free market" almost as if, as Verleger wrote (1990b), "it hadn't lost its virginity on August 6th with the United Nations vote". It was this confidence that determined the line of behaviour diligently taken by all the other industrialised countries, *i.e.* not taking any decision that might interfere with the market, waiting for events (on the supply side) to decide the outcomes (on prices) without intervening to try and point them as far as possible in the right direction, for example through a prompt and well-aimed management of the stocks to calm down prices, as was called for on many sides.[14] Though there was indeed no doubt that, if left to itself, the market would find a new equilibrium, it was also true that this equilibrium would bring higher prices the more substantial and long-lasting the actual deficit (or even the expected deficit) of crude oil and products. If the West had paid attention not only to the fate of its own economies, but also to those of poor countries, it should have intervened at the right moment to prevent prices allocating scarce supply and rationing demand.[15] This began to happen in the autumn of 1990 when the weakest links in the consumer chain (India, Pakistan, the Philippines, Brazil, etc.) began gradually to withdraw from the oil market. They were faced with the impossibility of paying prices two to three times higher than before, and were forced to cut domestic consumption, with only a few days stocks in reserve.[16] There was not the slightest doubt that this was yet another demonstration of the efficiency of the markets, just as there could be no doubt about the absurdity that this took place while the reserves of crude oil and products were, according to western governments, even "too" full to the brim!

In the first three meetings held by the Governing Board of the Paris Agency after the crisis had broken out (the first, incredibly, a month after on August 31st), the line that emerged as decision-making policy was *to decide nothing*: to allow things to take their course, despite the prospect of an increasingly "tense market in the winter months due to the higher seasonal demand, due to the reduced possibility of drawing upon the commercial stocks, and due to the saturation of refining capacity" (IAE/P, August 31st 1990). Whereas all the producing countries were "encouraged" to increase oil supply as much as possible if they had not already done so, the Agency rejected, without even deigning to mention it, OPEC's request to do the same by drawing upon the ample stocks of the OECD countries. This rejection and silence meant in fact that they did not wish to start any kind of dialogue with the producing countries, even in the exceptional circumstances, despite having decided, on that specific occasion, to increase production to guarantee "the stability of the market and the regular provision of oil to the consumers", thus completely isolating Saddam Hussein.

The highest level of political collaboration – both in the West, and in relations between the West and Arab countries – thus came up against the lowest possible collaboration on an economic level. Here the only understanding was between the United States and Saudi Arabia, establishing a huge investment commitment ($1.5 billion) in a very short time to increase Saudi productive capacity (news of which was to emerge only at the end of the crisis). This understanding also meant the accumulation (beforehand?) of additional stocks, actually managing together the organisations in which they held an unchallenged leadership, IAE and OPEC respectively. To conclude, it seems to us that, more than by the infrequent summits of Paris and Vienna, the development of the crisis was influenced by the many conversations between Washington and Riyadh on the "hotline" promptly set up between the U.S. Energy Department and the Saudi Oil Ministry ("Petroleum Economist", April 1991).

7.5. THE REASONS FOR SUCCESS

7.5.1. The aid provided by the Arab countries

At the end of August 1990 all the OPEC countries (including Libya, which was to go back on its initial decision) began to increase their production, suspending the "ceiling" of 22.49 million barrels a day fixed at the OPEC Geneva Conference a month before.

This was a decision of fundamental importance because around 90% of the incremental capacity of oil was located in those countries.[17] In order to face a supply deficit of 4.9 million barrels a day and a seasonal increase in consumption for an additional total requirement of 7.4 million barrels a day, it would have been necessary to force the pumping of oil from the wells to the maximum, beyond sustainable capacity levels, and at the same time, draw upon stocks to a steadily increasing extent. There were margins of manoeuvre on the supply side, therefore, but they were gradually more limited the longer the crisis lasted, and at any rate they were not such as to justify "wait-and-see" policies or great optimism. On the demand side, there was even less room for intervention, given: (a) that it was not easy to substitute oil (in the short term) with other sources; (b) the low elasticity of prices of final demand (because of the heavy percentage played by taxes); (c) that governments were reluctant to adopt measures of consumption restriction with their uncertain outcome, which were also counter-productive at the psychological level.

Table 7.7. *World oil production (mill. bbl/d)*

	1990				1991		
	July	August	November	December	January	February	March
Saudi Arabia	5.45	5.85	8.35	8.61	8.14	8.20	8.00
Iran	3.05	3.30	3.20	3.30	3.20	3.30	3.40
Iraq	3.40	1.00	0.43	0.43	0.25	0.00	0.00
Kuwait	1.85	0.10	0.08	0.08	0.05	0.00	0.00
UAE	2.00	1.60	2.35	2.35	2.45	2.48	2.50
Qatar	0.37	0.40	0.37	0.37	0.37	0.39	0.39
Venezuela	2.00	2.05	2.30	2.35	2.35	2.35	2.35
Nigeria	1.70	1.80	1.90	1.90	1.90	1.90	1.90
Indonesia	1.23	1.30	1.33	1.40	1.45	1.45	1.45
Libya	1.25	1.40	1.50	1.50	1.50	1.50	1.45
Algeria	0.75	0.75	0.80	0.80	0.80	0.80	0.80
Gabon	0.27	0.27	0.30	0.30	0.30	0.30	0.30
Ecuador	0.28	0.30	0.29	0.29	0.29	0.30	0.30
Condensates	2.17	1.98	1.98	1.98	1.98	1.98	1.98
OPEC total	*25.77*	*22.10*	*25.17*	*25.65*	*25.03*	*24.94*	*24.82*
USA	7.14	7.22	7.30	7.30	7.40	7.30	7.30
North Sea	3.31	3.30	3.86	3.58	3.69	3.89	3.90
Others	12.02	12.07	12.34	12.63	12.29	12.21	12.20
Condensates	2.73	2.73	2.75	2.75	2.75	2.75	2.75
Non-OPEC total	*25.20*	*25.31*	*26.25*	*26.26*	*26.12*	*26.15*	*26.15*
Total of Western countries	*50.97*	*47.41*	*51.42*	*51.91*	*51.15*	*51.09*	*50.97*
Former USSR	11.37	11.30	11.05	10.90	10.70	10.70	10.30
Others	3.06	3.11	3.13	3.13	3.10	3.10	3.10
World	*65.39*	*61.82*	*65.60*	*65.94*	*64.95*	*64.89*	*64.67*

Source: "Petroleum Intelligence Weekly".

The other OPEC countries began making up the initial deficit of 4.9 million barrels a day as early as August, continuing in the months to follow, until in December the maximum level of sustainable capacity was exceeded (at 4.4). This was due mainly to an increase in Saudi production of 3.05 million barrels a day (about 70% of the entire OPEC increase) to 8.35 million barrels a day, with peaks some days of over 9 million barrels a day, well over its peak capacity (tab. 7.7.). With overall OPEC production in November approaching pre-crisis levels, the world oil system seemed to have recovered at the level of global arithmetic (though not in the case of specific local or company situations) a position of substantial demand-supply equilibrium. The markets also reacted in part to the readjustment with a price fluctuation sliding from $35-40 a barrel from mid-September to mid-October to $30-35 a barrel at the end of 1990 and $20-35 in mid-January 1991.

At the beginning of winter, the oil system saw its productive capacity exploited over and above its margins of security, with a reserves potential only marginally touched upon. These were the only hope for dealing with the seasonal increase in demand, as well as and above all for facing the consequences of a war that seemed by then inevitable. The war would put at risk a productive capacity of up to 13 million barrels a day, if all those infrastructures were taken into account that could have been hit by medium and long-range Iraqi missiles (up to 900 km).[18] This was perhaps a remote possibility, but it caused those who hazarded predictions for the future to claim that the explosion of a conflict would make prices shoot up to "over $50 a barrel" according to the "Financial Times" (Montagnon, 1990), to $65 according to the World Bank and up to $100 according to Yamani ("Business Week", September 8th 1990; October 15th 1990). These forecasts were not so completely wide of the mark, given the way things had gone in previous crises, with market situations less critical than those which seemed imminent at this time.

7.5.2. The puzzle of the stocks

But what capacity for resistance could the available stocks actually give the West? Given the extreme confusion the issue had generated, this was a very important question. At the beginning of the crisis (tab. 7.8.), world stocks were valued at 5.3 billion barrels, a level that was to remain basically unchanged up to the eve of the war. It was senseless however to relate this quantity to the absent production of Iraq and Kuwait, or to the entire consumption, in order to calculate the capacity for resistance in the first case as 1,080 days, and in the second case 100 days, although this was normal practice, beginning with the IAE itself. If stocks were taken to mean "actual excess over current requirements", or in other words, quantities that cannot be utilised without affecting the regularity of supplies, we may see that on the eve of the war the actual margins for manoeuvre were a good deal narrower than any calculations of accounts might lead one to hope. If we consider that in the US system, the holder of around a third of all the world's stocks on land, 70% of those controlled by the (commercial) firms are considered by the Department of Energy as "minimum operative level" (below which deficiencies emerge in the distributive system and sales) and equal coefficients are applied (for type of product) to other areas, we may conclude that in January 1991 the companies were able to draw on *stocks for no more than nine days of actual consumption*. This figure was confirmed by Peter Holmes, President of Royal Dutch-Shell, who calculated (at operative level) eight days of surplus supplies ("Financial Times", October 10th 1990). To these could be added (tab. 7.9.) twelve days of oil in navigation, though mainly related to supplies already spoken for, and twelve days of strategic stocks, 2/3rds controlled by the United States and 1/3rd equally by Japan and Germany: countries who were certainly not willing to share them, where necessary, with others.

Table 7.8. *World oil stocks at month's end (mill. bbl)*

	1990						1991
	Lug	Ago	Set	Ott	Nov	Dic	Gen
Oil at sea	678	599	695	658	700	687	725
Producing countries' stocks	72	61	60	78	86	117	116
Independent companies' stocks	60	65	61	59	56	44	53
OECD commercial stocks	2,777	2,804	2,806	2,791	2,732	2,712	2,663
Non-OECD commercial stocks	961	983	980	965	993	998	1,007
Total non-governmental stocks	4,548	4,512	4,602	4,551	4,567	4,558	4,564
Governmental stocks	792	797	797	797	794	793	793
Total	5,340	5,309	5,399	5,348	5,361	5,351	5,357

Source: "Petroleum Intelligence Weekly".

Table 7.9. *Nominal and actual oil stocks as of 31 December 1990 (mill. bbl)*

	Nominal stocks	Actual stocks	Days of consumption
Oil at sea	725	725	12
Commercial stocks	3,839	575	9
Strategic stocks	793	793	12
Total	5,357	2,093	33

Source: "Petroleum Intelligence Weekly".

7.5.3. A "lucky" recession

Thus, on the eve of the war, the picture was much less reassuring than governments and international institutions struggled to make it appear. A decisive aspect of the outcome that might have been seen on the markets was (beside possible damage to Saudi plants) that of the *length* of the crisis. Although after the first hours of fighting America's huge superiority was clear, at the end of January 1991 the White House spokesman, Marlin Fitzwater, was still claiming that the US Administration "believed it had to think in terms, not of days or weeks, but months" ("Financial Times", January 26-27th 1991). This period was obviously quite beyond the capacity of resistance of the West.

How, then, can we explain the later course of events, which turned out better than the most optimistic expectations, with prices that fell $9-10 in the hours following the beginning of military operations (then settling at $19-20 a barrel), with flows of supply fully satisfying current requirements? A number of commentators have traced these results to the optimal functioning of markets and to the absolute rationality of the operators, capable of foreseeing everything, even that no Iraqi Scud would hit the huge Ras Tanura refinery or the port of Juaymah, or that (as was later revealed), the White House statements on the length of the conflict were propaganda rather than a reflection of their real beliefs. Markets and operators certainly gave an excellent demonstration of their worth, and credit should be given for this. But besides these motivations, primary importance must be given (even though it is normally neglected) to the slump in oil demand caused by a totally unforeseen economic recession, both in its extent and its effects, before the crisis. If we compare the demand expectations (tab. 7.10.) formulated by the Paris Agency in August 1990 with the data of the last quarter of 1990 and the first quarter of 1991, there is a positive difference in the two periods (also considering the additional requirements of the military forces in the Gulf) respectively of 2.3 and 1.1 million barrels a day (-4.3% and -2%). Still more substantial is the difference between forecast and reality in international trade, to which the change in prices responds more directly. American imports alone (about 30% of the global total) fell by 2.3 million barrels a day in the four weeks before the war (just as OPEC production was reaching its highest point of the last decade), after touching an anomalous peak of 8.2 million barrels a day in July 1990, before the start of the crisis (for reasons that it would be interesting to investigate further).

Without a recession, the international system would have found itself in a very different situation, needing to use up the slight surplus of operative stocks, and given their non-homogeneous qualities in some markets when compared with the seasonal qualities required.[19] A trend of this kind would inevitably drive companies to protect themselves on the market, generating a price rise that instead did not happen. The decision taken by IAE countries on January 11th 1991 to authorise member governments, in case of war, to draw upon stocks and take measures to limit consumption by 2.5 million barrels a day, with further support from the overwhelming military superiority of the allied forces, helped to bring prices down. The conclusion to be made is that if at the end of the crisis the optimists were proved right, this was mainly due to reasons other than – and even opposite to – those they had put forward, and among these the recession certainly played the major part.

7.6. AFTER THE CRISIS

Whatever the ideas about how things actually went, there is one thing that everyone can agree on: the Gulf crisis put the spotlight on the critical nature of the western world's reliability of access to oil resources in the Middle East. This is an issue that has already profoundly affected the international political scene and the history of these countries from the beginning of this century, and that today appears in a much more complex framework. There are two reasons for this: the full sovereignty these countries have managed to achieve over the past two decades, at the same time acquiring full control over the exploitation of their mineral resources; and the dominant role the Gulf will inevitably play in the future on the international scene, as its share of control of production

Table 7.10. *World oil supply and demand (mill. bbl/d)*

	IV 89	1989	I 90	II 90	III 90	IV 90 (2)	IV 90 (3)	1990	I 91 (2)	I 91 (3)	II 91 (4)	III 91 (4)	IV 91 (4)	1991 (4)
Demand[1]														
North America	19.8	19.3	18.9	18.7	19.1	18.7	19.8	18.9	18.5	19.7	18.4	18.7	19.0	18.7
Europe	13.3	12.5	12.9	12.4	12.7	12.6	13.4	12.7	13.2	13.2	12.3	12.6	13.3	12.9
Pacific	6.2	5.8	6.5	5.5	6.0	6.1	6.4	6.0	6.7	6.7	5.5	5.6	6.4	6.1
OECD	39.3	37.5	38.4	36.6	37.9	37.4	39.6	37.6	38.4	39.6	36.2	36.9	38.7	37.6
non-OECD	15.1	14.7	15.3	14.8	15.3	15.6	15.7	15.3	16.0	15.9	15.7	15.8	16.2	15.9
Total	*54.4*	*52.2*	*53.7*	*51.4*	*53.2*	*53.0*	*55.3*	*52.9*	*54.4*	*55.5*	*51.9*	*52.7*	*54.9*	*53.5*
Supply														
OECD	15.9	15.9	16.1	15.7	15.4	16.2	15.8	15.8	16.2	15.9	15.6	15.7	16.0	15.9
Developing countries	9.8	9.6	9.8	9.8	9.9	10.2	9.9	9.9	10.1	10.0	10.1	10.1	10.1	10.1
Exportations countries with centrally planned economy	1.8	1.9	1.4	1.9	2.1	1.4	1.7	1.7	1.2	1.4	1.5	1.5	1.2	1.4
Processing gains	1.3	1.3	1.3	1.3	1.3	1.3	1.3	1.3	1.3	1.3	1.3	1.3	1.3	1.3
OPEC NGLs	1.9	2	1.9	1.9	1.8	1.9	1.9	1.9	1.9	1.9	1.9	1.9	1.9	1.9
Non-OPEC total	*30.7*	*30.7*	*30.5*	*30.6*	*30.5*	*31.0*	*30.6*	*30.6*	*30.7*	*30.5*	*30.4*	*30.5*	*30.5*	*30.5*
Oil deficit	23.7	21.5	23.2	20.8	22.7	22.0	24.7	22.4	23.7	25.0	21.5	22.2	24.4	23.0
OPEC	*23.7*	*21.8*	*23.7*	*23.6*	*21.9*	*23.1*	*22.9*	*23.1*	*23.0*	*23.6*	*22.5*	*23.4*	*23.8*	*23.2*
Stock change	0	0.3	0.5	2.8	–0.8	1.1	–1.8	0.7	–0.7	–1.4	1.0	1.2	–0.6	0.2
Total supply	*54.4*	*52.2*	*53.7*	*51.4*	*53.2*	*53.0*	*55.3*	*52.9*	*54.4*	*55.5*	*51.9*	*52.7*	*54.9*	*53.5*
Normal stock change				–1.4	1	1.1	–0.7	–0.7	–1.4	–1.4	1	1.1	–0.7	0.1

1. Excluding countries with centrally planned economy. 2. Actual data. 3. Forecasts before crisis. 4. Forecasts at start of April 1991.
Source: IAE.

approaches the share of proven reserves. Basically, the Gulf controls almost all the incremental capacity of oil (especially its low cost capacity), and it is this that gives it most power (Mabro, 1990). In addition, the greater role of the Gulf will be accompanied by a growing dependence of the industrialised countries on imports, due to the combination of a fall in their domestic production with an increase, however moderate, of their consumption. In the absence of adequate corrective measures, these trends show that the reasons for the political and economic vulnerability that oil supplies have made so obvious in the past are destined in time to get worse.

To what extent, therefore, has the Gulf Crisis changed the terms of the "oil question"? And are western policies able to readjust their energy systems to satisfy their primary reliability as well as environmental needs? Some general and perforce brief reflections may be made:

Firstly, none of the reasons and contradictions leading to the Gulf Crisis seem to have been resolved. Indeed, the war is destined to scatter the seeds of future resentment and disturbance, (a) because of the new impetus given to Arab radicalism of a fundamentalist religious type, (b) because of the strong dissatisfaction within the moderate Arab regimes due to the non-appearance of those democratic reforms promised lavishly during the crisis; and (c) because of the political instability created within the delicate equilibrium of the national regimes in the Middle East, together with their international alliances.

Secondly, the great increase in energy consumption forecast for the next 15 years, around 50% from 8 to around 12 billion toe, poses problems of immense importance for the whole world, and in particular for the industrialised world. These problems affect (a) the commitment of financial, technological and industrial resources required by the supply side; (b) the possible effects on the environment in a situation that is in some respects already believed to be critical (these effects would require an additional commitment of resources, technologies and consistent state policies to be prevented); (c) the geopolitical location of additional flows of supply (in the case of oil arriving almost entirely from the Middle East) and of demand, to be found for more than 80% in low income areas of the world.

To deal with both problems (limiting dependence on foreign oil and directing the additional supply especially to poor countries, at what for them are accessible prices), industrialised countries have to make a further effort to re-convert their energy systems to guarantee greater efficiency in energy use, a development of renewable resources, and an improvement of the technological uses of coal and nuclear energy. Although the declared intentions of western governments seem to show full awareness of the need for policies to these ends, in practice these meet formidable obstacles in the form of low energy prices (incompatible with expected profitability levels from investment); in policies of limiting state spending, which apply almost everywhere; and in the increasingly widespread social opposition to energy investments. The inability of policies to correct "market failures" thus creates the foundations for price mechanisms to be left once again to allocate a supply forecast in the future to be small, to the detriment of poor countries. From these considerations, a major conclusion can be drawn. There is a real risk that where the domestic policies of western countries fail to bring about the hoped-for adjustments in their energy systems (lower demand, greater domestic production and less dependence on foreign sources), it will be left to foreign policy, and the power relations in which it is expressed, to guarantee the "international order" that the end of the Gulf Crisis has perhaps only momentarily re-established.

Notes

* This chapter is a reworking of my essay, "La crisi del Golfo Persico tra cronaca e prospettive", in Bruni F., Onida F., (ed.) *L'economia italiana nel mondo che cambia*, ISPI, Bologna: il Mulino 1991.

1. The possibility of, the reasons for, and methods of a conflict between Iraq and Israel were discussed in "The Economist" of April 22nd 1990 in the article *The elephant and the hawk.*

2. It has been calculated that between 1982 and 1989 Saudi Arabia, Kuwait and the Arab Emirates provided Iraq with $30 billion zero-interest loans and that it gained a further 15 billion from the sale of oil produced in the Neutral Zone (equally controlled by Kuwait and Saudi Arabia), the area that was ironically to become the Gulf's first scene of war.

3. "The country which has the greatest influence in the region, through the Arabian Gulf and its oil, will retain its superiority as a super-power, without rivals that could compete", declared Saddam Hussein on that occasion. "This means that if the peoples of the Gulf, together with all the Arabs, are not careful, the region will be governed by the will of the United States... Oil prices would be fixed according to one point of view: to foster American interests and ignore the interests of others" ("The Economist", September 29th 1990).

4. The reconstruction of "The Economist" of September 29th 1990 starts with the first signal of February 24th 1990, with the meeting of the Arab Cooperation Council quoted in the text, then passing to the second signal of April 2nd, when Saddam Hussein announced he had chemical weapons at his disposal which could destroy Israel. The third signal came on May 28th, with the above-mentioned attack on the United States at the meeting of the Arab League. The fourth was on July 17th, when Saddam Hussein gave advance notice of a possible invasion of Kuwait. The fifth and last was on July 25th when the Iraqi troops began to move towards Kuwait.

5. The quota system was re-introduced by OPEC from September 1st 1986 after the suspension coinciding with the oil counter-shock. However it was only on January 1st 1989 that there was full agreement between the 13 member countries on the distribution of the quotas, although the rich countries did not afterwards respect it.

6. Particularly: harsh weather, transport difficulties of refined products on the river Rhine, maintenance work on the North Sea oil rigs, and a crisis of Romanian exports of products with the fall of the Ceausescu regime in January 1990.

7. As well as the losses caused by the lower prices of crude oil, Saddam Hussein accused Kuwait of stealing crude oil, from 1980, produced in the oil fields of Rumaila, whose borders have been contested by Iraq since 1961, valued of $2.4 billion.

8. To the ingenuous question of an American senator as to why "the liberation of Kuwait was of such vital interest for us ... to make us ready to risk thousands of American lives if necessary" (The Economist", November 17th 1990), the best answer, by no coincidence, was provided by the editorialist of the influential magazine "Oil and Gas Journal" (January 14th 1991): "because the United States has chosen its role of leader country in the importation of oil. It has decided to import rather than to produce an essential resource for its economy. Thus it cannot ignore or neglect its oil interests in the Middle East... the choice not to produce energy involves the decision to fight to defend the oil of others ... for the United States the military and energy choices of the past can today involve war", as was in fact to happen a few days later.

9. For this aspect, and the main implications deriving from it, cf. Pinzani (1991).

10. The unused non-OPEC extraction capacity of crude oil was only 0.3-0.4 million barrels a day, the refining capacity not more than 1-2 million barrels a day. The US refineries had a utilisation rate in July of around 93%: very nearly the technical maximum.

11. Table 7.2. sums up this data in part. A lower exportation of crude oil for 3.95 million barrels a day emerges, equal to about 18% of the entire trade calculated on the markets in July 1990, and of products for 0.95 million barrels a day, 11% of product trade. From a qualitative perspective, the crisis removed medium-density

crude oil at 31° API gravity (though with values for some oilfields higher than 35° API) with a fairly high sulphur content, around 2.4%, but above all it removed high-quality medium-light products (naphtha, gasoline, diesel oil) obtained from the sophisticated Kuwaiti refineries, in no way replaceable, as regards specific and productive potential, by other refineries. The combination of an inferior quality of crudes to replace those lacking with an inferior quality of plant able to refine them in Japan and Europe, thus meant that to obtain the same amount of high-quality products required by the market, larger amounts of crude would have to be processed than before. The supply crisis, given the same demand, was therefore worse than its nominal value suggested.

12. Cf. Graham (1990, 1991). The calculation of this cost takes into account the losses suffered by countries such as Turkey and Egypt due to the embargo decided by the UN. Until mid-November 1990 the cost for Saudi Arabia stood at $22 billion. Japan also later agreed to contribute the sum of $9 billion towards the cost of the war.

13. Due to the reduced role that energy would play in the formation of revenue compared to the Seventies, and due to the reduced purchasing power of the dollar, so that the $20 of nominal increase recorded in the first eight weeks of the crisis were equal to $14.5 in 1979 and to $7.5 in 1973.

14. Among the principal supporters of this type of proposal were Verleger (1990b), who had made a closer examination of the decisive role played by the stocks in the 1978-79 crisis, and the Saudi former minister Ahmed Zaki Yamani (1990). On this subject also see Clô (1990b).

15. During the months of the crisis practically nobody worried about what was happening in the Third World. One of the few was Philippe de Latour (1990), who suggested transforming part of the useless financial aid to poor countries in the form of real aid and also in the form of oil supplies. This proposal was more extensively set out in the journal "Energia" (de Latour, 1991).

16. The cost of imports of crude oil in India increased in 1990 by 40% to $5 billion, corresponding to 1/3rd of all the currency expenditure of the country.

17. A capacity that was initially estimated at 3.6-4.4 million barrels a day by the well-informed PIW (August 27th 1990) for additional flows available for not more than 90 days, without in other words causing damage to the oil fields (sustainable capacity), and in 4.2-5.3 million barrels a day for peaks of production available for not more than 30 days (surge capacity). Other sources were more cautious, with estimates of the excess capacity at 3.7 million barrels a day according to the "Financial Times", (August 9th 1990) and of 3-4 million barrels a day according to the "Oil and Gas Journal" (August 13th 1990).

Estimated incremental oil capacity (mill. bbl/d)[1]

	Production July 1990	Incremental capacity Surge[2]	Substainable[3]
Saudi Arabia	5.3	2.6-2.8	2.1-2.30
Venezuela	2.0	0.4-0.5	0.5-0.75
Libya	1.25	0.3-0.4	0.2
Iran	3.05	0-0.4	0-0.2
Other OPEC countries	4.93	0.5-0.6	0.5
Total OPEC countries	16.53	3.8-4.7	3.3-3.95
Non-OPEC	16.42	0.4-0.6	0.3-0.45
Total	32.95	4.2-5.3	3.6-4.4

1. Estimate at start of August 1990. 2. Available for no longer than 30 days. 3. Available for 90 days.

Source: "Petroleum Intelligence Weekly".

18. According to "Le Monde", October 11th 1990, Saddam Hussein had a number of medium and long-

range missiles at his disposal, and perhaps one, the Tamouz-1, with a range of 2,000 km. For risks to the infrastructures, see PIW (August 27th 1990) and "The Economist" (January 12th 1991).

19. Two thirds of the 590 million barrels of strategic stocks of the United States were, for example, made up by heavy qualities virtually incompatible with processing plants and the structure of final consumption. See "Oil & Gas Journal" (August 13th 1990).

8 OIL RESOURCES: FACTS AND CONJECTURES

8.1. DOUBTS OLD AND NEW

On the threshold of the coming millennium what prospects exist for oil resources, their physical availability, and their economic and political accessibility? For a number of reasons this question has recently become of renewed interest. *First*: the fact that oil substitution processes, generated by relative price changes in the Seventies, have gradually died out, especially after the price counter-shock. As we will see, oil demand has recovered to the point that it is growing in parallel with the overall growth in energy demand. Hence, there is renewed pressure on the efficiency of mineral resource oil stocks. *Secondly*: oil retains and in the long-term will continue to retain its leadership amongst commercial energy sources being used world-wide. This is for two reasons: firstly, because it is the most popular (about 40% of the total), and secondly, because it is the cheapest source. The low relative costs of oil (both in investment and in production) and its greater supply elasticity confer on it a formidable competitive advantage compared to other sources of energy (Figg. 8.1. and 8.2.).[1] Obviously the cost advantage (or differentiated revenue) depends – in the competitive game – on price policies that are actually set up by various competitors[2] and by tax policies. In this game, however, oil does retain its role of main playmaker: on one hand being in a position to condition the entire energy price structure and on the other, the supply curve (accepting to produce at any price) of the other sources. Finally, there is a *third* reason, connected with the previous two, based on the renewed interest in oil resource matters. This is the renewed debate about the actual amount of resources, their adequacy for satisfying a demand once more on the rise, and the foreseeable duration in the long-term. "The end of cheap oil" was the alarmed cry of two geologists in "Scientific American" in May 1998 (Campbell and Laherrère, 1998). Pessimists and optimists are once more in conflict, even if only within the restricted professional circle.

In fact it may seem paradoxical to hint at the possibilities of oil shortage if we look at the oversupply flooding the markets nowadays, bringing prices down, or if we recall the "mistaken prophecies" of a not so distant past. It is equally wrong, however, to ignore the terms of this debate, for there can be no certainties regarding the future of the oil business and because politics often count more than economics.

The attitude of the scientific, political and business worlds towards the prospects of mineral resources has always varied between great euphoria and darkest pessimism; between the myth of abundance, sustained by technological optimism, and that of shortage overshadowed by ecological pessimism due to the expansion of population, consumption and production processes (Clô, 1998).

In the field of energy resources the situation has regularly recurred, both in the past and nowadays: in the early Seventies, when the more pessimistic visions prevailed, and today, when totally opposite sentiments have taken over. The explosion of oil prices, and the complex overlapping of events now famous as the "energy crisis", were generally interpreted not as a relatively simple problem, that of *insufficient production capacity and concentration of marginal production in the OPEC area*, but as the far more critical problem of *absolute shortage of traditional resources* hastily identified in the decreasing ratio between reserves and consumption.

It cannot be denied that in the vast chorus of consensus some dissenting voices were raised which admitted that natural resources could be substituted with other production factors. Thus the existence of any problem whatsoever was denied "since the world could do without natural resources and thus their extinction is an event and not a catastrophe" (Solow, 1974, p. 12). Some even admonished that "expressions such as 'energy crisis' or 'energy shortage' were fictitious. But to believe in fiction is a reality. It means that higher oil prices are more easily accepted by public opinion, as if they had been fixed by Nature, when they are really fixed by political strategy" (Adelman, 1972b, p. 73).

With the wisdom of hindsight and euphoria of the price collapse, nowadays many

Figure 8.1. *Investment cost of energy sources (doll./bbl)*[1]

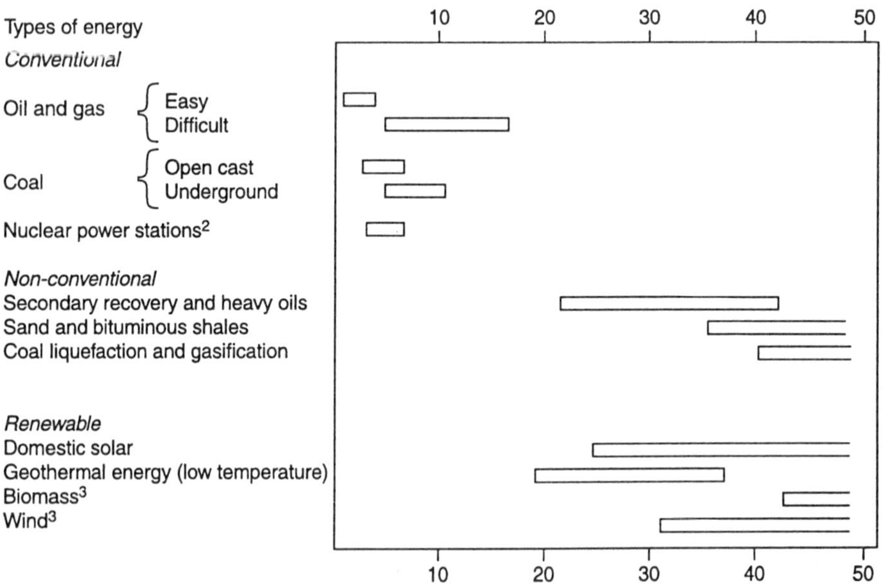

1. Rather favourable real costs and not potential mean costs.
2. Cost of equivalence of the fuel to produce electricity in the power stations.
3. Special costs and/or small quantities available in practice.

Source: J. Masseron (IFP).

Figure 8.2. Investment costs of energy sources (doll./bbl/y)

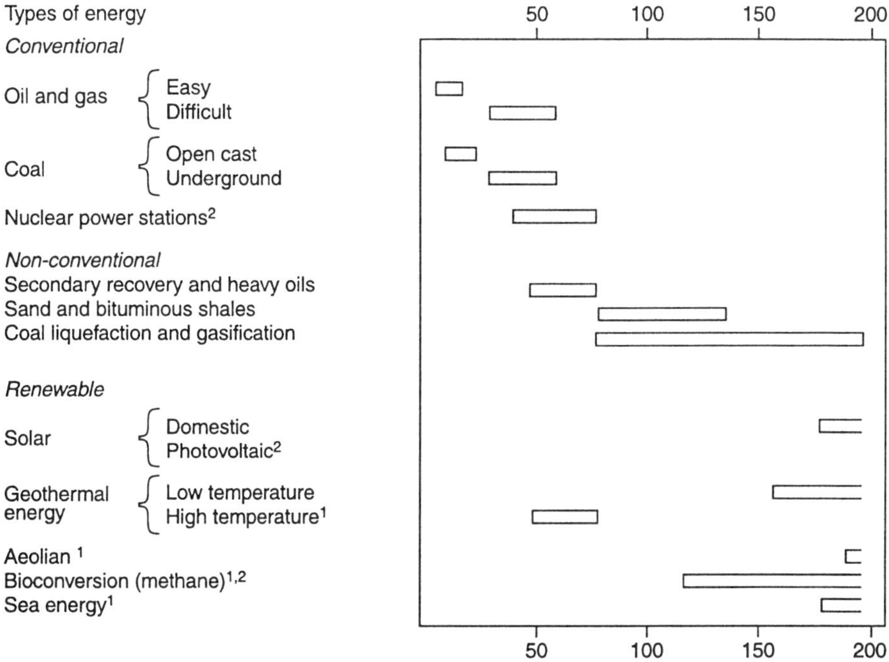

1. Special costs and/or small quantities available in practice.
2. Fairly insignificant at present.

Source: J. Masseron (IFP).

agree with Solow and Adelman, with the risk of perpetuating an inverted version of the errors of the past. It should not be thought that in the long-term the problem of the finiteness and political vulnerability of energy resources has been resolved. In reality the only problem to have been resolved is that which originated the price rise, i.e. in the insufficient production capacity available given the rate of demand.

It is not, in itself, particularly sensible to be complacent about the immense and increased physical horizons of fossil resources, which with today's depletion rate should last about another 1,000 years, or the even more extensive horizons promised by the mythical renewable resources. Simple resource data indicate nothing, in fact, about their economic *accessibility* (means, costs, time), about their quality *compatibility* with the scheme of end uses, about the level of their environmental *compatibility*. If we bear all of this in mind, we will realise how the margins of resources that we can truly rely on at any given historic moment are far narrower that in theory. Furthermore, if this was not the case we would have far greater difficulty explaining what occurred in the past, when oil prices exploded, despite the abundance of resources available, or what is happening now when the poor social acceptability of coal and nuclear power has effectively excluded these sources from the range of new investment decisions and additional incremental consumption: despite their having been seen as reliable (until quite recently) for the replacement of oil and for the diversification of supply sources.

8.2. FIGURES (AND THEIR LIMITS)

When discussing oil resources, fact and conjecture appear to overlap continually, especially if we look at today in order to plan for tomorrow. This includes the complexity of the matter, subject material, its intrinsic uncertainty, and the various interests at stake. In an ocean of figures and available analysis it is never easy to distinguish fact from conjecture: what efficiency really is. Since we are not able to do this, we will propose a simplified version of what we feel to be the principal "truths", the figures upon which they hinge, and the relative limits of their significance. Obviously we should begin with the central theme regarding the volume of proven reserves and ultimate oil resources (following the definitions set out at the beginning of this book).

8.2.1. Proven reserves

Proven oil reserves have grown without interruption since World War II and they are now twice current production (Fig. 8.3.). By 1950 cumulative world production was 60 billion barrels[3] with 4 billion barrels being extracted annually. Proven reserves were estimated at 77 billion barrels,[4] enough to last approximately 20 years at those depletion rates, but a lot less if consumption increases were borne in mind. Limited horizons do not seem to arouse any fear for the future, whether in political or professional circles. If anything, they are held to be largely suitable for supporting the massive penetration of oil in world-wide energy accounts that would soon be observed. (Paradoxically, quite the opposite of what was to occur two decades later in the face of far more comforting estimates of reserves).

Figure 8.3 *Oil: world reserves and production (bill. bbl)*

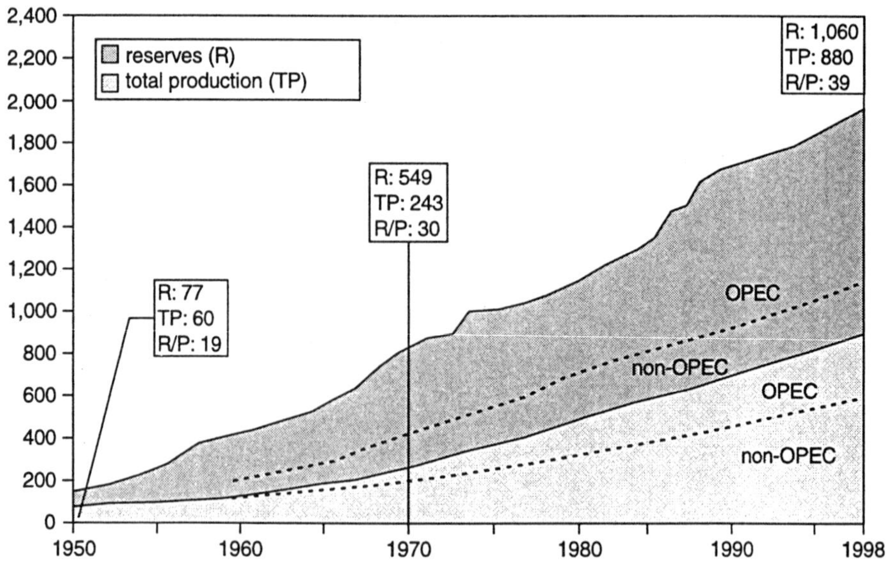

In retrospect it could be said that this vision was to demonstrate great foresight (Tab.8.1.). In the following half century – 1950-80 – , in fact, 11 times the oil estimated below ground at the start of the period was extracted and, despite this, today's reserves are estimated at 14 times greater than those of that time (1,060 billion barrels as opposed to 77 billion barrels). In brief: for the entire period the *gross* increase in proven reserves has been in excess of 1,800 billion barrels, whilst the net increase has touched 1,000 billion barrels, with a reserve/production ratio that has doubled in about 40 years. Between 1970 and 1998 the proven reserves did not fade out, as predicted by the "prophets of doom", gathered together in the Rome Club, but in fact increased by 511 billion barrels, despite world consumption for the same period being 638 billion barrels. "In spite of the succession of traumas in the oil market", wrote the great geologist Peter Odell, who was amongst the optimists from the start, "processes of discovery and development of reserves have continued in a way which justifies the conclusion that the *world is running into oil, rather than out of it*" (1994, p. 92 our italics).

These extraordinary and unforeseeable results are – *in today's consensus of opinion* – facts that justify feelings of optimism for the future. Despite the inevitability that at some point in the future total oil production will reach its peak and consequently start to decline,[5] there is little empirical evidence that this moment is imminent: if anything, figures suggest the opposite. According to the *American Petroleum Institute* "oil resources are far more abundant today than they have ever been in the past" (API, 1995, p. 1).

This conclusion appears even more realistic if we bear in mind not only the oil reservoirs already discovered and efficient in today's techno-economic conditions – which it has been stated define the boundaries of proven reserves – but also those that could be so tomorrow.

In other words, we take a dynamic and not a static standpoint regarding oil exploitation. From this standpoint there are two aggregates to bear in mind: (a) resources as yet undiscovered or identified which, when added to proven reserves, define the extreme boundary of last resources; (b) so-called "non-conventional" oil, which unlike the conventional sort we have discussed so far, cannot be found as such in nature and has to be transformed industrially from other raw materials (asphalt sands, extra-heavy oils, bituminous shale).

8.2.2. Ultimate resources

Statistically, the ultimate resources limit can be obtained by adding the following four items to proven reserves:

- *reserves to be discovered*: whose existence may be presumed on the basis of acquired knowledge;
- *extension of current reserves*: thanks to additional drilling;
- *re-evaluation of current reserves*: thanks to better knowledge or better techniques;
- *increase in recovery rate* of local reserves through improved techniques.

If these aspects are also borne in mind, the quantity of oil that could be extracted in the future can be estimated as vastly superior to that expressed by proven reserves. The estimate of last resources, as was said at the start of this volume (section 1.2.2), is certainly the most subjective, uncertain and disputable, in the oil world. The variability of the quantities formulated over the past fifty years is proof of this (Tab. 2 in Statistical

Table 8.1. *Synoptic table of proved oil reserve progress (bill. bbl)*

Year	Reserves	Production		Reserve variation([1])			Reserves: current production (years)	Distribution of reserves					
		Current	Accumulated	Gross ([2])	Net			Non-OPEC	%	OPEC	%	Middle East	%
1950	77	4	60	-	-		19	33	43	44	57	32	42
1955	158	6	84	+105	+ 81		26	46	29	112	71	98	62
1960	291	8	118	+167	+133		36	73	33	218	67	183	63
1965	354	12	168	+113	+ 63		30	94	27	260	73	215	61
1970	549	18	242	+269	+195		30	150	27	399	73	336	61
1975	648	20	343	+200	+ 99		32	185	29	463	71	387	60
1980	653	23	458	+120	+ 5		28	219	33	434	66	363	56
1985	766	21	565	+220	+113		36	230	30	536	70	432	56
1990	1.001	24	680	+350	+235		42	230	23	771	77	662	66
1995	1.028	25	801	+148	+ 27		41	243	24	785	76	665	65
1997	1.045	26	853	+69	+ 17		40	240	23	805	77	677	65
1998([3])	1.060	27	880	+42	+ 15		39	252	24	808	76	672	63
						%							
1950-70	+472	-	+182	+654	+472	100	-	+117	25	+355	75	+304	64
1970-85	+217	-	+323	+540	+217	100	-	+80	37	+137	63	+96	44
1985-98	+294	-	+315	+609	+294	100	-	+22	7	+272	93	+240	82
1950-98	+983	-	+820	+1.803	+983	100	-	+219	22	+764	78	+640	65

N.B. Rounded values. 1. Compared to the previous period. 2. Gross of accumulated production. 3. Our estimates.

Sources: Statistical Enclosures (Tab. 3 and Tab. 5).

Enclosures). This variability has, however, tended to shrink over time – thanks to the larger amount and better quality of information available – in a range between 1,700 to 2,200 billion barrels. A similar conclusion is also to be found in Colitti and Simeoni's excellent 1996 study.

Table 8.2. *Estimate of last oil resources (bill. bbl)*

	1981	1986	1993
Reserves discovered (a)	1,091	1,223	1,698
Reserves already produced (b)	449	526	699
Proven reserves residue (c) = (a) – (b)	642	697	999
Reserves to be discovered (d)	445	412	387
Re-evaluation and extension (e)	202	180	128
Recovery left over (f)	360	286	177
Total last resources (g) = (c)+(d)+(e)+(f)	1,649	1,575	1,690

Source: Colitti and Simeoni (1996, page 8).

If we use cautious methods and hypotheses, they give an estimate of last oil resources (on which we can rely until 2030) of 1,690 billion barrels. 1,000 have already been discovered, another 387 are still to be discovered and the remaining 305 billion barrels derive from re-evaluations, extensions, and increased recovery rate. In *static* terms, this reserve could last – at a constant efficient rate of current production (relative to 1998) – for about another 60 years,[6] that is to say until about the middle of the next century. Despite being routine, this calculation is however misleading, both because demand will not remain constant over time and because oil reservoirs decline during the second half of their lives and production gradually diminishes, until they are completely exhausted. All the factors that influence total oil reserves are of a *dynamic* nature and therefore vary constantly. A reservoir's reserves cannot be known exactly until it no longer produces.

8.2.3. Hubbert's curves

The geologist Hubbert studied the exploitation of the oldest American oil provinces. In 1956 he devised a model that sketched a profile, through time, of the *production* of each reservoir. A "bell-shaped curve" (Fig. 8.4.) emerged that can be split into three phases: (a) *rapid growth*: when a growing number of wells are running; (b) *stabilisation*: when the exhaustion of the older wells is barely compensated by new wells of equal capacity; (c) *gradual decline*: when the exhaustion of resources dominates substitution investments. The value of the integral in the area beneath the curve represents the last resources that may be recovered, that is to say production accumulated over time expressed as a logistic curve.

The supply curve does not vary, therefore, at a constant rate for "n" years – as indicated by the reservoir/production ratio – and suddenly stop. It is instead the function of investments and discoveries that occur throughout the exploitation of the reservoir. When

Figure 8.4. *Current production and total production of an oilfield*

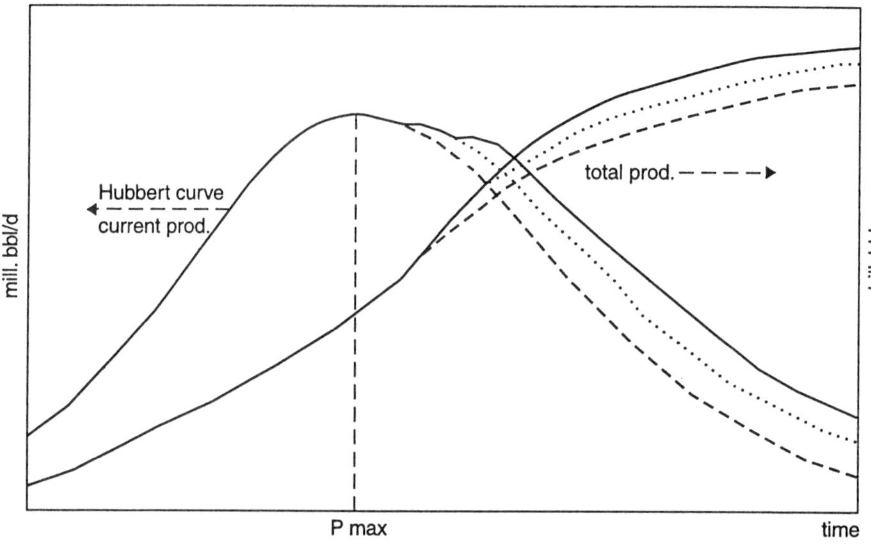

basic resources then begin to decline, maximum production will occur and hence the phase of decline may possibily extend for quite a long time, depending on how far investments and technical progress allow even just one part of the extracted barrels to be replaced.

Hubbert's model maps out over time the profile both of *production curves* and *discovery curves*, which is the determining priority of the former. In this case the curve is bell-shaped so that initially the growth rate is limited by sporadic discoveries of limited size, then accelerated by productivity increments of the more interesting discoveries, and finally the decline when these gradually reduce. Both curves may refer to the life of a single reservoir or to all the reservoirs that identify an oil province or, by a progressive totalling, a global context.

Though Hubbert's model is geological and not economic, it takes for granted that the market is perfectly competitive: with the cheaper resources extracted initially, costs gradually increasing as the discoveries become more difficult, prices that follow cost trends, and demand that reacts as a consequence.

It was during their consideration of the possible exploitation profile of all the reservoirs – hypothesising that each one produces at "maximum efficient rate" (MER)[7] – that Colitti and Simeoni estimated that the oil production curve on a world-wide scale could reach its (theoretical) peak in 2016, then see a gradual decline towards the middle of the next century.[8] In other words: on the basis of resources pinpointed to-date and leaving aside rational behaviour from a strictly technical-economic aspect, *crude oil production could encounter a turning point – according to Hubbert's curves – in the second decade of the next millennium.* The interest of this estimate does not derive solely from the reliability of the method with which it has been obtained, but also from its substantial similarity to conclusions reached by other authors using different systems (Drollas, 1986, Hubbert, 1956, Pauwels, 1997). Even this estimate, however, cannot be seen as a *forecast*: it is more of a reflection on the maximum production that is technically viable, given resources and technology. The effective course of events depends on many variables – demand, technological progress, prices, interest rates, and investment cycles – and hence results noticeably different even from the technical optimum.

Non-conventional oil resources must also be added to the conventional. Estimates of their entity are even more problematic than those we have seen for conventional oil. There are several reasons for this: the diversity of the criteria for defining these resources,[9] the still embryonic phase of their industrial exploitation, and the consequent impossibility of drawing long-term possible supply curves that do not take into account credible technological routes. In conclusion, the reasons for uncertainty are too significant for the quantity evaluations calculated so far to be held reliable and significant.[10] Since we are basically talking about a prospect still to come, their importance becomes secondary.

For the purposes of our argument, it seems to be more important to highlight quality aspects upon which rest the exploitation prospects of these resources. There are three in particular. First: *the physical entity of non-conventional oil resources can be held to be almost unlimited*: since it is dependent on the use of precise technical, economic and commercial prerequisites. Secondly: *the principal "barrier" to their entry on the market is represented by current "limit prices" of conventional oil, given their current production costs*. Third: the development of mineral technologies is, in any case, moving towards *the long-term of the use of these resources* and allows them to be included amongst those on which the human race may realistically rely. We will fall back on them when conventional oil yield undergoes an obvious decline or when political reasons – as in the Seventies – prevail over those of economy.

What conclusions can we draw from the prospects disclosed by oil resource estimates? To conclude our discussion we could repeat those conclusions reached by Colitti and Simeoni in their study (our italics): (a) "world-wide oil production has not yet reached its historic peak: *the petroleum industry has not reached its 'maturity'* and is not yet near to meeting its decline" (ditto, p. 34); (b) "*only two areas have already entered their final oil production phase: the United States and Europe*. That is to say: the first systematic oil extraction area and a marginal area where important reserves were not discovered until the end of the Sixties. The other areas are still relatively 'young', and at least one extremely large area, the Middle East, is very young" (ditto, p. 41); (c) *the petroleum industry has before it a "relatively extensive long-term and may plan its future for almost a century ahead"*. The era of hydrocarbons is not near its close. "So there is every possible reason for its producers to adopt long-term strategies for their production and for their markets" (ditto, p. 41).

It was also mentioned earlier that these conclusions reflect today's consensus of opinion. Whilst this comforts us totally as far as oil's *theoretical* availability is concerned, nothing can tell us however, what its *actual* conditions of access are. In other words: as in the past, the oil resource question in the future will not be a problem of absolute restriction due to physical entity, but rather *a problem of economic accessibility, political reliability, and environmental tolerability*.

8.3. INVESTMENTS, TECHNOLOGY, AND "RELATIVE SHORTAGES"

At the root of the pessimism and erroneous prophecies of the past regarding the "imminent exhaustion of oil resources" there has always been an underestimation of or lack of faith in the force of technological progress. These are perhaps inevitable consequences of "our ignorance both of the geological basis to rely upon, and of the future *patterns* of technology, of consumption, and of the competitive energy supply" (Adelman, 1972a, p. 41). Economic theory and history teach us that in reality the processes of technological substitution and innovation create *relative* situations of shortage, which at first sight may appear to be *absolute*. "The category of relative scarcities, the trigger mechanisms for technological innovation due to relative scarcities and the differential revenue are all components of an economic theory applicable not only to the case of 'declining fertility' but also of 'growing fertility', due to technical progress" (Quadrio Curzio, 1988, p. 40).

Models of either type of situation can be encountered in the last fifty years of the petroleum industry's mineral history, related to *three periods* in which distinctive elements are displayed such as: investments, technology and efficiency. In the paragraphs that follow we will examine in brief how efficiency has varied between the period that preceded the first oil crisis (1950-1970), that which followed (1970-1986) and that which came after the prices counter-shock (1986-1998). For the first two periods we will make use of valuable statistical sources, sadly interrupted in 1985, of investments monitored per geographic area and per production phase by the *Chase Manhattan Bank* (CMB). For the last period we will use the more aggregated data of the *Institut Français du Pétrole* and *Arthur Andersen*.

8.3.1. The old and the new worlds: boom and decline (1950-1970)

This was undoubtedly the most exciting period. In fact, the majority of the 600 oil "provinces" and so-called "giant" and "supergiant" reservoirs identified across the world date back to those years:[11] over 170 of the 360 listed in 1995.[12] Although these reservoirs represent only 1% of all the fields discovered in the world, they account for two thirds of current global production and contain over half of proven reserves.[13] It was chiefly thanks to these exceptional discoveries – mainly localised in the Middle East and in the ex-Soviet Union – that world-wide reserves encountered the strong upward swing from 77 billion barrels in 1950, to 550 in 1970, despite the exponential increase of production. At that time, only one barrel of oil was extracted for an average 2.5 barrels discovered.

Overall mineral activity yield (exploration, development and production) is high and on the uprise in the "new" oil world, whilst a new decreasing curve in the "old" American world begins to be apparent. It was therefore logical that investments started to move in that direction, even if they continued to remain mainly concentrated in the less efficient and less economic oil provinces of North America. Of the approximately 8,000 new reservoirs explored *annually* in the Sixties, only 40 were located in the Middle East, against approximately 7,200 in the United States and Canada.

The unit costs of oil discovering in the United States grew in an exponential fashion, whilst the number of new discoveries registered a drastic decline.[14] In 1968, the discovery of the "supergiant" Prudhoe Bay reservoir in Alaska seems to have started a new phase in the history of the American petroleum industry, almost a century since its birth. Nonetheless, it was to serve only to delay the negative "turning point" of that production that had inexorably begun – despite the immense financial resources absorbed – back in 1973.

Paradoxically the great mineral successes of the Fifties and Sixties were premonitory for the subsequent crises. The possibility of developing enormous resources at low cost and with little investment, their adequacy for fully satisfying growing demand, the emergence of an actual surplus of production capacity, the arrival of non-integrated competitive newcomers, the pressure to drop market prices against list prices, and the consequent drop in productivity of the petroleum majors,[15] are all reasons that explain the consolidation, *at the end of the Fifties*, of a new set of advantages for international industry and the emergence within it of new behaviour and new trends. In particular: (a) the start of a negative phase in the mineral investment cycle and its progressive spread to non-OPEC[16] areas; (b) poor incentives for improving mineral technologies: both for exploration and for extraction. The recovery rate from reservoirs did not exceed 20%-25%; (c) a tendency of businesses to designate all incremental investments to the *downstream* oil cycle phases and upstream chemical sector. The mineral investment quota, which hit peaks in excess of 60% in the Fifties, dropped below 35% in about 1970. The fall in mineral investments, the amount of time lag required to bring to fruition the new "provinces" in Alaska and in the North Sea, and the temporary disappearance of spare oil capacity on international markets, were to originate the dramatic "turning point" of the Seventies (chapter 5).

Table 8.3. *Investments in the world oil industry: absolute values (bill. doll. 1986) and percentages*

Years	Total	Annual	Mineral[2]	%	Transport	%	Refining	%	Other	%
1950-59	306	31	189	62	39	13	44	14	34	11
1960-69	446	45	218	49	60	14	91	20	77	17
1970-79	843	84	416	49	163	19	184	22	80	10
1980-86	823	118	570	69	56	7	146	18	51	6

1. Excluding Countries with Economic Planning (PEP). 2. Includes exploration, development and production.

Source: For the years between 1950 and 1984: CMB; 1985-86: our estimates.

8.3.2. Decreasing yields (1970-1986)

The increase in prices that began in 1970 is at the root of the new, positive circle of investments. It allowed companies to make strong productivity increases and hence they were able to self-finance. After the drop and only partial recovery in the Sixties, annual *investments encountered an exponential growth*, and reached the historic 120 billion dollars peaks of the early Eighties (Tab. 8.3.). Overall, the petroleum industry world-wide (excluding countries with economic planning) invested a constant 1,665 billion dollars (1986) in these 16 years compared to 600 in the previous 16 years.[17]

The response of the petroleum companies – and of the western countries of which they were the expression – in years scarred by the serious recession brought on by the Seventies oil crisis, was an extraordinary absorption of wealth. In fact, the chief intention was to re-build lost or abandoned production capacity outside of OPEC "walls". Incremental investments were indeed on the whole (70%) destined to mineral activity.

The results achieved were certainly positive if we look at the preparation of new oil production capacity as an alternative to that of the OPEC.[18] Far less if the analysis is aimed at *decreasing marginal yields* of investments in terms of proven reserves. Their real increase of 2.5 times actually corresponds to a net increment in reserves – considering a time lapse of three years – of about 8 times less (Table 8.4.).[19]

The conclusion is that an investment of about 1,000 billion dollars (constant) between 1970 and 1986 in the mineral phase was only just able to recover the drop in reserves. Their world total – net of OPEC – at the end of the period, is only a fraction above the peak reached in 1975.

The *annual discovery rate for oil* – given by the variation in reserves plus production – reached a maximum of 7 billion tons in the period 1965-69 (Table 8.5.) and has now halved.[20] This clear drop is even more clearly accompanied by the *dimension of new reservoirs*. Of the approximately 30,000 reservoirs discovered in the mid-Eighties only about 35 (mainly located in the Middle East) can be categorised as "supergiant" (with over 5 billion barrels of reserves) representing about half of all the reserves at that time. Of these, only 2 were discovered after 1970, compared to 23 in the two previous decades. A smaller size determines, *coeteris paribus*, a greater unit cost, lower productivity and fewer possibilities of future re-evaluations of extent. New reservoirs, apart from increasingly rare exceptions, have tended to become smaller and of more difficult accessibility, frequently situated in deep seas, far from centres of consumption and with production costs of 15-20 dollars/barrel as opposed to the 5-10 dollars/barrel of the most expensive of the period.

At that time the amount spent in investments for new discoveries was in the region of between 500 and 1,000 million dollars. The effect of this was to lower the revenue per single reservoir, when of smaller dimensions; their greater sensitivity to price variation in the selling of oil; and the growing difficulty that businesses of relatively small dimensions had in resisting on the market or of gaining access to it. Acquiring reserves of oil, even buying up the businesses that owned them, was more economical in unitary terms in those years – on an approximately 1:2 ratio – than seeking and discovering. So in the early Eighties there was a *great wave of mergers and acquisitions* that mainly involved the United States. Between 1980 and 1985 we count 21 – including the take-over of *Gulf Oil* by its "sister" *Chevron* – for a total of 80 billion dollars (of the period), with the passage of ownership of about quarter of all American reserves (but without those investments contributing to an increase of even a single barrel).[21]

If, with Barnett and Morse, we accept as a resource's economic indicator of shortage the trend of its real cost of extraction per product unit, data will lead to the conclusion that *the Seventies marked the appearance of a growing scarcity of oil resources*, coinciding with price increases, although the scarcity was not at the origin of these (Barnett and Morse, 1963, Barnett, 1979).

To speak of relative oil scarcity when its price show a downward spiral – from circa 36 dollars/barrel in 1980 to 27.5 in 1985 and then 13 in 1986[22] – may seem nonsense. However this is not so if: (a) we abandon the idea that the short-term dynamics of the oil market are able, always and in any case, to offset to their best advantage long-term dynamics and interests; (b) we bear in mind that decreasing yields were only relative to the non-OPEC area.

Table 8.4. *Mineral investments and variation in proved oil reserves per geographic area*[1]

	Investments	(billion dollars 1986)			Var. reserves (billion tons)			
	1950-59	1960-69	1970-79	1980-84	1953-62	1963-72	1973-82	1983-87
United States	138	142	214	239	+0.5	+0.9	–0.9	–0.1
Canada	12	19	37	31	+0.3	–0.1	–0.1	–
W. Europe	4	7	57	55	+0.2	+1.7	+1.7	–0.1
Africa	4	15	24	23	+1.3	+6.3	–0.4	–0.4
Middle East	6	9	27	29	+16.7	+23.4	+1.0	+4.4
Other	25	25	57	78	+2.9	+1.5	+1.1	+1.3
World total	189	217	416	455	+21.9	+34.4	+2.4	+5.1
– Developed nations	154	168	308	325	+1.0	+3.1	+0.7	–0.2
– Developing nations	35	49	108	130	+20.9	+31.3	+1,7	+5,3

1. Excluding PEP.

Source: investments: CMB; reserves: API "Basic Petroleum Data Book".

Table 8.5. *Mineral investment performances in the petroleum industry world-wide*[1]

	Annual investments (billion $ 86) (A)	Rate discovered[2] (billion tons/y) (B)	(A) : (B) ($ 86/ tons)	Reservoir dimensions			
				N.	Giant billion tons	N.	Supergiant billion tons
1950-54	15.3	2.0	7.7	45	7.7	11	16.8
1955-59	22.5	1.2	18.8				
1960-64	20.7	2.4	8.6				
1965-69	22.8	6.9	3.3	69	10.8	12	15.8
1970-74	29.7	5.5	5.4				
1975-79	53.4	2.9	18.4	40	4.7	2	12.0
1980-84	91.0	3.8	23.9	n.a.		n.a.	

1. Excluding PEP. 2. Obtained from the sum of the annual variation of proven reserves and the annual production.

Source: investments: CMB; reserves: API "Basic Petroleum Data Book".

The traditional way of reasoning, which sets forth the economic theory of natural resources – the necessity of falling back on marginal areas once the comparatively richer areas are exhausted – is not of great help with oil. This scheme can, in fact, be adapted to a world in which it is the economy rather than politics that direct decision-maker behaviour and where government intrusion really is at a minimum. At the origin of decreasing margins on mineral costs, we actually have inefficient geographic distribution as a consequence of the priority objective – in some respects necessary – of petroleum companies to re-create outside the Middle East the control margins of supply that they had lost within that area. "From 1973 onwards – wrote Adelman with force – water has begun to flow uphill: drilling activity quadruplicated in the United States until 1985, whilst in Saudi Arabia it dropped by two thirds" (Adelman, 1986).

The "split" between the Anglo-American petroleum industry and producer governments has meant that mineral investments were not directed where they would appear to have the greatest chance of success, lower costs, and better crude oil quality, but where businesses had greater chances of being able to operate and greater certainty of maintaining full control of their investments, profits, and supply flow. Where, in substance, the political risk appeared lower then the host governments would modify to their advantage contractual terms and rights of property. "Geological and political, rather than economic and managerial factors have principally been involved in determining the global reserves' development process" (Odell, 1994, p. 96).

Over 70% of all mineral investments were thus concentrated in the "marginal" areas of North America and of Northern Europe, holders of only 8% of proven reserves, whilst as little as 15% was invested in the 13 OPEC countries, despite 77% of reserves being situated there, with discovery costs 10-15 times lower (Table 8.6.).

Table 8.6. *Distribution of reserves and investments and investment unit costs: 1970-1984*

	Reserves (%)		Investments	Unit costs
	Proven	To be discovered	%	(dollars 1980/tons)
North America	5	12	60	45
Europe	3	3	13	71
OPEC	77	50	15	5
Other[1]	15	35	12	24
Total	100	100	100	

1. Excluding PEP.

Source: Investments: CMB; unit costs: Schneider-Schulz (1986).

The economically inefficient method in which investments are distributed is well summed up by the comparison between costs and reserve variation in the various areas. The 633 billion dollars invested in developed countries has achieved a net increase in their reserves of just 0.5 billion tons, whilst the 238 billion dollars spent in developing countries has increased reserves by 14 times (Tab. 8.4.).

These discomfiting comparisons demonstrate the enormous waste of resources that the world-wide economy has had to support because of the enduring political instability in those areas richest in oil, and the antagonism between them and consumer/importers, for the demise of control structures, which in the past, despite everything, the petroleum industry had exploited to ensure conditions of price stability and orderly, more efficient development of investments.

8.3.3. Technological innovation and growing yield (1986-1998)

The petroleum counter-shock in 1986 rapidly dragged down the volume of investments in the entire chain of petroleum value. Even mineral activity has been heavily hit, further aggravated by the fact that whilst in other phases the capacity achieved resists across time, in the minerals sector it has to be continually "renovated" with respect to the extracted crude oil. Between the early and mid Eighties mineral investments dropped twice over and then even further towards the end of the decade (Fig. 8.5.). There were three reasons for this: (a) businesses had lower financial resources available; (b) the asymmetric trend between oil prices (dropping) and its incremental costs (rising); (c) the prospect that oil prices could begin to recover when marginal productions left the market.

Investments for maintaining or expanding production capacity do not take place – or are heavily reduced – if incremental costs are in excess of anticipated prices. The same thing happens if the best information available projects even higher prices for the future than those currently existing. In this case it is opportune to refrain from investing in reserves: to keep them for future, more profitable use. "The higher the cost, the lower current profit will be and the greater the advantage from the postponement. To express this concept in other words, the lower the cost of finding new reserves, the lower the costs of the opportunity ('use costs') of producing today rather than tomorrow" (Adelman, 1986).

If we know the ratio for each area between output and incremental costs it will then be possible to determine – for each price level – a maximum limit of achievable production. In other words: of constructing a long-term supply curve. Or, alternatively, of determining a minimum price threshold below which mineral activity will diminish, thus compromising supply capacity in the long-term. This double functional relationship, *"prices-reserves-production capacity"*, on one hand, and *"prices-demand"* on the other, is the mechanism that regulates the market and determines long-term equilibrium. Its functioning in the short- to medium-term is, however, strongly distorted by inertia and unforeseeable time lags (both on the supply side and on the demand side), which may give rise to depression cycles that alternate with growth cycles: with reciprocal damage to producers and consumers. Stability, as we have mentioned many times, is precious to everyone.

Whilst price reductions promptly affect investment cycles[23] – and therefore future supply capacity – they produce limited effects on current production unless they

Figure 8.5 *Mineral investments in the oil industry (bill. doll.)*

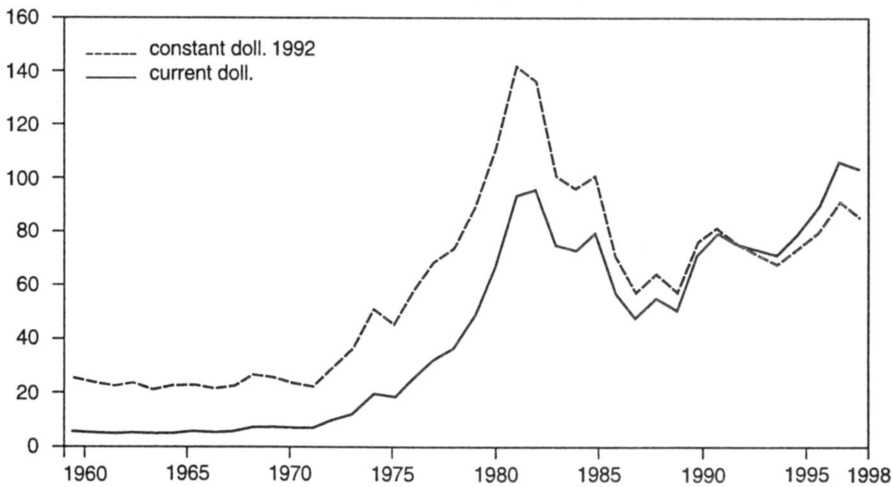

Source: 1960-84: CMB; 1985-98: IFP.

drop (permanently) below direct costs. For this to happen – as Saudi Arabia had hoped when it began the "price war" early in 1986 – prices would have to have dropped below the 8-10 dollars/barrel minimum touched in July of that year.[24] Even a very low price would only have made a marginal reduction to current production, whilst investing in additional capacity would not have become very profitable. The natural exhaustion rate of reservoirs would not have been covered by new oil so that in time production capacity, and thus production, would have tended to contract. There would have been inevitable effects on equilibrium prices, which would have been even more evident as the recovery in demand gradually consolidated. This was the consensus scenario sustained at that time by geologists, economists and businessmen. They were neither stupid, nor naive, nor excessively prudent. Rather more prosaically, they were interpreting the data and knowledge available at that time.

This knowledge did not incorporate *the "breaking" point in technological progress that was to be seen in research, development and mineral production techniques.* Three aspects were involved: (a) *seismic activity*: through the more advanced application of information technologies in processes of acquisition, accumulation, and processing of data; (b) *equipment and drilling techniques*: applied to small diameter bores and "derived" drilling that over recent years has seen increasingly complex well architecture; (c) *polyphase production*: pumping techniques that offer great efficiency potential for deep sea discoveries.

Of these three "families" of innovations the most extraordinary – for the effect it has had – is the first: substantially deriving from the *information technology revolution.* Basically this has made it possible to multiply and improve in an exponential fashion the geological knowledge of reservoirs: with the possibility of thus reducing the uncertainties, the risk, and the effort required for their exploitation. In particular, *3-D seismic design* technology has made it possible – thanks to the use of in-depth sensors – to construct the high definition image of a reservoir[25] on a 3-D basis (width, height, and depth). It would not be an exaggeration – given the advantages derived – to say that the application of I.T. to mineral cultivation is comparable to that of fertilisers used at the beginning of the century for agriculture.[26]

With less effort, far greater results are achieved in all the phases of a reservoir's life. In particular:

a) *increased capability for identifying the location, the profile and the dimension of reservoirs*: with an enormous reduction in mineral costs and risks. In the exploration phase the success rate in the number of wells drilled has gone from 1 in 10, twenty years ago, to 1 in 5 today, whilst in the next decade there are hopes of dropping to 1 in 3 (Babusiaux and Boy de La Tour 1999, p. 19);
b) *the accessibility level of oil resources is increased*: with the possibility of "attacking" mineral themes, border areas, and single objectives that are at the same time technically and economically not viable. In fact, the base of resources that can be relied upon has effectively increased. Add to this the fact that as mineral knowledge increases, so do estimated reserves, *coeteris paribus*;
c) *mineral operation times have been reduced*: due to the greater success rate of drillings, the halving of the time that was necessary from the acquisition of the seismic datum and its interpretation, and the increased possibility of calibrating the type of intervention to specific geological requirements. Cutting down time has three important implications: (i) increased elasticity of the supply in relation to

upward price variations; (ii) reduced costs because unproductive capital remains tied up for less time; (iii) reduced investment risks when prices oscillate. The time required to render operative a reservoir is thus cut down from 5-7 to 2-3 years;

d) allows better *optimisation of the overall reservoir exploitation curve*. The adoption of a 4-D seismic technique – that adds time to the other dimensions – means that a reservoir's useful life can be defined with greater precision, integrating geology and mineral geophysics with production engineering.

Technological innovations in seismic activity have been synergically coupled by *horizontal and directed perforation*: this makes it possible to bore horizontally in mineralised zones, with crude oil recovery exceeding that achieved only with vertical bores.[27] The greater drainage capacity simultaneously allows the increase of production levels, conformity of reservoir efficiency, and reduction in the number of wells to drill. The costs of horizontal drilling are "20%-40% higher compared to conventional drilling, but productivity is 2 to 5 times greater and the recovery rate is noticeably higher with an increase in recoverable reserves, in some cases of 30%-40%" (Babusiaux and Boy de La Tour, 1999, p. 21).

In general, new technologies have made it possible "attack" the new frontier of oil exploitation: *in deep waters*. These waters are over 200 metres deep with seabeds not easily reached by man for installation operations or equipment maintenance. These areas cover 55 million km^2 of the earth's surface, compared to 15 of traditional offshore, and only in 2.3 of them has exploration activity started: in the Gulf of Mexico, in Angola, and in Brazil (Martin, 1999, p. 70). The rate of technological improvements in this field is astounding and largely attributable to new concepts in drilling platforms. In 1988 Shell Oil installed the Bullwinkle platform, in the Gulf of Mexico. This is the world's tallest, anchored at a depth of 400 metres (80m higher than the Eiffel Tower) compared to 300m reached in 1978 and 6m in 1947. Five years later the same company developed the Auger reservoir, at a depth of 870 metres, using a fixed floating platform (*tension leg platform*) kept in place by cases of ballast located on the seabed. Subsequently, with the Mensa reservoir, south east of New Orleans, a record 1,640 metres was reached, whilst at King's Peak it will soon be possible to exceed a momentous 2,000 metres below sea level. Only a decade ago such triumphs would have been unthinkable, when in the aftermath of the petroleum counter-shock the dismantling of offshore platforms progressively exceeded that of new installations.[28] Today over a third of the world's oil production is extracted from offshore reservoirs and their development absorbs half of all investments in the mineral upstream (OCSE-AIE 1996).

The "break" in traditional technological *patterns* in mineral activity has produced extraordinary results and its diffusion and evolution is far from being defined as exhausted. The result has been an increase in productivity, an increase in oil resource boundaries, and a reduction production times and costs. It is however not easy to quantify with precision since it is difficult to separate the reduction of structural cost from others tied to economic trend dynamics in the instrumental sectors.[29] According to the *Institute Français du Pétrole* in the early Nineties *unitary technical costs of production*[30] *have dropped by an average of 1 dollar/barrel per annum*. Those of non-OPEC production – where more new technologies were applied – passed from values between 12 and over 20 dollars/barrel, to average values below 10 dollars/barrel. This has made it possible to reduce the differential disadvantage compared to the great Middle Eastern reservoirs: from a 10 to 1 ratio to a 3-4 to 1 ratio between 1985 and 1995 (Appert, 1997, pp. 39-40).

Fig. 8.6 *Five-year average of discovery costs (excluding revisions) (doll./bbl)*

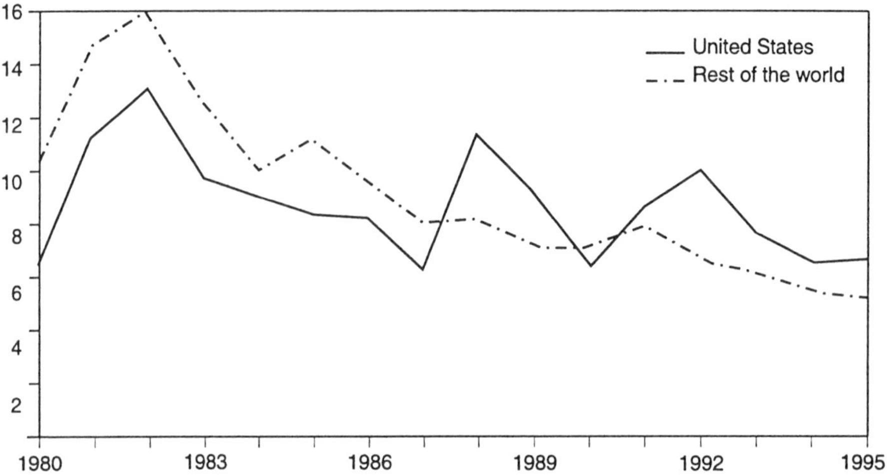

Source: Andersen Consulting, *Oil and Gas Disclosures*, Houston, Texas, annual report.

As far as oil discovery costs are concerned – which influence investment decisions given their relationship to market prices – a reduction of values approaching 13 dollars/barrel down to 7 was observed in the United States, whilst for the rest of the world the reduction was even more evident: from an average of 16 dollars/barrel to under 6 (Fig. 8.6.).

The end result of all this is that technological progress has begun an *increasing efficiency* phase in oil production that has made it possible to: (a) *face the price drop positively*: reducing the break-even point for existing production and at the same time allowing replacement and expansion with new capacity, despite the not irrelevant drop in real investments; (b) *render the price drop itself structural*: not so much because it was generated by short-term competitive dynamics on the market, but rather due to changes in technology and the reduction of structural costs that the dynamics brought with them; (c) *expand the level of technical and economic accessibility of oil resources*: expressed not only in the increase of proven reserves stock, but chiefly in the progress of technologies used for production. Such is this progress, that it *more than compensates* for the reductive effects on reserves stock that would otherwise have produced the price drop of crude oil and the drop in mineral investments. Obviously this positive net result on reserves would be even more significant – in terms of quantity – if investments deriving from progress had not been limited chiefly to "marginal" areas outside of the Middle East.

Between 1985 and 1998 proven reserves increased in *net* terms by almost 300 billion barrels, which added to the 300 or more extracted in the period, gave a gross increment in reserves that exceeded 600 billion barrels. This amount was equivalent to that obtained in the "fabulous" twenty years of the post-war boom. The comforting reassurance that comes to us through the force of technological progress, the return to an abundance in current supply and that which data project onto the resources, and the structural drop in costs and prices are indisputably positive *facts*. They do not, however, erase the complexities and problems that continue to hold good for the future for economic and oil politics.

8.4. OIL RESOURCES AND GEOPOLITICS: THE "PASSAGE EAST"

These reasons are, in substance, rooted in two equally indisputable facts: (a) *oil resources remain concentrated chiefly in the Middle East*; (b) *the Middle East is inevitably destined to acquire a role of absolute leadership on the world-wide petroleum scenario.* Neither of the two facts actually prefigures a scenario that is *necessarily* pessimistic. But nevertheless they lead us to believe that events and behaviour of a political nature may interact – more in the future than they have in the past – with facts of an economic nature: conditioning according to casual combinations that cannot be foreseen. In other words the geopolitical distribution of oil resources does not allow us to believe that it was univocally market mechanisms or parameters of economic rationality to determine conditions of equilibrium on oil markets or to guide choices of economic agents. In the future, more than in the past, economic models or theory will not be able to offer us very much help.

The reason can be summed up in a few facts. Today the Middle East "controls" – as has actually been the case since the Fifties (Tab. 8.1. and Tab. 8.7.) – 63% of the world's oil reserves and over 50% of ultimate resources. On the other hand, its contribution to world-wide production does not exceed 30%. Obviously, since *today's* reserves and resources are *tomorrow's* production, the Middle East's quota of this latter will inevitably grow in a measure that is twice that of today. In other words, from 30% it should progressively push towards 65%. The time required to close these "scissors" may be delayed by the intensive exploitation of marginal areas outside of the Middle East: but such a possibility has an economic limit that is greater than the physical limit.

Table 8.7. *Geographic distribution of last oil resources (billion barrels)*

Regions	Proven reserves[1]	(%)	Reserves to discover	Re-evaluations & extensions[2]	Advanced recovery[2]	Last resources[3]	(%)
Europe	18.8	1.9	15	6	4	43.8	2.6
North America	28.8	2.9	58	19	20	125.8	7.4
Latin America	125.0	12.5	56	13	15	209.0	12.4
Middle East	662.9	66.3	99	40	79	880.9	52.1
Africa	62.0	6.2	26	14	17	119.0	7.0
Russia[4]	57.0	5.7	91	24	32	204.0	12.1
Far East	44.6	4.5	42	12	9	107.6	6.4
World total	999.1	100.0	387	128	176	1,690.1	100.0

1. End 1993. 2. In new and old fields on the basis of technologies applicable up to 2030. 3. Obtained from the sum of the previous columns. 4. Includes other CIS.

Source: Colitti and Simeoni (1996).

Furthermore, their drawing closer is increasingly evident from the asymmetric positioning of various petroleum areas along Hubbert's curves. Whilst, in fact, the Middle East – in the case both of "discovery curves" and that of "production curves" – can be placed in the ascending phase, the United States, the world's number two producer, has been settled for some time now on the descending phase: with reserves and production that have decreased, respectively, by 40% and 35% compared to the levels of the early Seventies.[31] For its part, Europe encountered the peak of its reserves towards the mid-Seventies – when it set off to exploit the North Sea – but then met a gradual erosion down to today's levels of less than 35%. This situation should not vary in direction: resource stock estimates of what is still to be discovered are substantially close to those already discovered. On the other hand the production curve has not yet met a negative turning point, with levels stabilising at 6-7 million barrels/day over the last few years.

The drop of proven reserves in the United States and in Europe has just been compensated – always to the exclusion of the Middle East – by additional oil strikes in Latin America (especially in Mexico and Venezuela) and in Africa (Libya, Nigeria, Angola, Gabon), Western Europe's reserves have remained substantially stable, also because of the generalised mineral activity crisis. The conclusion of all this is that oil reserve stocks outside of the Middle East – and even more so outside of OPEC – after the relative increase registered in the Seventies, remains unvaried at about 200-250 billion barrels, in correspondence with increased production from 1980 to the present time, from 35 to 43 million barrels/day, with a reserves/production ratio reduced to 16 years. The gross increment in reserves – obtained thanks to new, though minor, discoveries and technological progress – has barely served to replace growing production, but has not modified the size of initial reserves.

From this we may conclude that the doubling of world reserves that occurred between 1970 and 1998 – for an absolute value of something in the region of 500 billion barrels – is attributable to over 400 billion barrels (80%) of the increases registered in OPEC countries and in particular – for 336 billion barrels (70%) – in the Middle East. Despite a huge commitment of financial, technological and industrial resources, the *natural* order of things has in no way been modified: since the geopolitical distribution of reserves known to us is exactly that of half a century ago.

The latest wave of pessimistic visions of oil's short-term future is based on the awareness that the increase in world-wide proven reserves is mainly derived not from new strikes but rather from a revision of official OPEC government estimates (Campbell, 1997; Campbell and Laherrère, 1998). These authors argue that the estimates are not very reliable and not homogeneous compared to more correct methodologies, and too conditioned by political reasons: that governments are interested in over-estimating data for real reserves to condition, in their favour, the redistribution of production quotas within OPEC.

This over-estimation error of reserves is perhaps in part true, as that of under-estimating was true in the past: when data were tightly controlled by the majors, for they served as a reference basis for their contractual relationships with the producing States. Asymmetric information and opportunist behaviour surely conditioned those times. Apart from all this there is another dual fact that encourages us to be optimistic: (a) OPEC's official estimates are "taken seriously" by all of the petroleum industry: as proved by their inclusion in the statistics distributed by the "Oil & Gas Journal" or BP Statistical Review of World Energy; (b) the extraordinary progress of technology that has occurred over the last decade has been applied only marginally to the OPEC universe. It is legitimate to hope that if this had occurred the results – in terms of reserves – in such fertile "terrain" would have been even more favourable than that which has been seen in marginal areas.

Figura 8.7. *Non-OPEC reserves and production*

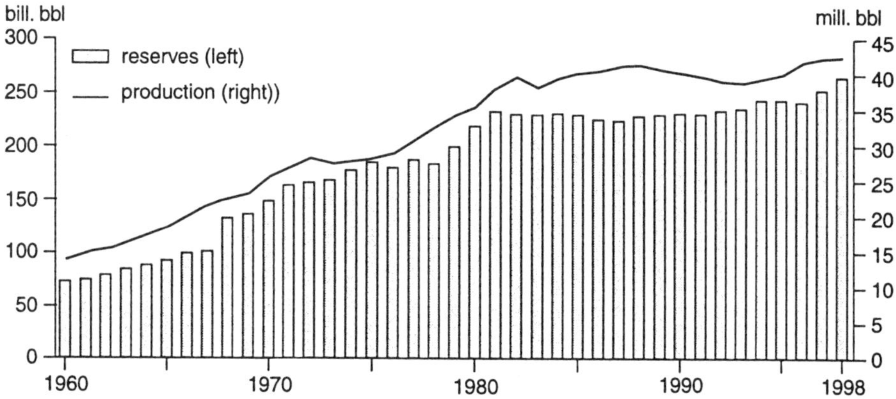

More than a new debate on the improbable exhaustion of resources, we prefer to highlight here – as a final indication – what we feel is really the great challenge and the great change in the future of oil: its *"passage east"*. For three contextual and inevitable reasons. First: the weakening of the Western world's domination, especially that of the United States. Second: the "return of the Middle East", which it has been documented will achieve a specific weight noticeably greater that of the past.[32] Third: the international oil game sees the advent of central Asia and the Caucasian region.

Table 8.8. *The ten largest holding countries of oil reserves*[1]

		Reserves (billion barrels)[2]				Production (million barrels/day)				P:R
No.	Country	1998	%	1970	Δ 98-70	1998	%	1970	Δ 98-70	1998 years
1	Saudi Arabia	259	25	129	130	8.0	12	3.4	4.6	89
2	Iraq	113	11	32	81	2.1	3	1.5	0.6	147
3	Kuwait	94	9	67	27	1.8	3	2.7	-0.9	143
4	Abu Dhabi	92	9	12	80	1.9	3	0.6	1.3	133
5	Iran	90	9	70	20	3.6	5	3.7	-0.1	69
6	Venezuela	73	7	14	59	3.1	5	3.6	-0.5	65
7	Russia	49	5	77	-28	5.9	9	7.3	-1.4	23
8	Mexico	48	5	3	45	3.0	5	0.4	2.6	45
9	Libya	30	3	29	1	1.4	2	3.4	-2.0	60
10	China	24	2	20	4	3.2	5	1.2	2.0	21
1-10	Subtotal	872	84	453	419	34.0	51	27.8	6.2	70
11	Rest of world	162	16	158	4	32.3	49	17.1	15.2	14
1-11	World total	1,034	100	611	423	66.3	100	44.9	21.4	43
11	-OPEC	800	77	399	401	27.7	42	22.4	5.3	80
	-Non OPEC	234	33	212	22	38.6	58	22.5	16.1	17

1. Rounded values. 2. Year-end data.

Source: "Oil & Gas Journal" (various numbers).

Table 8.9. *The largest oil reservoirs*

	Field	Countries	Year of discovery	Period	Estimate of final recovery (bill. bbl)
1.	Grawar	Saudi Arabia	1948	Jurassic	83
2.	Burgan	Kuwait	1938	Cretaceous	72
3.	Bolivar Coastal	Venezuela	1917	Miocene-Eocene	32
4.	Safaniya-Khafji	Saudi Arabia/ Neutral zone	1951	Cretaceous	30
5.	Rumaila	Iraq	1953	Cretaceous	20
6.	Ahwaz	Iran	1958	Oligocene-Miocene-Cretaceous	17.5
7.	Kirkuk	Iraq	1927	Oligocene-Eocene-Cretaceous	16
8.	Marun	Iran	1964	Oligocene-Miocene	16
9.	Gach Saran	Iran	1928	Oligocene-Miocene-Cretaceous	15.5
10.	Agha Jari	Iran	1938	Oligocene-Miocene-Cretaceous	14
11.	Semouor	USSR	1966	Cretaceous	16
12.	Abqaiq	Saudi Arabia	1940	Jurassic	12.5
13.	Romashkino	USSR	1948	Carboniferous-Devonian	14.3
14.	Chicontepec	Mexico	1978	Eocene-Cretaceous	12.3
15.	Berri	Saudi Arabia	1964	Jurassic	12
16.	Zakum	Abu Dhabi	1964	Cretaceous	12
17.	Manifa	Saudi Arabia	1957	Cretaceous- Jurassic	11
18.	Fereidoon-Marjan	Iran/Saudi Arabia	1966	Cretaceous- Jurassic	10
19	Prudhoe Bay	USA	1968	Cretaceous-Triassic	9.6
20	Bu-Hasa	Abu Dhabi	1962	Cretaceous	9
21	Qatif	Saudi Arabia	1945	Jurassic	9
22.	Khurais	Saudi Arabia	1957	Jurassic	8.5
23.	Zuluf	Saudi Arabia	1965	Cretaceous	8.5
24.	Cantarell	Mexico	1976	Palaeocene	8
25	Raudhatain	Kuwait	1955	Cretaceous	7.7
26.	Sarir	Libya	1961	Cretaceous	7.5
27.	Hassi Messaoud	Algeria	1956	Carboniferous-Ordovician	7.5
28.	Shaybah	Saudi Arabia	1968	Cretaceous	7
29.	Abu Sa'fah	Saudi Arabia/Bahrain	1963	Jurassic	6.6
30.	Minas	Indonesia	1944	Miocene	7
31.	Al Bermudez	Mexico	1973	Jurassic	6.5
32.	Asab	Abu Dhabi	1965	Cretaceous	6
33.	Bab	Abu Dhabi	1954	Cretaceous	6
34.	Taching	China	1959	Cretaceous	6
35.	East Texas	USA	1930	Cretaceous	6.8
36	Umm Shaif	Abu Dhabi	1958	Jurassic	5

8.5. NEW PLAYERS

It is not clear whether the oil industry was first founded in Pennsylvania or in the Caucasian area. What is certain is that even without going back as far as the pages where Marco Polo describes flourishing oil commerce in Baku in 1273, the shores of the Caspian Sea have always been very important in the events of this industry. In fact it was the discovery of mineral deposits near Baku, today's capital of Azerbaidzhan, that gave enormous thrust in the 1880s to the newly emerging Russian oil industry. Thanks to those deposits and to transport systems that moved crude oil to the Black Sea coast, Russia came level with and, for several years, overtook American production in the period immediately preceding World War I. Foreign capital, especially European, flowed copiously and with the favour of the tsarist administration. Important innovations occurred in the mineral upstream, as in the case of the first offshore well, explored in 1923 in the waters of the Caspian Sea. When the Russian Revolution exploded in 1917, the consequent nationalisation of industry and the preference of soviet organisations to develop, after World War II, the great reservoirs in the area between the Volga and the Urals, decreed the decline of Caucasian oil.

The development potential that emerged after the collapse of the USSR in 1989 consequently indicated a "return" to the petroleum scenario after half a century or more. Despite great obstacles posed by legislation that is still inadequate for a market economy, western businesses have begun to invest in exploration activity. The conjectural character of scarce accumulated information and still marginal production do not make it possible to estimate production potential with relative certainty, as occurs in mature areas. The various estimates available place possible and proven reserves stock in a range, respectively, of 8-40 and 70-150 billion barrels, overall for the "four tigers" of the Caspian Sea: Azerbaidzhan, Kazakhstan, Turkmenistan and Uzbekistan.

Table 8.10. *Oil reserves in Caucasian countries (billion barrels)*

Country	Proved			Possible	Total
	EIA	BP	OGJ	EIA	EIA
Azerbaidzhan	3.6-11	7	1.2	27	30.6-38
Kazakhstan	10-16	8	5.4	85	95-101
Turkmenistan	1.4-1.5	-	0.5	32	33.4-33.5
Uzbekistan	0.2-0.3	0.6	0.6	1	1.2-1.3
Total	15.2-28.8	15.6	7.7	145	160.2-173.8

Source: EIA (1997), BP (1998), OGJ (1999).

Such potential is in any case of great interest and it encouraged researchers, in the most cautious of hypotheses, to compare it to a new North Sea (16 billion barrels) or, in the more optimist, even to Saudi Arabia. In any case, the Caucasian area was destined to play a role of great importance: not so much under a quantitative aspect – with exports estimated as growing from 0.4 million barrels in 1997 to 3-4 in 2010[33] – as under a geopolitical profile.

There is, in fact, a section of interesting continuity between past and present in the role that companies and western capitals assign to the Caucasian area. Thus, as at the end

Figure 8.8. *Distribution map of oil resources*

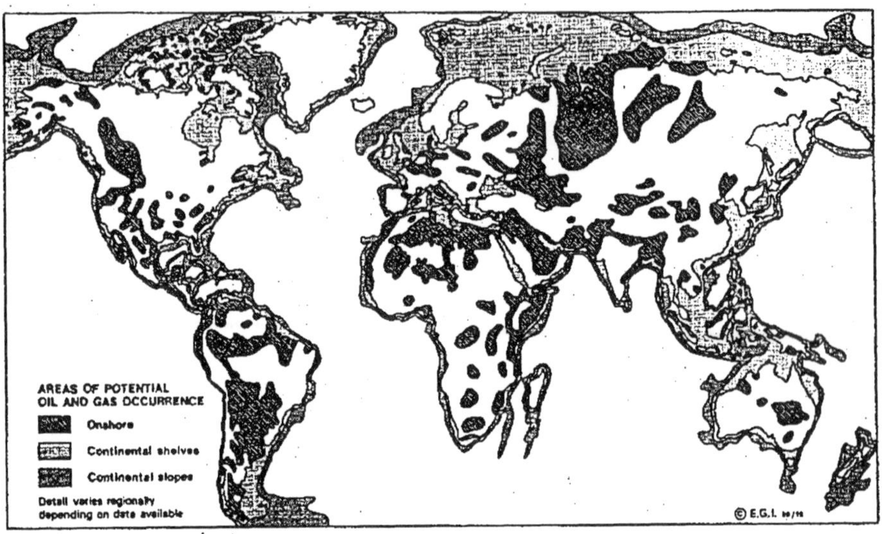

Source: Odell (1999).

of the last century, when European interest was to ensure itself from that production an alternative source to the American supplies monopolised by Rockefeller's Standard Oil Company, today's interest is all concentrated in the possibility of attenuating with the new central Asian and Caucasian areas the Middle East's otherwise dominant role in the future supply of oil (and in part that of Russia in natural gas). This explains why the entire game of resource exploitation is being played on the transport routes (overland and maritime) under construction for use by the final markets of eastern Asia and Europe. An intricate match, once again, where political reasons will count for more that those of the economy (Andreatta, 1998, Stauffer, 1998).

Notes

1. Data – processed by Prof. Jean Masseron of IFP in Paris – date back to the mid-Eighties. Since then there has been undoubted improvement in the cost of several renewable sources (photovoltaic, Aeolian). We do feel, however, that related distances for oil have not been greatly modified, at least as far as size is concerned.

2. In the Seventies the differential advantage of oil on the *cost side* was more than cancelled by the *price side* which was pushed way beyond levels of indifference with other fossil and nuclear sources. Hence its substitution with these sources.

3. Compare API 1995 p. 20.

4. The historic series about proved reserves – and relative sources – are contained in Table 3 of Statistics Enclosures.

5. Theoretically, the peak of production is reached when accumulated production exceeds half of ultimate resources.

6. The datum is obtained by comparing the estimate of last resources by Colitti-Simeoni (1,690 billion barrels) to 1998 production (26.7 billion barrels as in Tab. 5 of Statistical Enclosures) after deducting from the former the production accumulated in the period 1993-1998 (also deduced from Tab. 5 of Statistical Enclosures).

7. By MER we mean maximum production rate achievable from each oil field without causing irreversible damage to the reserves that can be obtained from that field.

8. The estimate is obtained from a sample of 1,750 oil fields of various ages (over 30,000 existing in the world): representative of 91% of all proven reserves and 92% of world-wide production.

9. According to the Paris Agency (OCSE-AIE, 1998) non-conventional oil includes: oil shale, oil sands, liquefied carbon, biomass fuels, gas oil. Overall, these components contributed to world oil balance in 1996 for circa 1.2 million barrels/day over a total production of 72 million barrels/day. According to other sources, non-conventional oil should also include oil found at extreme sea depths, even though it does not undergo any industrial transformation (as the more precise definition of non-conventional oil would require).

10. According to Perrondon A. et al. in the article *The World's Non-Conventional Oil and Gas, Hydrocarbons of last resource* in "Petroleum Economist" March 1998, page 98, the resources of non-conventional oil may account for between 300 and 2,000 billion barrels, with an estimated 400 at oil prices in excess of 5 dollars/barrel. According to Rainbow R. in *Oil Outlook 2020* at "World Energy Council" Madrid 1992 they amount to 3,000 billion barrels for only extra-heavy oils and asphalt sands (in the oil price range of 35-38 dollars/barrel) and 2,500 billion barrels of bituminous shale (at 58 dollars/barrel). According to this author last oil resources (conventional and non) hit about 9,000 billion barrels. Finally, according to the United States Department of Energy in "International Energy Outlook 1998, with Projections Through 2020", April 1998, page 38, the non-conventional resources could amount to 5,000 billion barrels.

11. An oil field qualifies as a "giant" when it holds over 500 million barrels of recoverable oil; "supergiant" has over 5,000 million barrels.

12. Of the 600 provinces discovered in the mid Nineties, 450 have been moderately or intensively explored. Of these, 150 were found to be sterile, *i.e.* unsuitable for producing commercial quantities. About 90% of oil recovered is localised in the 20 largest provinces that also contribute to the greater part of current production; another 8 provinces contain another 5%, and the other 272 the remaining 5%. Information taken from Colitti and Simeoni (1996, page 12).

13. Compare IFP (1996, p. 151 and ff).

14. Compare Adelman (1972a) p. 31.

15. Compare section 4.7.

16. *Mineral investments in the petroleum industry (millions of dollars)*[1]

Year	United States	Other Countries	Total Min.Inv.	Min.Inv./ Total Inv.(%)
1947	1,750	475	2,225	55
1950	2,100	475	2,575	59
1955	4,100	1,110	5,210	61
1957	4,525	2,180	6,705	59
1960	3,735	1,875	5,610	53
1963	3,650	1,520	5,170	42
1965	3,760	2,025	5,785	44
1970	4,335	2,895	7,230	36

1 World excluding PEP. Current values.
Source: CMB.

17. The analysis of the period 1970-1986 is taken largely from my *Instabilità e interdipendenza dei sistemi energetici* of 1988.

18. Refer to section 6.3 and, in particular, to Table 6.4.

19. Annual investments rose from 24 billion $ in the period 1960-69 to 62 in the period 1970-84 compared to variations of annual reserves in the two periods, respectively, of 3.8 and 0.5 billion tons.

20. About half derived from re-evaluation of old oil fields discovered in a chain in the Fifties and Sixties.

21. The M&A wave also pushed many businesses to adopt costly defence measures as potential *raiders* through "*stock repurchases*". In the early Eighties in the United States this accounted for 25 billion current dollars in the group of businesses monitored by CMB.

22. Compare Tab. 1 in Statistical Enclosures.

23. See Fig. 8.1.

24. Evaluation of direct costs is en extremely uncertain subject: it depends on a number of assumptions. The data in the table below are estimates of the Texas Eastern Corporation published by the influential "Petroleum Intelligence Weekly" of 9 December 1985.

Direct costs of oil production

Costs (1) (dollars/barrel)	Production (mil. barrels/g)	Capacity (mil. barrels/g)	Areas of production
< 2	8	15	Middle East (M.E.)
2-4	12	15	M.E, Africa, North Sea (N.S.), South America
4-6	10	10	N.S., USA, Africa, South America
6-8	6	6	N.S., USA, Canada, South America, Asia
8-10	4	4	USA, Canada, Europe, Asia
10-12	2	2	USA, Canada, Europe
12-14	2	2	USA, Canada, Europe
14-24	1	2	USA, Canada, Europe
Total	45	56	

1 Direct costs of production (costs excluding depletion, depreciation, amortisation, interest, capital cost).
For Adelman (1986), direct costs are lower: in the North Sea an average of about 2 dollars/barrel. (with 60% of production <2 and 80% <4 dollars/barrel) whilst in the USA 65% of production is <4 and 30% <8 dollars/barrel.

25. Exxon was the first business to develop, as early as 1972, the "3D-Seismic". For an interesting and simple description of this technology and what its results have brought to Exxon Compare *Exxon's Deep Sea Extraction*, in "Argomenti Esso", no. 33, December 1998, p. 28.

26. Thanks to Giovanni Ceccaroni for this interesting analogy.

27. "The deviated well technique is based on the use of turbines moved by pressurised mud. Drilling direction, first vertical, then gradually deviated a few degrees at intervals of every few meters of drilling. Horizontal crossing of the production level is normally very extended, reaching hundreds of meters" (SSEM, 1998, p. 127). In September 1996 BP reached a record of 8,700 meters of deviated and horizontal wells at the Wytch Farm (Great Britain) site when exploring an offshore oil field that had begun of a land platform (Appert, 1996, p. 39).

28. To realise how import offshore is, we should remember that ten years ago North Sea production was forecast for 2000 at 2.5 mil. barrels/g, whilst it can really be placed nearer to 6. The great hopes placed in the new offshore "province" of the Gulf of Mexico – the only one able to staunch declining American production – seems to be weakening. Today that province produces over 1 million barrels/day, from concessions of which half are to be found below 900 meters.
Compare IFP, *Panorama 1998*, especially the report *Gulf of Mexico: the American Oil Recovery*, pages 63-70 and of the President of Royal Dutch Petrol. Co. C.A. Herkstroter *A Continuing contribution – Oil and*

Gas in the 21st Century, Paper 1996. Also compare Boy de la Tour, *Technologies pétrolière: les nouvelles frontières*, in "Revue de l'Energie", no. 456, February 1994 and in the review "Petroleum Economist": *New drilling technology is better than cost-saving*, April 1996, p. 52 and *Deeper ever deeper – special supplement on new technology*, December 1995, pp. 2-5.

29. The 1986 price drop and investments have brought about a heavy capacity surplus in the mineral industry sector suppliers, especially of drilling systems. Relative tariffs dropped – because of competition between suppliers – by 70% between 1983 and 1993. These reductions were then recovered with capacity use increases.

30. Total unit costs are obtained from the sum of exploration costs plus operational costs plus amortisation.

31. Compare data tables 3-6 in the Statistical Enclosures.

32. The Middle East's historical peak quota in world-wide production was in 1974 with 37%. In the future, as has already been stated, it will tend towards 66%.

33. The lower estimate is taken from the *International Energy Forecast Software of Houston*; the larger from Shenoy et al., *Caspian Oil Export choices clouded by geopolitics, project economics*, in "Oil & Gas Journal", 19 April 1999, pp. 29-35.

9 FROM MONOPOLISTIC STABILITY TO COMPETITIVE INSTABILITY

9.1. OIL AS A COMMODITY?

The industry and the oil market have emerged radically modified from the second petroleum crisis. A new phase, the *fifth*, has begun, in which the free forces of the market and the competition once and for all seem to prevail over any dominating position of the great multinationals, over every form of their oligopolistic co-ordination and over every cartel policy. Price formation is regulated by competition and the market. An increase in the number of figures dealing with each other daily, the disappearance of long-term contractual relations, the organisational destructuring of large integrated groups, cause an absolute unpredictability and volatility in prices in a highly uncertain framework. As was the case during the first phases of the take-off and development of the industry, variations often in double percentage figures are observed daily. "When there is much supply and demand, and both are uncertain and subject to vast seasonal or annual fluctuations, the conditions are created in which the forces of demand and supply on the free markets change in continuation. This play of uncertain economic forces produces constant fluctuations in the prices, thus giving rise to the creation and success of the futures markets" (Hieronymus, 1982, p. 21). New forms of transactions, of the paper type and typically financial, appear alongside physical transactions and there is a gradual increase in the volumes traded. The level of information available on the market increases, as well as its transparency and efficiency. The combination of the physical and economic globalisation of the markets with their financialisation has helped to accentuate price instability. Anything that happens in a Texas refinery or an oil platform in the North Sea, even the slightest change in climate or political tension tends to impact, in real time, on the whole world price structure (of crude oil and products) and on all contracts in both the physical and paper markets. Oil seems therefore to have lost its specificity and become like any other commodity, such as grain, copper or bacon (handled on the New York futures market). Thus it follows that public policies have also lost much of their reason for existing, unless the reason is to reduce their influence to make the markets even more efficient.

There is much truth in all this. However, as in all the "opinions of consensus" emerging through time, these positions run the risk of being reductive and misleading about the *complexity* of the problems interacting on the oil market, as well as reflecting temporary and not structural long-term conditions; underestimating the close functional relations that continue to link economic variables and political variables; giving the reasons for market failure as finally overcome. Oil still retains its own *specificity* which prevents it from being assimilated, *sic et simpliciter*, with any other commodity, although the nature and the conditions of the market are today very different from those of the past. In Marcello Colitti's words: "Oil is a political commodity, because it is vital to the modern economy, and to anyone who wants to create one; and because it is mostly produced [*and will be increasingly so in the future,* Author's note] in countries which for a long time had little will of their own, and whose role was at best that of a pawn on a chess board, in which the strong pieces were in the hands of powerful corporations and colonial governments" (Colitti and Simeoni, 1996, p.119). On the other hand, the Gulf crisis of 1990 had been the most obvious and dramatic confirmation of the political specificity of the oil commodity. Here it is important to look back at the main structural transformations affecting the oil industry in the Eighties, in order to outline its new distinctive traits and the subsequent reactions.

9.2. NEW STRUCTURAL ORGANISATIONS

The most important structural changes concerned the supply of oil as regards control, degree of concentration, property rights, management methods, types of commercial transactions, type of operators, and criteria of price formation. The main consequence of the expulsion of the majors from almost all the OPEC producer areas was the *vertical and horizontal disintegration* of the international industry, initially dominated by them.

It is of course true that other instrumental conditions in its life cycle which had once boosted integration had gradually been modified. These included the extension of the market, the specialisation of certain stages, and the greater completeness of the contracts. There is no doubt however that the speed and intensity of the changes observed was initially to be attributed to the change in the property rights in OPEC areas and to the subsequent loss of control of the mining resources by the majors. Their expulsion, coming as it did at the same time as the entry into the market of a number of new producers (of different size and in different parts of the world), led to two types of interconnected effects:

(a) a considerable *reduction in the degree of concentration* of crude production (calculated on the market quotas of single countries) with respect to the peaks reached during the Seventies: in order to form a positive relation between concentration and prices (Verleger, 1991) (Fig. 9.1.);
(b) the *disappearance of co-ordinated supply management* through the consortiums, long-term contracts, and information feed-backs allowed by vertical integration.

Just a few figures (Tab. 5.7.) are needed to illustrate the transformations occurring. In 1973 the majors controlled 30 mill. bbl/d of crude oil, equal to 65% of the world production, of which 80% was from OPEC countries. As regards usage, 75% went to their own refineries and 25% to third parties through long-term contracts. In 1998 the situation was radically different. The production controlled by the majors has fallen by about 4 times in absolute terms, from 30 to 9 mill. bbl/d, and by about 5 times in percentage of the world production, from 65% to 14%. The drop in the availability of crude – entirely attributable to the OPEC areas – was followed by downsizing also in the downstream stages, with a drastic redimensioning of refining capacities (also due

Figure 9.1. *Concentration index of oil supply and oil real prices*

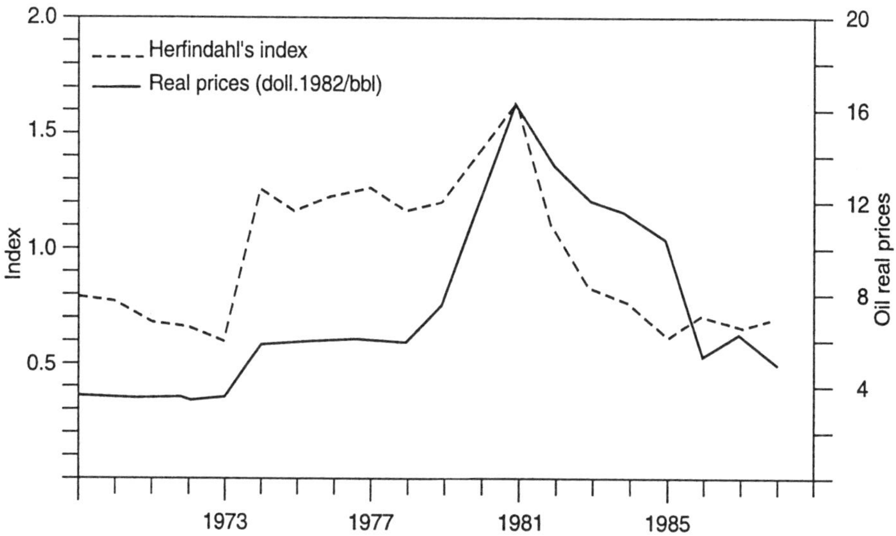

Source: Verleger (1991).

to the low or negative margins which could be gained) and of final sales. The 24.3 mill. bbl/d of 1973, with a market share of 55%, had fallen in 1985 to just 18.1 mill. bbl/d and then risen to 25.3 mill. bbl/d in 1998: equivalent to the 40% of the market.

The degree of vertical integration of the majors – given by the relation between flows of availability of oil and sale flows of products – fell from values around 1.0 throughout the years of their dominion (within a situation of almost perfect balance) to values in a range of between 0.3-0.4. There was thus a huge surplus in the quantities sold downstream with respect to those available upstream. Instead the opposite is true for public companies in the producing countries which control the largest market shares today. Figure 9.2. shows the positioning of the world's 15 largest oil companies with respect to a perfectly balanced line of equivalence of activities or of integration. The greater the distance from that line, the greater the degree of vertical disintegration of the individual firms and of the industry as a whole, depending on their degree of representativity. The data is from 1995. As we can see, the oil industry is clearly split down the middle: on one side the companies of the leading OPEC countries in the supply of crude oil (to the left or above the line of equivalence); on the other, the majors who hold the greatest control of the final market, though the market shares are much smaller than in the past (to the right and below the line). Though the strategy of the major firms in both groups was to strengthen their degree of vertical integration, it is very far from values of equilibrium. The same applies to Saudi Aramco (Saudi Arabia) and NIOC (Iran), the world's first and second largest crude oil producing firms, and on the downstream side, R.D. Shell, Exxon, BP or Mobil. A greater level of equilibrium is reached by the smaller firms, due to their firm control over the domestic markets (the Kuwaiti KPC or the Venezuelan PDVSA), or to upstream integration strategies (the Italian ENI and the American Amoco).

As well as the figures, however impressive they may be, which have marked the breakdown of the dominion of the great majors though not their overall economic force, it is important to analyse the effect of this on the configuration of the international market, especially as regards its degree of competitiveness and price dynamics. The expulsion of the majors was not indeed only due to the drastic cut in their size, their integrated circuits, and their geographical diversification. An equally important element was the change brought about in the relations between producer countries and firms, which previously could be placed within a framework, in a theoretic line, in a diagram not far from that principle-agent. Between countries and firms there has always been a close relation of bi-univocal interest. The reasons for the conflict could be attributed to the methods of sharing out the profits gained, or at least, this was the case until the end of the Sixties. These relations were later to remain for at least a decade, though a large section of the ownership rights had been taken away from the companies. On the mining side, what had been taken away from them was in fact almost completely restored to them on the commercial side, through preferential contracts. Thus the handling of the volumes of supply and control of the market remained firmly in the hands of the majors. It was only with the second crisis, plus the breaking of the preferential contracts, that

Figure 9.2. *Crude production and product sales of the first 15 world oil companies (mill. bbl/d)*

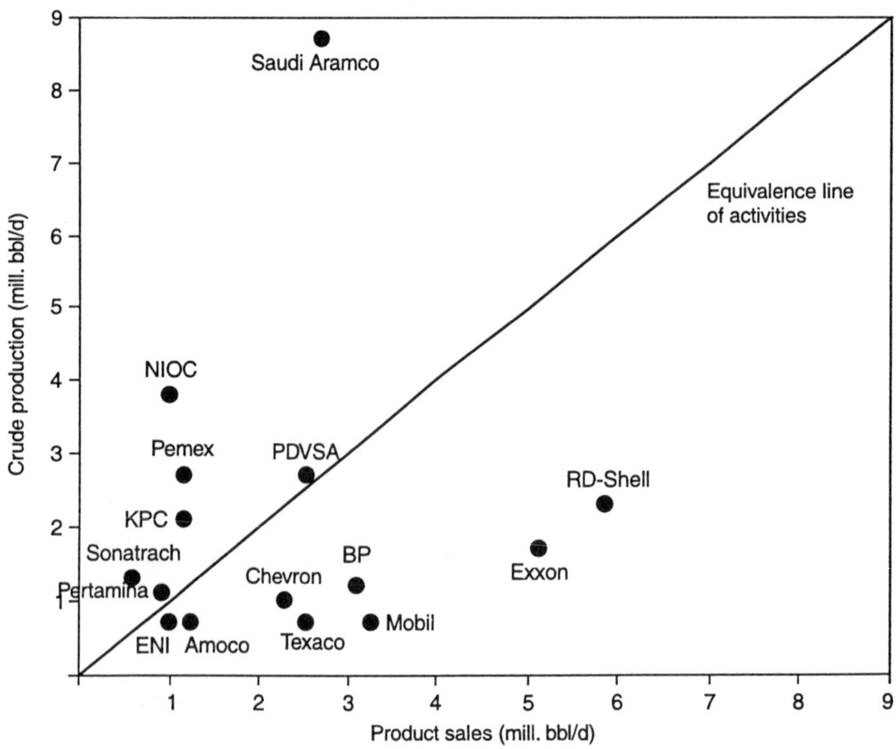

their central role disappeared. Indeed, it was only then that the market fully substituted integration or quasi-integration as a system of allocation of resources and as a system of fixing prices. The relations between countries and firms changed from *co-operative* to *conflicting*. Instead of being agents of the countries, the companies became their counterparts. This reversal of positions is also due to the fact that instead of net sellers of oil, the majors were now net buyers, and thus interested in exploiting the opportunities offered by the competition between producing states as best as possible.

9.3. NEW REACTIONS

The fact that competitive conditions (though not perfect competition in the strict sense) prevailed on the oil market is the *combined* outcome of two phenomena. On one hand, the emergence of new structural layouts, as we have seen above. On the other, the development of new types of by the main players: firms, producer states and consumer states. This reaction in all three cases was orientated towards giving the market, however imperfect, a more important role. We have already discussed the companies. The processes of organisational destructuring and their different positioning on the market could not fail to push them towards full acceptance of the challenge of competition and globalisation of the markets.

The consumer states saw the dismantling of all the apparatus which had once been used by public policies. In no country[1] was there any trace of policies for administrative control of oil prices to be found, or legislation or policies to protect national companies, reserves of activities attributed to the states,[2] or obstacles to trading. Even the long and in a certain sense extraordinary experience of the public companies seemed to be running out, as all or part of their share capital[3] was placed on the market.

Encouraged also by the return to conditions of abundance of oil supply, what used to be known as energy policies have largely disappeared, sharing the conviction that the virtues of the market are sufficient on their own to guarantee conditions of lasting equilibrium. The only instruments used by public policies today are environmental regulations, which gradually raise the quality standard of the products, and the system of taxation on the consumption of oil products: with the revenue for the seven largest industrialised countries equal to twice that collected by all the OPEC countries put together.[4]

Finally, the new reactions of the producer countries should be explained. After the unhappy and unsustainable experience of fixing high prices for oil – with compatible volumes of supply – and after the countershock of 1986, OPEC countries adopted a policy of fixing the quantities according to prices determined exogenously by the market. From price-makers, they became price-takers. Though perhaps unavoidable, this choice certainly helps to defend their long-term interests, especially those of Saudi Arabia.[5] "The largest OPEC exporter has embraced a 'long-term policy', that is, a policy of depletion of its own reserves which takes into account the need to keep its market share, and to maintain the market for its oil 'in being' for a long time to come; and has more or less left to a highly speculative futures market the task of fixing the price. To put it more simply, the largest exporter has decided that it is useless to try to regulate both quantities and prices, and has settled for the first – that is, for securing for itself a large share of the market at prices at which the cheaper producer gets his rent, but very little goes to the dearer one, and none to costlier energy sources" (Colitti and Simeoni, 1996, p. 123).

To conclude, new layouts and new reactions have encouraged the consolidation of *conditions of greater competition* on the oil market, although its features are still very remote from those of a perfectly competitive market. If this were so, the prices should tend towards the margin costs of the last unit necessary to balance demand and supply: which certainly cannot be said to be the case with a marginal costs and prices ratio of up to 1:40. In fact, if we use *Lerner's index* [(price – marginal cost)/price] as a criterion to measure the "degree of monopoly" of the industry, in other words of the monopolistic power of the companies (seen as their capacity to fix prices higher than those which would occur in a competitive equilibrium), we would see how this tends towards its maximum value, the unit[6] (Lerner, 1934). The fact that this power does not mean high prices is only apparently contradictory. This is because, as we have already said, the careful monopolist does not always find it convenient to raise his prices and profits in the short term, but instead prefers to preserve his market position in the long term. The leading OPEC countries seem to have learnt from the mistakes of yesterday and the results obtained show they were right.

9.4. THE MARKET PENDULUM

9.4.1. The renewal of demand

Beneath the cover of a comforting abundance in oil supply and of the return of its prices in real terms to the values of 1973-74, in the decade between 1985-95 the oil markets saw changes of great import, the combined outcome of more than one phenomenon. As regards *demand*: its positive reaction to the lower prices (although the strong increase of fiscal pressure has done much to soften this effect); the economic growth in a percentage of two figures for the new industrialised countries in Asia; the net increase in the world population at a rate of 240 thousand people a day (with a total in the decade of 876 million to 5,732 million;[7] the slowing down in the processes of replacement of oil, due to the work that had already been done, to the low economic convenience of renewable resources, to the growing environmental opposition to nuclear power and coal, and to the disinterest of energy policies. In short, economic and political reasons have once more began to operate in favour of oil.

Between 1985 and 1995 the world demand increased by 10 mill. bbl/d to 70 mill. bbl/d, abundantly exceeding the previous peak of 1979 (Tab. 9.1.). In terms of percentage, the growth was less than that of the past, but much less so if we look at the absolute values which take on even greater significance since this is a non-renewable resource. The increase would have been much greater if the crude oil consumption in the former Soviet bloc had not been practically *halved* after 1990, and what is more, at the same time as the collapse of domestic production. Outside this area, there are three trends of importance. The first is the strong and constant growth of consumption in *developing countries*. In ten years this has increased by about 9 mill. bbl/d to 23.4, doubling their share of world consumption from 16% in 1973 to 33% in 1995. To give an idea, it is as if two industrialised countries of the size of Japan and Germany had appeared simultaneously on the oil market. This trend is destined in the future to expand, with

Table 9.1. *World oil demand by area: 1970-2010*[1] *(mill. bbl/d)*

	1970	1973	1979	1985	1990	1995	Forecast	Absolute var.	
							2010[2]	1995-85	2010-95
OECD	*34.0*	*40.8*	*42.1*	*34.7*	*38.1*	*40.4*	*45.4*	*+5.7*	*+5.0*
North America	16.2	19.7	20.9	17.6	18.9	19.8	22.7	+2.2	+2.9
Europe	12.5	14.9	14.8	12.0	13.0	13.9	15.1	+1.9	+1.2
Others	5.3	6.2	6.4	5.1	6.2	6.7	7.6	+1.6	+0.9
Non-OECD	*12.4*	*16.6*	*23.4*	*25.3*	*28.4*	*29.9*	*46.2*	*+4.5*	*+16.4*
Former USSR	5.3	6.3	8.7	9.0	8.4	4.7	7.1	–4.3	+2.4
China	0.6	1.0	1.8	1.9	2.3	3.3	6.6	+1.4	+3.3
Others-Asia	1.4	2.2	3.2	3.7	5.6	8.0	14.0	+4.3	+6.0
Others	5.1	7.1	9.7	10.7	12.1	13.9	18.5	+3.1	+4.6
Total	*46.4*	*57.4*	*65.5*	*60.0*	*66.5*	*70.3*	*91.6*	*+10.2*	*+21.4*
Annual variation (mill. bbl/d)		+3.7	+1.4	–0.9	+1.3	+0.8[3]	+1.4		

1. Peaks in oil consumption in the Seventies were reached in 1973 and 1979. 2. IAE 1995, "Energy saving case". 3. +1.5 excluding former USSR.

Source: IAE

annual increases exceeding those most recently recorded (+1.1 mill. bbl/d). The second trend is the revival of demand in *industrialised countries*, though at rates lower than previously.

Between 1985 and 1995 it increased by 5.7 mill. bbl/d, recovering 80% of the drop that followed the second oil crisis. The third trend is the new role that oil has in the coverage of incremental energy consumption, thus inverting the situation of displacement that occurred after the increase of its relative prices. Oil in fact made the largest contribution (36%) to the increase in world energy consumption of about 1.2 bill. toe between 1985 and 1995, to 8.1 bill. toe, ahead of methane gas (34%), nuclear power (18%), coal (9%) and renewable resources (3%). This trend too is destined in the future to be accentuated, if the current structure of relative prices of the various sources remains and given the opposition in industrialised countries to the additional use of nuclear energy and coal. According to the Paris Agency, over the period 1995-2010 oil consumption could increase on a world scale by 21.4 mill. bbl/d (to 91.6), mainly as a consequence of the greater demand from developing countries; an increase in 15 years of the same proportion of that recorded over the last 25 years (Tab. 9.1.).

9.4.2. The return of OPEC

As regards the oil supply, these appear to be the most important trends in the decade from 1985-95 (Tab. 9.2.):

(a) the huge increase in OPEC production: from 17.6 to 27.5 mill. bbl/d, thus recuperating more than 7/10 of the previous drop;

(b) the additional OPEC production satisfied more than 90% of the increase in world demand: with total non-OPEC flows remaining substantially stable;

(c) this stability is in turn the result of very different trends in the different areas. In particular:

- the gradual decline of the US production (from 10.6 mill. bbl/d in 1985 to 8.3 in 1995), due to the irreversible halt in the activity of thousands of stripper wells no longer competitive at current price levels;[8]
- 40% drop in the production of the former USSR at levels difficult to recuperate without making huge investments;
- the very positive trend of North Sea production (Great Britain and Norway) with the entry into production of new second-generation oil fields, often of small dimensions. This unexpected increase is due to the extraordinary results achieved by the companies in developing new mining technologies with huge increases in productivity and reduction of extraction costs;
- positive trend of the production in "Other" countries, especially in developing countries, increasingly diversified geographically: Vietnam, Malaysia, Syria, Yemen, Angola, Congo, Colombia, Brazil, etc.

The trends observed in the last decade are destined to be consolidated in the future. Indeed, the expected increase in the demand cannot fail to be satisfied mainly by OPEC countries and particularly by the Middle Eastern ones.

The "passage to the Middle East" does not *in itself imply* a negative scenario, though there is no doubt that the entirely political question of the reliability of the supplies will arise again because of it.

However, we may learn from past crises that it is the dynamics of the economy

Table 9.2. *World oil supply by area: 1970-2010 (mill. bbl/d)*[1]

	1970	1973	1979	1985	1990	1995	Forecast 2010	Absolute var. 1995-85	Absolute var. 2010-95
OECD	*13.2*	*14.0*	*14.9*	*17.1*	*15.9*	*18.0*	*13.9*	*+0.9*	*–4.1*
North America	12.8	13.1	11.9	12.5	11.0	11.0	8.0	–1.5	–3.0
Europe	0.4	0.4	2.4	4.0	4.3	6.3	5.3	+2.3	–1.0
Others	–	0.5	0.6	0.6	0.6	0.7	0.6	+0.1	–0.1
Former USSR	*7.5*	*8.6*	*11.7*	*12.0*	*11.5*	*7.1*	*9.7*	*–4.9*	*+2.6*
OPEC[2]	*23.2*	*31.2*	*31.5*	*17.6*	*25.1*	*27.5*	*49.4*	*+9.9*	*+21.9*
Others	*4.7*	*5.1*	*8.5*	*12.5*	*14.4*	*17.3*	*19.0*	*+4.8*	*+1.7*
Total	*48.6*	*58.9*	*66.6*	*59.2*	*66.9*	*69.9*	*92.0*	*+10.7*	*+22.1*

1. Includes natural gas liquids and processing gains. The difference in demand level is due to stocks variation. 2. Includes Ecuador until 1990, then included in "Others" (1995: 0.4 mill. bbl/d).

Source: IAE

which give critical strength and great force to political reasons, and not vice versa. Some conditions are needed to prevent this happening:

- the dynamics of the prices should be able to provide the financial resources necessary, both to producer countries and companies, to expand the production capacity in line with future requirements. Price falls in the short term – even for minimal conjunctural excess in supply – would certainly discourage investments, thus creating the premises for future situations of crisis;
- the conditions must be created for a return of the capital, of the technologies, and of the managerial skills of the Western companies to producing countries, especially in the Middle East, on the basis of contractual relations unlike those of the past. The unused capacity in the OPEC countries, equal to 5-7 mill. bbl/d (including Iraq), is able to guarantee coverage for only 3-5 years of growth of world consumption (IFP, 1996, p. 96). In current conditions, it is unlikely that it will be able to achieve the additional necessary capacity of 15 mill. bbl/d within 10-15 years;
- the third condition is that the extraordinary technological progress (especially in off-shore areas) seen in the last few years is able to continue in the non-OPEC area, in order to compensate for the natural decline of existing capacities (1.5-2 mill. bbl/d) and if possible increase them. This returns us, once more, to the central question of prices and to the possibility that their current instability can generate conditions of market equilibrium in the future.

9.5. CRITERIA OF PRICE FORMATION

The fixing of oil prices today depends entirely on the free play of the market forces, *i.e.* the daily comparison of demand and supply. Besides the criterion of a general nature, it is important to analyse just how this fixing is done. From this we can judge the efficiency of the markets in terms of transparency, informative capacity, and significance of the values handled. In the mid-Nineties, the approximately 72 mill. bbl/d produced on a world scale is evaluated on the basis of the quotations expressed daily on the spot or futures markets of five crude oils taken as reference or market benchmarks: Brent (UK), Alaskan North Slope and West Texas Intermediate (United States), Dubai and Tapis (Far East). Basically each of them acts as a focal point for fixing the prices of the commercial flows from the rest of the world towards their membership area or inside their own area. As an example, the *fob* prices of Arabian Light towards Europe are fixed backwards starting from the Brent spot price, after deducting the actual cost of transport and the quality differential. The price considered is usually the price relative to the moment of landing at the port of destination, calculated by applying a factor of delay in the formula.

Thus in January 1995 Arabian Light was quoted at 16.01 doll./bbl with a differential of 0.95 with respect to the benchmark Brent crude (Tab. 9.3.). Price levels lower than this threshold of equivalence indicate competition mainly with the productions of the North Sea. Figure 9.3. shows the three macro-areas of destination of the commercial flows (America, Europe and Asia) and the relative benchmark crudes (West Texas Intermediate (WTI) and Alaskan Northern Slope (ANS), Brent, Dubai and Tapis respectively). Of the 72 mill. bbl/d produced overall in 1996, about 15 (21%) are evaluated with formulae fixed to Brent (Tab 9.4.). These are the volumes produced

Table 9.3. *Price formulae at January 1995*

Crude destination	Benchmark crude	Price basis	Time factor	Adjustment factor (doll./bbl)	Final price (doll./bbl)
EUROPE					
Arabian Light 34°	Brent	Fob	40 days post load	–0.95	16.01
Iranian Light 34°	Brent	Cif	unloading date	–0.25	16.63
Libyan Es Sider 37°	Brent	Fob	unloading date	+0.15	17.13
USA					
Arabian Light 34°	WTI	Fob	50 days post load	–2.60	16.34
Nigerian Bonny Light 37°	Brent	Fob	15 days post load	+0.32	17.08
FAR EAST					
Arabian Light 34°	(Oman + Dubai)	Fob	monthly average	+1.90	16.84
Kuwait 31°	(Oman + Dubai)	Fob	monthly average	+0.30	15.71

Source: "Petroleum Intelligence Weekly"

Figure 9.3. *World oil market: benchmark crudes*

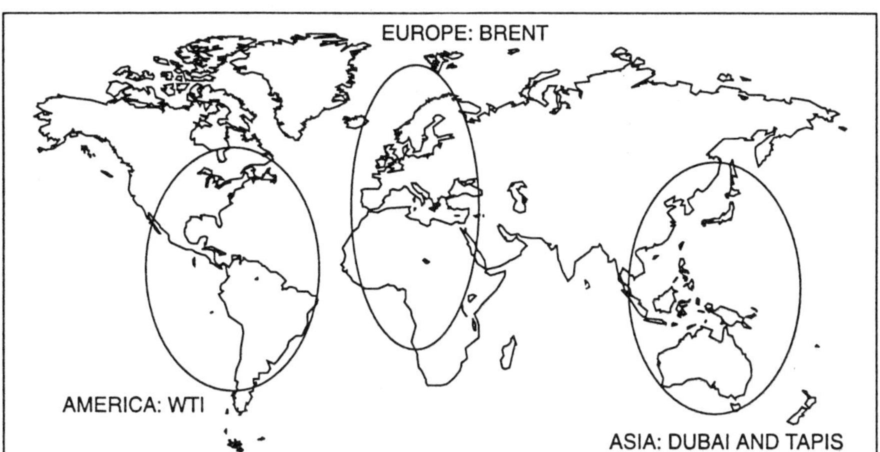

Table 9.4. *World oil production and reference price 1996*

Area/Producer	Description	Reference prices	Volumes (mill. bbl/d)
Europe	Local production	Brent	6.5
North Africa	Exports to Europe	Brent	2.0
West Africa	Exports to Europe	Brent	0.9
Former USSR	Exports to Europe	Brent	2.1
Middle East	Exports to Europe	Brent	3.3
Total Brent			*14.8*
North Africa	Exports to USA	WTI	0.3
West Africa	Exports to USA	WTI	1.6
Middle East	Exports to USA	WTI	1.7
North America	Intra-area production and export	WTI	8.5
South America	Exports to USA	WTI	1.9
Other areas	Exports to USA	WTI	4.6
Total WTI			*18.5*
Alaska	Production and export	ANS	1.8
Total ANS			*1.8*
Middle East	Export to Far East	Dubai	10.8
Total Dubai			*10.8*
Asia	Intra-area export and production	Tapis	5.5
Total Tapis			*5.5*
Consumptions of producing areas	Not valued at market prices		19.2
Processing gains	Not valued at market prices		1.5
World oil supply			*72.1*

Source: our calculations from various sources.

inside Europe (and mainly consumed here) or imported to Europe. The WTI price is the benchmark in the evaluation of 19 mill. bbl/d (26%) produced in or destined for the North American market. Also in North America, about 1.8 mill. bbl/d (3%) are fixed to the price of Alaskan North Slope. In the Far East the strong dependence on Middle Eastern imports results in volumes fixed to the price of Dubai for about 11 mill. bbl/d (15%), while locally Tapis crude appears in the evaluation of about 6 mill. bbl/d (8%) produced and consumed in the area. The remaining 19 mill. bbl/d (26%) of world production is not evaluated at market prices since it is destined for internal consumption in non-market economies.

This system of fixing the prices follows, in its logic, that of the import parity applied for a very long time when the multiple base point system was used. The methods of forming the base prices and the degree of adhesion are however radically different. While, in fact, in the past list prices were fixed administratively by the majors (at levels which were often divergent from the marginal market transactions), today's benchmark prices are determined by the market, through which pass the majority of the commercial flows and to which the flows inside the vertically integrated companies are linked.

There is also a different degree of adhesion to this system of prices by the competitors. In the past this was maximised by the oligopolistic co-ordination of the seven majors, while today it is forced to be flexible for several reasons. The first of these is the relatively low degree of concentration of the offer. Secondly, strong competition, especially in a situation, like the current one, of a "buyer's market". Thirdly, the objective difficulty for every producer to run differentiated price policies for each area of destination in a situation of continuous variation of the values of reference.

9.6. EFFICIENCY AND TRANSPARENCY

How should we judge the degree of efficiency and transparency of the oil markets? There are two limits which emerge most clearly: poor representativity of the benchmark crudes and poor transparency of spot quotations. The overall production of the five "pivotal" crudes amounted in 1996 to just 3.5 mill. bbl/d compared with 51 mill. bbl/d for physical market transactions and about 126 mill. bbl/d in paper transactions connected to them in the formulae of price fixing (Tab. 9.5.). Thus there is a ratio of 1:15 and 1:36 in the global values and up to peaks of 1:145 in trade on the American markets linked to West Texas Intermediate. In this last case, compared with a daily production of 0.75 mill. bbl there are transactions on the physical market worth 18.5 mill. bbl and 90,000 paper transactions worth 90 mill. bbl. The small volumes of benchmark crudes involves two types of risk: firstly, an excessive conditioning of the total transactions by entirely marginal productions; secondly, risks of manipulation of the benchmark prices.

Table 9.5. *Oil spot markets and linked markets: 1996*

	Crude oil					Total
	Brent	WTI	ANS	Dubai	Tapis	
Location	North Sea	Oklahoma (USA)	Alaska (USA)	Dubai	Malaysia	
Physical production (mill. bbl/d)	0.76	0.75	1.40	0.26	0.35	3.52
Linked markets (mill. bbl/d)	14.8	18.5	1.8	10.8	5.5	51.4
Name	Brent *forward* IPE *forward*	Nymex WTI *forward*	–	Dubai *forward*	Tapis *forward*	
Minimum lot (bbl)	500,000[1] 1,000[2]	1,000[3] n.a.	–	250,000	100,000	
Daily transactions (no.)	2[1] 34,653[2]	90,000[3] n.a.	–	1.8[4]	0.3[5]	
Daily transactions (mill. bbl/d)	0.5[1] 35.0[2]	90.0[3] n.a.	– –	0.4	0.003	125.93

1. For Brent forward. 2. For IPE futures. 3. For Nimex. 4. 44 contracts/month. 5. 7 contracts/month.

A second limit is given by the very significance of the spot prices taken as reference. In this context the situation is far from transparent and efficient. In fact, it is only for prices on the futures markets or those with deferred delivery that the markets have reached a high level of reliability as regards liquidity and transparency. The same cannot be said for quotations handled on the spot markets of crude and even less so for those of oil products. These latter are published by various specialised newsletters, the most important of which is "Platt's Oilgram Price Service" published by McGraw Hill of New York. It is possible to read the daily prices of the principal products "handled" the day before on the markets of Rotterdam, Mediterranean, North West Europe, Persian Gulf, Singapore, US West Coast, US Gulf Coast, New York and the Caribbean. Through these areas the majority of the 8 mill. bbl/d of oil products traded yearly transit on an international level (with respect to 28 mill. crude and 72 mill. total consumption). Platt's define the prices that they themselves report as "an estimate made by the internal staff on the basis of market transactions and information from sources considered reliable and active on the market". Every day the managing director telephones a close circle of multinationals, traders, brokers and large consumers for information on the transactions closed over the previous twenty-four hours (and even about their intentions to sell or buy). On the basis of these, through an entirely personal and unknown procedure, he processes and publishes the estimates of the quotations.

Their significance and transparency is therefore highly restricted and questionable. For several reasons: (a) the quotations do not reflect actual values traded, as on any stock exchange, nor represent the weighted average of these on the quantity; (b) there are no forms or possibilities of *ex-post* control of the actual conclusion of the negotiations; (c) the differences between published prices and actual prices are found (from the few surveys carried out)[9] often to be very great, especially when the markets are particularly unstable; (d) the quotations do not consider adequately the differences in quality and quantity of the lots traded, in terms of payment; (e) there are ample margins and risks of manipulation of the prices by a limited number of people working to estimate them, although the invisibility of these people leads most to take this estimate, wrongly, for that of Adam Smith.

Despite all this, Platt's quotations gained extraordinary significance in the Eighties. The great majority of international trading of oil products refer to these quotations, as in the case of transactions on the crude oil market, evaluated with formulae of the net-back type; even the final prices (net of taxes) in the consumer markets, where the game of competition (when fully operative) pushes the domestic prices into line with international prices.[10] Platt's prices are also used as reference for evaluating other energy sources competing with oil. This occurs in particular in long-term contracts of import-export of natural gas, where prices are normally indexed to those of the competing products in the various segments of consumption.[11]

Considering its limits, it may certainly be said that the role of what we could call "Platt's price system" is completely out of proportion. But it is used and in the absence of anything better, it continues to be indicated, more out of convention than conviction, as the expression of the "market" for oil products.

9.7. FUTURES MARKETS

The consolidation of the market and the competition as systems for the formation of oil prices, as well as the reappearance of normal conditions of volatility and uncertainty about their dynamics, have again brought out the *risk factor* as the central element in the decisional processes and the destiny of the oil companies. Their operating results in the short term and their survival in the long term depend on their ability to handle this factor. This is particularly true given the extraordinary nature of some of the investment choices, concerning new and unexplored areas of research and production in the upstream sector; new process and product technologies in refining; and new areas of consumption in distribution.

The challenge of the market, of the competition, and of the risk has forced the companies to make radical changes in their business culture, in the organisational layouts and in the use of new contractual instruments. In the new competitive context, it is not so much the size of the company that counts but rather its capacity to interpret and anticipate the market dynamics, to adopt the right decision of sale/purchase rapidly, to make full use of the relative convenience of the prices of crude and products. Flexibility has become more important than dimension and speed of decision more important than completeness of information. In this context, new instruments of negotiation have developed as well as new methods of transaction. Of these, deferred delivery markets and the futures markets have become particularly significant.

9.7.1. Price and efficiency of the markets

A market brings together various operators for the trading of goods in exchange for money; each single transaction involves and requires the conveyance of information. Thus the market as a whole is also a necessary instrument for conveying and utilising the information relative to it (Menger, 1981). The most important information is the price. Since it is determined by the agreed action of all the participants on the market, it represents a highly significant element not only for two single operators but also for any subject whose activity is linked to the asset being negotiated. The higher the number of transactions, the better the quality of the information and thus the efficiency of the market.

Each economic decision produces effects which spread over a varying period of time. In other words, there is an interval between the decision and the appearance of its result. This means that there could be a change in the conditions justifying that decision before the effect desired by the operator appears. In other words, there is the risk that a choice might later turn out to be wrong. Every operator makes forecasts to evaluate the current and expected conditions of a group of factors that can influence the price. In highly competitive markets, as soon as an operator finds out about the slightest change which might influence the price, he decides to sell or buy, thus changing the decision into a force which in fact affects the price.

The topic of the significance of the price is linked to the concept of the *liquidity* of an asset which refers to the degree of certainty with which its relative value may be evaluated (Burns, 1983). This depends on the existence of an active market which constantly expresses evaluations of that asset. Shares have high liquidity, since their relative values are almost continuously expressed, whereas a country house is an asset which is much less liquid, since there is no market on which it is continuously traded and it will be much more difficult to find out what it is worth. On markets where there is a large offer, supported by the holding of stocks, as well as a vast demand, the degree of liquidity is more accentuated. In general, the liquidity of a market depends on how easy it is to evaluate the quality and quantity of an asset (homogeneity of the asset); on there being a huge demand that must be satisfied with considerable urgency, to prevent certain supplies not finding buyers; on the possibility of holding stocks; and on low costs of transferring the asset from seller to buyer.

9.7.2. Risk, stocks and market of deferred delivery

Before making a decision, the operator analyses the present and future conditions of the market according to the information available to him. The risk will be much lower for those who check and make use of this information in good time and those who interpret it most rationally than for those who do not receive information or make bad use of it. Adam Smith finds the first reason for the existence of monopolistic behaviour in the control of information. Those who know about crucial aspects of the market will try to make sure that others do not get to know about them, thus trying to maximise their utility to the detriment of other operators, while at the same time reducing their risk.

As well as the most efficient use of information, there are other techniques for better risk management. Holding stocks may be seen as an expedient for limiting the consequences of unpredictable events, while on some markets it is also possible to resort to forms of insurance.[12] Another instrument is the conclusion of contracts whose terms (quantity, quality, place of delivery and above all price) are fixed *in advance* with respect to the physical exchange. The delivery is, in this case, deferred. A refiner may, for example, sell his products several months in advance, considering the price applied at that moment to be satisfactory for him. In this way he forgoes possible future increases but also protects himself from possible falls in price. In the competitive markets of commodities, *deferred delivery transactions* are a consolidated practice, and an increasingly extensive use of this may lead to a rapid expansion of so-called paper transactions.

Between the conclusion of a contract with deferred delivery and the physical passage of property an interval of varying length passes during which the seller is obliged to deliver and the buyer is obliged to collect the asset through the exchange of money. This dual obligation disappears at the moment of delivery. There is, however, another possibility. Before the expected expiry of delivery, the two counterparts may agree to conclude an *opposite* contract to the initial one, though identical as regards the terms of the transaction, thus neutralising the reciprocal obligations arising at the time of the first contract. The only element varying between the first and the second transaction is the price, determining a profit or a loss for the two operators. For example, if an operator A sells B 100 tons of diesel oil at the price of 20 for delivery after 3 months, after a few weeks, the two operators can agree on a contract where B sells the same quantities and qualities with the same expiry, thus perfectly offsetting the initial commitment. If the price goes up in the meantime, A will pay B the difference and vice versa if the price has fallen.

In very liquid markets the second contract does not have to be concluded between the two initial counterparts, although this might happen with a third operator. A, who intends to release himself from the initial commitment towards B with delivery at three months, may decide to purchase the same quantities with the same expiry from C, thus offsetting his commitments and substituting operator C in the initial sale to B. Since these transactions do not give rise to any physical consignment of the asset, they are known as "paper transactions". *Futures markets are formed by the multiplication of these transactions and their institutionalisation.*

9.7.3. Institutional characteristics and informative role

For the operators, futures markets are an efficient means of handling uncertainty, since their main function is to allow the reduction of the risk connected with the variability of the price of the asset handled on the physical market. The futures market has four fundamental characteristics:

Standardisation. Futures transactions are done according to contracts whose terms of quality, quantity, date and place of delivery are standardised and unchanging. Each contract gives the right to receive the assets, if purchased, or to deliver, if sold, at a future date.

Rarity of the deliveries. Only between 1% and 3% of the total futures transactions actually ends in delivery of the goods. In the majority of cases in fact opposite contracts are then opened so as to balance the commitments perfectly.

Liquidity. A high number of operators gives the futures markets a great liquidity with benefits on the degree of significance of the price.

Parallel movement of the prices. The only contractual term which is not standardised is the price. The difference between the price at the time of opening a contract and that of its closure determines the result of the operation. This value determines credit or debit to the operator. However price movements are not random since they follow the price trend formed on the physical market thanks to operations of arbitrage between the two markets made practicable by the possibility of actual delivery. Although operators rarely resort to this, the possibility of delivery is still an excellent link between the futures market and the physical market.

Since they refer to contracts whose delivery (rare but potentially always possible) is assigned to a time after their formulation, futures prices may be described as an *estimate of the price* in the future. Here we should dispel a common myth. The informative utility of these prices should not be evaluated on the basis of the capacity to predict the future price, which almost always differs due to the uncertainty and unpredictability of the markets. Futures contracts exist in fact for the purpose of those assets characterised by the high variability of their spot prices and thus the unpredictability of their trend. If futures prices were totally predictable, there would be little variability of spot prices, and soon this form of contracting would be eliminated. Indeed, we have already stressed that the efficiency of the operations of coverage (the first reason for the existence of futures markets) does not depend on the capacity to predict but on the constant relation between prices. The futures price is always an estimate, which may be considered as the most accurate possible given the information available at the time it is made. When new information on the conditions (current and future) of the market arrives, it is promptly taken into consideration by a large number of operators (including speculators) and variations in the futures price are made with the subsequent decisions of purchase or sale.

Examples of coverage on futures markets

Let us take the imaginary case of a London-based trading company which in mid-October purchases 100 tons of diesel oil which it intends to hold in stocks until January to satisfy possible demands from their clients. The agreed price is the spot price published that day by the "Petroleum Argus" of 140 doll./ton. Some days later, however, the choice turns out to be rash since news has arrived which suggests that prices are going to fall. The company could deal with this risk in three ways:

(a) immediately sell the 100 tons of diesel oil at 140 dollars and then buy it back at the beginning of January at lower prices. This decision avoids the risk of the possible depreciation, but leaves the door open to new ones: in fact it would be necessary to intervene first with a sale then with a purchase on the physical market, where the heterogeneity of the requirements of each individual operator is accentuated;

(b) conclude a sale with a deferred delivery at the beginning of January at the October spot price with a discount of 5 dollars/ton, *i.e.* 135 dollars. The potential purchaser would probably want a bigger discount if huge price falls are forecast. On the other hand there would be the same problems of heterogeneity as in the previous case. Another trader with exactly the opposite expectations would have to be found, if possible;

(c) contact a Commodities Exchange agent from whom to order a futures sale of a contract (100 tons of diesel oil) with expiry in January.

The futures operation opens on 15th October at 148 dollars. The stocks of diesel oil acquired on the physical market are not touched, and left to be sold in January. This sale goes through on 5th January at the spot price of 117 dollars/ton quoted by the "Petroleum Argus". The forecast of the falling prices is thus confirmed and the stocks have depreciated by a total of 23 dollars/ton (with a total loss of 2,300 dollars). The same day however, the operator orders the closure of the position opened in October on the futures. Through the purchase of a contract expiring in the same month, January, at the price of 117 dollars, the profit on the futures market (including the agent's commission) is 31 dollars/ton (148-117, with a profit of 3,100 dollars). The loss on the spot market is thus more than compensated by the gain on the futures market. The following diagram (where "–" means purchase and "+" sale) summarises the operation with values in dollars/ton.

Diagram 1

	15th October	5th January	
Market type			
physical	-140	+117	-23
futures	+148	-117	+31
(expiry January)			
		net gain	+8

Long hedging

In January a German diesel oil trader concludes a contract of sale with delivery at the end of March at 130 doll./ton. Being unable to hold stocks, only in March will he purchase the quantity to fulfil his commitments, thus risking having to pay a spot price higher than that already gained on the concluded sale. In order to attenuate this risk, the operator decides to make use of a futures market by long hedging (equal to the total that he has to deliver) with expiry in March at 128 doll./ton. At the end of March, as feared, prices go up and the trader purchases a batch of diesel oil at 139 doll./ton against a selling price fixed at 130. Thus the loss is of 9 doll./ton. At the same time, however, he orders his agent on the Commodities Exchange to close the position, opened three months before, on the futures market with a sale of the same contract expiring in March which quotes 140 doll./ton for that day, determining a gain of 12 doll./ton. The operation may be summarised by the following diagram:

Diagram 2

	10th January	31st March	
Market type			
physical	+130	+139	-9
futures	-128	-140	+12
(expiry March)			
		net gain	+3

Actual delivery

The previous examples have been chosen to illustrate the utility of the coverage: *understood as taking on an opposite commitment on the futures market to what the operator has on the physical market*. In both cases the operation is concluded with a net gain: deriving from a relatively favourable trend in futures prices. For a positive final result to emerge, the basis (difference between the futures price and spot price) must tend to zero as the expiry date draws near. When the relation between the prices has not been advantageous, as the expiry date of the open contract approaches, there is an easy way out: *actual delivery*. This rarely takes place, though the possibility of opting for this solution exists and must exist. We may see why in the following example: in mid-January an operator decides to cover his own stocks of diesel oil just purchased at the spot price of 133 doll./ton with a sale on the futures market at 134 doll. expiry March. At the end of March however, instead of the prices falling as the operator feared, they rise: spot prices to 135 dollars and futures contracts to 140. If he decides to close the sale position on the futures market opened in January (with expiry March) at 134 doll. with a purchase at 140, the operation would turn out as follows:

Diagram 3

	15th January	28th March	
Market type			
physical	-133	+135	-2
futures	-134	-140	+6
(expiry March)			
		net gain	+4

In this way there would be a loss in the futures (-6) greater than the re-evaluation of the stocks (+2), reflected in an increase of the basis. Instead of proceeding with the second futures operation, the operator prefers to resort to the possibility of performing the physical delivery on the futures contract at the price fixed in January of 134 dollars. In this way he will gain 1 dollar less than the 2 that he would have been able to gain by not covering (*i.e.* operating solely on the spot), but limiting the net loss he would have suffered if the futures operation had been concluded, started on the basis of expectations which later turned out to be mistaken. Actual delivery is not only important for those who cover to enable them to limit losses due to bad forecasts. Even more important is *the fact that it leads to a close relation between the futures price and the spot price: ensuring that the basis tends towards zero as the expiry date approaches*. Going back to the previous example where a spot price of 135 dollars is quoted in March and a futures contract expiring in the same month is quoted at 140 dollars, operations of arbitrage are triggered between the two markets in order to re-establish the balance between the two prices (the situation illustrated in the diagram should therefore be temporary).

9.7.4. Development of the oil futures markets

The first futures markets in oil transactions emerged in the United States at the end of the 1860s, in response to annual variations of 40%-60% in prices. A spot market based on pipeline certificates appeared in January 1871 at Titusville and later in about twenty towns from Chicago to New York, where the Petroleum Exchange began to operate in 1877. In the 1880s Wall Street became the most important futures market, especially after the formation of the New York Mercantile Exchange, known by the acronym Nimex, from the merging of the Mining Stock and Petroleum Exchanges. With characteristics very similar to those of today – negotiations from 10 am to 5 pm, the same dimension of lots (1,000 bbl) and products handled (futures, options etc.) – it can trade quantities up to 1 mill. bbl against a physical production of 0.1. With the consolidation of the near monopoly of Standard Oil these first experiences of the futures market came to a halt, only to re-emerge strongly about a century later, when the price curve began to fluctuate in ways not dissimilar to those of the past.

There are two main stock exchanges: New York, organised in the New York Mercantile Exchange (Nimex), and London, organised in the International Petroleum Exchange (IPE). Nimex handles contracts for diesel oil, unleaded petrol, fuel oil low in sulphur and particularly for oil with reference to West Texas Intermediate (WTI) and delivery in the Cushing plants in Oklahoma. The IPE handles contracts for diesel oil and crude with reference to Brent.[13]

Table 9.6. clearly shows the extraordinary development of the futures market on both sides of the Atlantic. In a decade, the trading of crude on the Nimex has increased about 100 times, in 1994 exceeding 106 mill. bbl/d (with a total of 27 million contracts), compared with a physical production of the same benchmark crude, WTI, of around 1 mill. bbl/d. Crude trading on London's IPE is no less important: 40 mill. bbl/d compared with a Brent crude production of around 1 mill. bbl/d. The vast and continuing importance of the trading is the best proof of the success of these markets, of the high liquidity that they are able to guarantee and the trust that is placed in them by the operators. The range and number of companies accredited on New York and London are wide and varied. Two thirds are companies involved in different sectors of the oil industry (integrated, refiners, distributors, producers, traders, etc.) and one third are pure speculators, a percentage which increases significantly in times of the greatest tension. What effects has the enormous growth of the futures market had on the oil market? The question is subject to great controversy today between those who stress the capacity of these markets to reduce the risk for the operators and those who highlight the negative outcome on the actual stability of the prices. One fact is certain: the development of these markets has greatly reduced the barriers to entry and cost of speculation. The dimension of a cargo of crude purchased on the international market was once not less than 600 thousand barrels, with a cost which could vary from 6 to 20 million dollars. In addition, the majority of companies were reluctant to deal with suppliers or buyers outside their environment and thus the market was not very liquid. Futures markets have reduced the minimum quantity negotiable to just 1,000 barrels and removed the need for a well-established reputation (Verleger, 1987, p. 182). All that is needed is a cash deposit of 5,000 dollars to gain one.

Table 9.6. *Transactions in crude and product futures markets*[1] *(mill. bbl/d handled)*

	Nimex market			IPE market	
	Crude (WTI)	Gasoil	Unleaded gasoline	Crude (Brent)	Gasoil
1979	–	0.13	–	–	–
1980	–	0.95	–	–	–
1981	–	3.95	–	–	0.61
1982	–	1.75	–	–	1.84
1983	1.69	7.41	–	0.10	1.80
1984	1.36	8.37	0.00	0.01	1.58
1985	15.92	8.83	0.53	0.02	1.51
1986	33.25	13.10	1.76	0.01	2.77
1987	57.86	17.04	8.16	0.00	3.26
1988	74.84	19.58	13.86	2.18	4.58
1989	81.81	22.87	17.87	6.63	5.79
1990	94.37	25.41	20.74	16.14	7.67
1991	83.03	26.40	21.78	21.00	8.47
1992	83.77	31.77	26.49	24.30	10.19
1993	99.33	33.35	29.16	34.99	10.70
1994	105.60	36.44	30.73	40.01	11.25
1995	96.49	32.41	29.27	39.60	15.46

1. Data for commercial year of 252 days. For trade on IPE market less than 252 days were considered for 1981 for gasoil and up to 1988 for Brent crude.

9.8. COMPETITION, INSTABILITY, UNCERTAINTY

To conclude, vertical disintegration, lower degree of concentration of the supply, fragmentation of the decisional units, and no or weak (OPEC) co-ordination of the management of volumes of supply are the main reasons why conditions of full competition prevail on the oil market. This outcome has been brought about by other facts that are

Figure 9.4. *Oil sales in the international markets (%)*

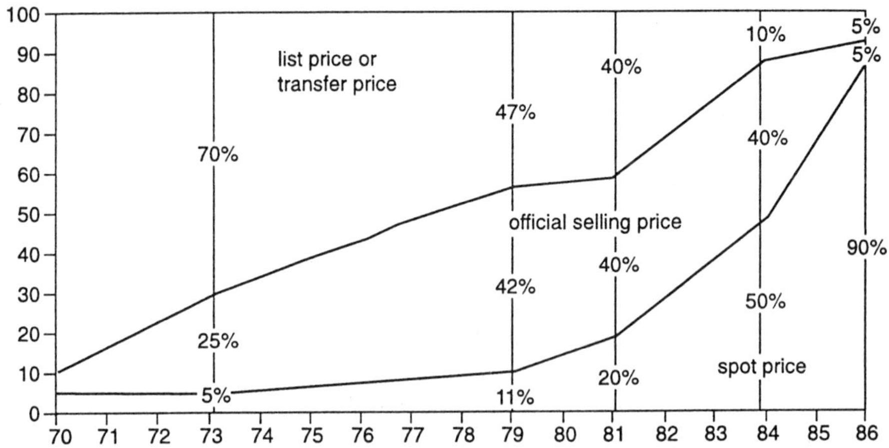

worth examining. *Firstly*, fragmentation of the decisional units also on the side of demand: due to the widening of the basin of consumers, the entry into the market of new subjects (intermediaries, large consumers, final distributors), and the downsizing of the largest companies. *Secondly*, greater geographical dispersion of the production and export units with subsequent greater difficulty in handling the so-called spatial competition. *Thirdly*, globalisation of the markets, due to the disappearance of any regulatory or tariff-based impediment which still prevented – as in the American case – the free circulation of crudes until the early Eighties. Today the oil market truly is a single world market.

Each of these reasons has contributed, singly or in combination, to increasing the degree of competition in the oil market and in determining – as is typical in this form of market – a closer short-term adherence between prices and variations in demand. From here comes an accentuated volatility of the prices which may be attributed also to the increasing instability and unpredictability of the levels of final consumption (total and for type of product); to the natural difficulty in programming (in good time) the levels of supply of crude; and to the differentiation of quality with respect to the changed needs of the consumers (for environmental reasons or due to changes in refining technologies). Besides these and other specific reasons, the volatility of the prices may theoretically be attributed to the conditions of indeterminacy and instability which are naturally part of the oligopolistic markets, in the absence of any form (implicit or explicit) of decisional co-ordination, as we examined in the second chapter.

Lastly, it is the type of contracts through which commercial transactions take place which increase the volatility of the prices. Once, these mainly happened with long-term contracts at list prices (Fig. 9.4.). Variations of these were shifted to the exchange values of the existing contracts with specific pre-fixed methods. Today the situation is the reverse. Almost all the transactions in fact take place with *spot contracts* at prices fixed daily. What are the reasons for this drastic change and what are its effects? There is basically one reason: in conditions of uncertainty and volatility of the prices, no operator, seller or buyer, is willing to commit himself unless it is in the short term and for single lots. Long-term contracts would be too risky or difficult to handle. As for the outcomes, two in particular are worth mentioning. First of all, the nature of the contracts now accentuates the very instability of the prices from which it drew its motivation. A slightest disturbance on the crude market is transmitted immediately to all existing and subsequent transactions: at the same time shifting from crudes to products (or vice versa, if the disturbance was caused by even just one of these) and requiring a certain time before the structure of the relative prices of the crudes and the products regains an internal coherence. The advantages that speculators draw from these continuous alterations do not correspond to identical advantages for the industrial operators who find themselves having to decide and plan with no inkling of what will happen the next day. Secondly, the instability of oil prices reverberates not only on all of the transactions on this market, but also on substitute markets (particularly the methane gas market) and on secondary markets (electricity) where the evaluation of the goods and services exchanged is widely correlated to oil prices.

As shown in Figure 9.5., in the decade 1986-96 oil prices were subject to wide and continuous fluctuations in an average range of 12-18 doll./bbl for the benchmark crudes Brent and Dubai: with minimum levels close to 10 doll./bbl and peaks (during the Gulf

Figure 9.5. *Weekly spot prices for Brent and Dubai (doll./bbl)*

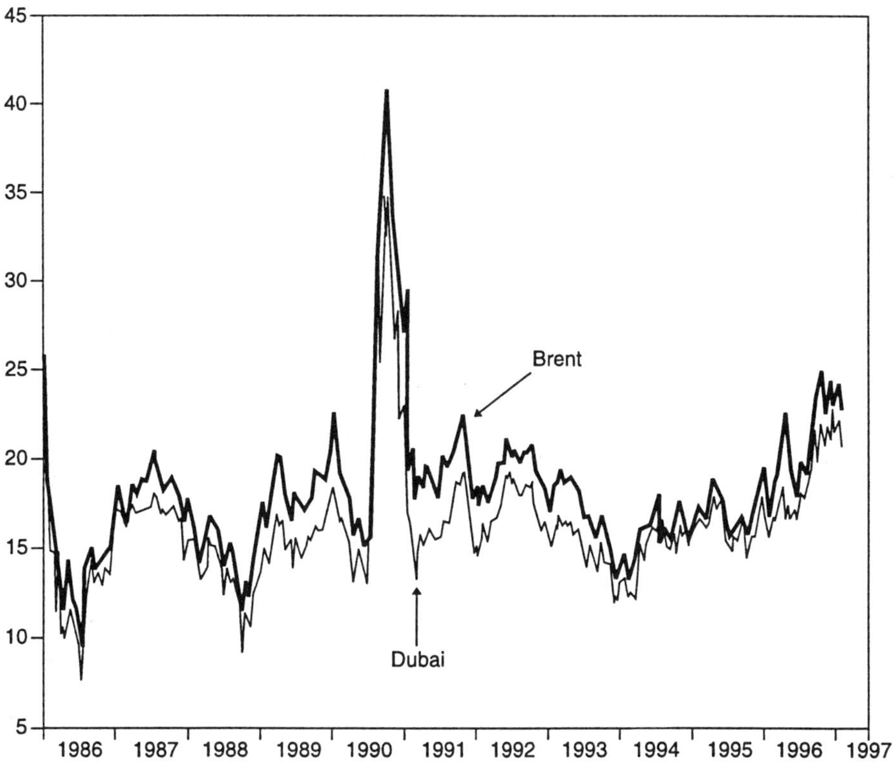

crisis) exceeding 40 doll./bbl. The interpolating line appears to be rising with end-of-period values exceeding 20 doll./bbl. Over the ten-year period, as many as eight had two-figure annual percentage fluctuations, of which six had higher than 20%.

Even more significant is the instability – in terms of frequency and amplitude of the fluctuations – in the prices of oil products. These markets in fact are subject to the tensions of the economic trend both of the consumer markets (downstream) and of the crude market (upstream): with price fluctuations that in turn have a retroactive effect, though in a more attenuated way, on crude quotations.

In the period in question, petrol quotations have fluctuated from 160-240 doll./ton, those of diesel oil from 130-220, those of fuel oil with low sulphur content from 65-110 doll./ton. The greater competition in the products markets, also due to the excess of refining capacity; the high costs of the environmental regulations, which can not always be transferred onto the final prices; the gradual reduction of the price differentials between light fractions and heavy fractions, on which the reasons of convenience of investments in upgrading of the refineries were based, are some of the reasons for the high instability and the depression of refining margins, and thus for the structural crisis which has hit this segment. Calculated for a complex refinery, relative to the processing of Brent crude, they have fluctuated from 0-4.0 doll./bbl.

Uncertainty, indeterminacy and instability of prices are the structural datum which must be faced each day by the players on the oil scene, under a competitive pressure that constantly modifies the reasons of convenience of the long-term choices. The markets are increasingly competitive with entries – in many areas of activity – at low or no investment. The incumbents thus find themselves having to bear the costs of continuity and sunken costs which are difficult to eliminate or alienate (it is more expensive to close a refinery than to keep it active, even though not profitable), while the newcomers take advantage of their presence with "hit and run" actions enabled by company and sectorial destructuring.

Thus the price dynamic often appears not to be tied by any expectations or by market fundamentals themselves, though still making reference to them. It sometimes happens that these are deduced from the first and not vice versa. The traditional interpretations often do not help to understand what is happening today or what could happen tomorrow. The *financialisation of the markets* has partly modified the nature and the mechanisms of their operation, exaggerating the speed and the significance of their reaction even to the smallest event. Exogenous variables, different from the normal fundamentals, can thus – in certain circumstances – influence the trends of the markets more than these. The reasons driving financial investors are not to be attributed to the events and the expectations on the physical market, though it is to this that they refer. In a market which appears *virtual* , the values of exchange on the futures market have a retroaction on the values of exchange on the physical market which in turn influence the former, in an interactive process where it is not easy to understand "who and what determines what".

9.9. OLD AND NEW CHALLENGES

Uncertainty is the great challenge facing the oil world. The oil companies, more than any other players, are fully aware of this: in the knowledge that deciding amidst uncertainty is highly risky, while doing so with the illusion of certainty can be – in an industry at such a high intensity of capital – absolutely ruinous. The dominating aspect of company choices, both operative and strategic, has become the management of risks, in a market where the external competitive conflict has widely replaced the hierarchical order within the companies. The question that has crossed the entire history of the oil industry – of how to reconcile competition (short-term) with stability (long-term) of the markets – is re-emerging today to its full extent.

The chances of success will greatly depend on the companies' ability to take up the new challenges and the new opportunities of the markets: reinforcing themselves on a level of technological innovation, production efficiency, organisational flexibility, and entrepreneurial strategies. Only in this way will they be able to liberate the immense financial resources that are a necessary requirment for the next 15 years – a constant 180 bill. dollars per year – to guarantee a harmonious development of the world oil and methane systems. The possibility of managing to do this is not however only linked to the forces of technique and the mechanisms of economics but also to political decisions.

The duty of politics is above all to pursue with tenacity the co-operation and the search for common interests with the producer countries, in order to allow, in a framework of greater certainty, the necessary return of foreign capital from where the oil reserves are mainly located. The whole world and the West in particular has an increasing need for Middle Eastern oil. The cost-opportunity of its exclusion would be unsustainable. The dissuasion of current investments and the shrinking of incremental supply is a policy destined to give market fundamentals some good extra reasons for a rise in prices. It is imperative that policies pursue the path of dialogue with producing countries. Only this will provide a stable solution to the regional problems of reliability, of economic growth, and of political stability: the first condition for a harmonious development of foreign investments. The mechanisms of international interdependence tie rich and poor countries, consumer and producer countries, increasingly closer together, with the result that the opposite position for everyone is costly and unsuccessful. The contradictions which are still a feature of the international energy scene may be overcome by the development of technology and industries, but no less by politics, and particularly the politics of international co-operation and dialogue.

Notes

1. We are referring to countries with market economies. It is a different question for developing countries and for the former USSR states in transition towards the market.

2. The most significant case was the suppression in France of the 1928 law which gave the state the monopoly of the importation of oil and products.

3. The most significant cases are: British Petroleum Co. entirely privatized in three stages in 1979 (5% with profits for £551m), in 1983 (7% for £566m) and in 1987 (36.8% for £7,240m); YPF in Argentina collocated for 45.3% in 1993 for more than $3bill.; ENI in Italy with two collocations: the first in 1995 (about 15% for Lit. 6,300 bill.), the second in 1996 (about 18% for Lit. 8,500 bill.); Repsol in Spain was privatized for 19% in 1995 and for 11% in 1996, thus reducing state participation to 10%.

4. The figure refers to 1994 with a fiscal revenue for the G7 of $240 bill. compared with $120 bill. profits from oil exports from the OPEC countries.

5. Saudi Arabia tried in vain, especially during the second crisis, to get a long-term strategy adopted by the OPEC countries.

6. Being in competition $p = CMg$ the "degree of monopoly" is equal to zero. In a monopoly there are no univocal values of Lerner's index, since there are no situations given by the elasticity of the demand. Its maximum value is the unit.

7. The data comes from Brown L. et al. (1996), *Vital Signs 1996*, from the Worldwatch Institute.

8. Domestic investments in the United States in the upstream oil sector fell from annual values of around $99 bill. in 1980-82 to values in 1994 of around $50 bill. Exploratory wells fell from 15,500 to 2,800; the total number of wells from 82,000 to 19,500.

9. Paradoxically, studies and interest on this matter have always been very limited. Only a few surveys carried out by the Commission of Brussels in the Seventies are available.

10. Internal ex-refinery prices higher than international prices would obviously lead to bringing in purchases from abroad and vice versa if they were lower.

11. Particularly: fuel oil with low sulphur content for thermoelectric and industrial uses, diesel fuel for domestic use.

12. In these cases, against the payment of a premium, the insurer takes on the risk of an event occurring which is negative for the insured party; a typical example is the farmer who fears weather damage to his crops.

13. In this contract physical delivery is not contemplated – in the case of non-closure of the initial obligations – but the liquidation of the positions through resorting to an index calculated on the basis of the forward delivery price for the first expiry of the Brent.

The draft of this chapter (particularly section 9.7. and relative sub-sections) was done with the collaboration of Davide Tabarelli.

Tables of statistics

Caution

The data given in the historic series may sometimes differ from those of the text due to the different statistic source and the different data processing methods used.

Table 1 – *World crude oil prices, 1859-1998*[1]

Year	Price (doll./bbl)	Change doll./bbl	Change %	Year	Price (doll./bbl)	Change doll./bbl	Change %	Year	Price (doll./bbl)	Change doll./bbl	Change %
1859	16.00			1909	0.70	0.0	−2.8	1959	1.92	−0.2	−7.7
1860	9.59	−6.4	−40.1	1910	0.61	−0.1	−12.9	1960	1.86	−0.1	−3.1
1861	0.49	−9.1	−94.9	1911	0.61	0.0	0.0	1961	1.80	−0.1	−3.2
1862	1.05	0.6	114.3	1912	0.74	0.1	21.3	1962	1.80	0.0	0.0
1863	3.15	2.1	200.0	1913	0.95	0.2	28.4	1963	1.80	0.0	0.0
1864	8.06	4.9	155.9	1914	0.81	−0.1	−14.7	1964	1.80	0.0	0.0
1865	6.59	−1.5	−18.2	1915	0.64	−0.2	−21.0	1965	1.80	0.0	0.0
1866	3.74	−2.9	−43.2	1916	1.10	0.5	71.9	1966	1.80	0.0	0.0
1867	2.41	−1.3	−35.6	1917	1.56	0.5	41.8	1967	1.80	0.0	0.0
1868	3.62	1.2	50.2	1918	1.98	0.4	26.9	1968	1.80	0.0	0.0
1869	5.64	2.0	55.8	1919	2.01	0.0	1.5	1969	1.80	0.0	0.0
1870	3.86	−1.8	−31.6	1920	3.07	1.1	52.7	1970	1.80	0.0	0.0
1871	4.34	0.5	12.4	1921	1.73	−1.3	−43.6	1971	2.13	0.3	18.3
1872	3.64	−0.7	−16.1	1922	1.61	−0.1	−6.9	1972	2.48	0.4	16.4
1873	1.83	−1.8	−49.7	1923	1.34	−0.3	−16.8	1973	2.81	0.3	13.3
1874	1.17	−0.7	−36.1	1924	1.43	0.1	6.7	1974	10.98	8.2	290.7
1875	1.35	0.2	15.4	1925	1.68	0.3	17.5	1975	10.71	−0.3	−2.5
1876	2.52	1.2	86.7	1926	1.88	0.2	11.9	1976	11.63	0.9	8.6
1877	2.38	−0.1	−5.6	1927	1.30	−0.6	−30.9	1977	12.57	0.9	8.1
1878	1.17	−1.2	−50.8	1928	1.17	−0.1	−10.0	1978	12.91	0.3	2.7
1879	0.86	−0.3	−26.5	1929	1.27	0.1	8.5	1979	29.82	16.9	131.0
1880	0.94	0.1	9.3	1930	1.19	−0.1	−6.3	1980	35.85	6.0	20.2
1881	0.92	0.0	−2.1	1931	0.65	−0.5	−45.4	1981	34.29	−1.6	−4.4
1882	0.78	−0.1	−15.2	1932	0.87	0.2	33.8	1982	31.76	−2.5	−7.4
1883	1.10	0.3	41.0	1933	0.67	−0.2	−23.0	1983	28.72	−3.0	−9.6
1884	0.85	−0.3	−22.7	1934	1.00	0.3	49.3	1984	28.08	−0.6	−2.2
1885	0.88	0.0	3.5	1935	0.97	0.0	−3.0	1985	27.52	−0.6	−2.0
1886	0.71	−0.2	−19.3	1936	1.09	0.1	12.4	1986	13.84	−13.7	−49.7
1887	0.67	0.0	−5.6	1937	1.18	0.1	8.3	1987	17.29	3.5	24.9
1888	0.65	0.0	−3.0	1938	1.13	−0.1	−4.2	1988	13.49	−3.8	−22.0
1889	0.77	0.1	18.5	1939	1.02	−0.1	−9.7	1989	16.34	2.9	21.1
1890	0.77	0.0	0.0	1940	1.02	0.0	0.0	1990	21.11	4.8	29.2
1891	0.56	−0.2	−27.3	1941	1.14	0.1	11.8	1991	17.38	−3.7	−17.7
1892	0.51	−0.1	−8.9	1942	1.19	0.1	4.4	1992	17.60	0.2	1.3
1893	0.60	0.1	17.6	1943	1.20	0.0	0.8	1993	15.20	−2.4	−13.6
1894	0.72	0.1	20.0	1944	1.21	0.0	0.8	1994	14.90	−0.3	−2.0
1895	1.09	0.4	51.4	1945	1.05	−0.2	−13.2	1995	16.32	1.4	9.5
1896	0.96	−0.1	−11.9	1946	1.20	0.2	14.3	1996	19.56	3.2	19.9
1897	0.68	−0.3	−29.2	1947	2.05	0.9	70.8	1997	18.77	−0.8	−4.0
1898	0.80	0.1	17.6	1948	1.99	−0.1	−2.9	1998	12.25	−6.5	−34.7
1899	1.13	0.3	41.3	1949	1.80	−0.2	−9.5				
1900	1.19	0.1	5.3	1950	1.75	−0.1	−2.8				
1901	0.96	−0.2	−19.3	1951	1.75	0.0	0.0				
1902	0.80	−0.2	−16.7	1952	1.75	0.0	0.0				
1903	0.94	0.1	17.5	1953	1.93	0.2	10.3				
1904	0.86	−0.1	−8.5	1954	1.93	0.0	0.0				
1905	0.62	−0.2	−27.9	1955	1.93	0.0	0.0				
1906	0.73	0.1	17.7	1956	1.93	0.0	0.0				
1907	0.72	0.0	−1.4	1957	2.01	0.1	4.1				
1908	0.72	0.0	0.0	1958	2.08	0.1	3.5				

1. 1859-1944: United States market prices; 1945-1972: posted prices of Arabian Light; 1973-1984: official selling prices of Arabian Light; 1985-1998: spot prices of Arabian Light.

Source: R.I.E. Database.

Table 2 – *Estimate of ultimate recoverable reserves of conventional oil in the world*

Year	Author	bill. bbl
1942	Pratt, Weeks & Stebinger	600
1946	Duke	400
1946	Pogue	555
1948	Weeks	610
1949	Levorsen	1,500
1949	Weeks	1,010
1953	Mac Naughton	1,000
1956	Hubbert	1,250
1958	Weeks	1,500
1959	Weeks	2,000
1965	Hendricks (USGS)	2,480
1967	Ryman (Esso)	2,090
1968	Shell	1,800
1968	Weeks	3,550
1969	Hubbert	1,350-2,100
1970	L,G, Weeks (National Academy of Science)	2,200
1970	Moody (Mobil)	1,800
1971	Weeks	2,290
1971	Warman (BP)	1,200-2,000
1972	Moody & Emmerich (Mobil)	1,800-1,900
1972	Bauquis, Brasseur & Masseron (IFP)	1,950
1972	Weeks	3,560
1972	Linden (Inst, Gas Technology)	2,945
1973	Schweinfurth (USGS)	2,964
1974	Bonillas (Socal)	2,000
1974	Howitt (BP)	1,800
1975	Odell	3,575-4,233
1975	Moody (Mobil)	2,000
1975	Grossling (USGS)	2,600-6,500
1977	World Energy Conference «Consensus»	2,265
1977	Nelson (Socal)	2,000
1978	De Bruyne (Shell)	1,600
1978	Klemme (Weeks)	1,750
1978	Nehring (Rand)	1,700-2,300
1979	Nehring (Rand)	1,600-2,000
1979	Halbouty & Moody	2,130
1979	Meyrhoff	2,190
1979	Roorda (Shell Oil)	2,400
1980	World Energy Conference	2,595
1981	Halbouty	2,275
1981	Strickland (Conoco)	2,071
1981	Colitti (Agip)	2,082
1982	Exxon	1,800-3,000
1982	Nehring (Rand)	2,400
1983	Masters & Root (USGS) Dietzman (AIE)	1,722
1984	Martin (BP) (al 2020)	1,685
1984	Ivanhoe	1,700
1984	Desprairies e altri (IFP)	2,050
1987	Masters (USGS)	1,744
1996	Colitti-Simeoni (ENI)	1,690
1998	US Geological Survey	2,200
1998	Perrodon, Laherrère, Campbell	2,700

Table 3 – *Estimated world proven crude oil reserves, 1936-1998 (bill. bbl)*

Year	North America [1]	Latin America	Western Europe	Eastern Europe [2]	Middle East	Africa	Asia and Oceania	World total	OPEC [3]
1936	10.6	1.4	n.d.	2.8	4.7	n.d.	n.d.	22.0	
1948	21.6	10.1	0.1	6.5	28.6	0.1	1.3	68.3	
1949	23.8	10.9	0.2	4.6	32.6	0.1	1.4	73.6	
1950	25.8	11.4	0.3	4.7	32.4	0.2	1.6	76.5	
1951	26.5	11.5	0.4	8.0	41.6	0.2	1.8	89.9	
1952	28.9	12.7	0.5	7.9	51.3	0.2	2.0	103.4	
1953	29.7	11.9	0.5	9.4	64.8	0.2	2.1	118.6	
1954	30.7	13.2	0.6	9.5	78.2	0.2	2.6	135.0	
1955	31.8	14.4	1.0	10.0	97.5	0.1	2.7	157.5	
1956	32.5	15.5	1.3	10.8	126.3	0.2	3.0	189.6	
1957	33.2	17.5	1.3	25.0	144.5	0.3	6.1	228.0	
1958	33.2	21.0	1.3	26.2	169.6	0.8	8.6	260.6	
1959	33.7	21.8	1.4	27.7	174.0	4.1	9.6	272.3	
1960	36.6	25.1	1.8	24.8	183.1	8.0	11.8	291.2	218.1
1961	36.8	24.7	1.8	24.7	188.1	9.6	11.6	297.3	223.6
1962	36.0	24.2	1.9	30.2	193.8	12.3	12.3	310.7	231.9
1963	36.6	24.3	2.3	33.2	207.0	16.3	12.6	332.4	248.0
1964	36.3	25.5	2.8	34.2	211.4	19.5	12.8	342.5	254.7
1965	38.9	25.2	2.6	36.2	214.8	23.2	13.0	353.8	260.4
1966	39.7	27.1	2.4	39.0	234.6	32.6	14.0	389.4	290.2
1967	40.2	26.9	2.8	36.9	248.5	42.4	14.0	411.6	310.4
1968	40.6	28.8	2.6	60.9	270.1	44.6	16.6	464.1	331.6
1969	38.4	29.2	2.4	61.0	332.8	48.4	15.7	528.0	391.6
1970	49.8	26.2	6.9	61.0	336.2	51.1	17.5	548.7	399.4
1971	46.5	30.5	9.6	58.8	346.4	54.6	28.8	575.2	411.0
1972	46.5	31.4	10.6	55.9	350.0	59.5	29.0	582.9	417.2
1973	44.7	29.5	17.5	52.3	350.0	61.6	31.7	587.2	419.3
1974	43.7	30.5	25.7	59.3	403.4	60.5	36.0	659.1	482.2
1975	39.8	36.1	25.6	61.9	387.1	59.1	38.6	648.1	463.1
1976	37.1	32.6	24.9	62.9	380.2	55.0	41.1	633.8	453.6
1977	35.5	40.2	27.6	62.0	376.8	53.7	40.4	636.2	447.9
1978	33.8	41.5	24.6	61.4	370.7	52.9	41.8	626.8	443.1
1979	36.6	57.1	24.1	62.9	362.7	53.2	39.0	635.6	435.6
1980	36.2	74.0	22.9	65.8	362.9	52.5	39.0	653.3	434.6
1981	36.7	84.3	22.5	65.7	365.2	56.4	39.3	670.1	438.3
1982	34.9	89.7	21.5	65.4	388.6	57.9	39.0	697.0	467.4
1983	34.3	90.8	22.1	65.3	397.1	57.3	38.7	705.5	475.3
1984	34.4	92.8	22.7	64.8	431.0	56.3	38.2	740.2	510.0
1985	33.9	118.6	22.2	64.4	431.9	56.2	38.5	765.7	535.8
1986	33.7	119.1	20.0	62.3	537.7	57.3	38.6	868.7	643.0
1987	34.1	121.7	17.4	60.6	567.8	57.1	39.2	897.7	674.0
1988	33.6	121.1	18.0	60.0	653.4	57.8	45.1	989.1	760.5
1989	33.3	120.6	17.0	59.8	663.3	58.8	47.5	1000.3	770.6
1990	32.7	122.2	17.0	58.8	662.0	59.9	48.3	1001.0	771.4
1991	30.8	125.1	17.1	58.7	662.9	62.5	48.5	1005.5	776.7
1992	29.0	129.8	17.7	59.1	663.4	63.9	44.4	1007.4	773.8
1993	31.0	130.9	18.0	59.1	663.3	63.6	43.8	1009.8	774.4
1994	31.6	131.3	19.9	59.1	665.8	68.5	43.3	1019.5	777.5
1995	31.7	131.6	21.1	59.1	665.1	75.5	44.2	1028.4	785.1
1996	31.3	138.8	18.7	59.1	676.0	75.4	43.5	1042.9	802.0
1997	31.2	140.2	18.1	59.2	676.8	75.5	43.5	1044.5	804.9
1998	32.0	152.7	18.7	59.0	672.0	81.3	44.3	1060.0	808.0

1. United States and Canada. 2. Former Soviet Union and Eastern Europe. 3. Effective Member Countries.

Source: 1936:A.E. Dustan, *Science of Petroleum*; 1948-1959: API, *Basic Petroleum Data Book*, 1996; 1960-1997: OPEC, "Annual Statistical Bulletin", various years; 1998: R.I.E. estimates.

Table 4 – *Estimated world proven crude oil reserves, 1936-1998 (%)*

Year	Nord America	Latin America	Western Europe	Eastern Europe	Middle East	Africa	Asia and Oceania	World total	OPEC
1936	48.2	6.4	n.d.	12.7	21.4	n.d.	n.d.	100.0	
1948	31.6	14.8	0.1	9.5	41.9	0.1	1.9	100.0	
1949	32.3	14.8	0.3	6.3	44.3	0.1	1.9	100.0	
1950	33.7	14.9	0.4	6.1	42.4	0.3	2.1	100.0	
1951	29.5	12.8	0.4	8.9	46.3	0.2	2.0	100.0	
1952	27.9	12.3	0.5	7.6	49.6	0.2	1.9	100.0	
1953	25.0	10.0	0.4	7.9	54.6	0.2	1.8	100.0	
1954	22.7	9.8	0.4	7.0	57.9	0.1	1.9	100.0	
1955	20.2	9.1	0.6	6.3	61.9	0.1	1.7	100.0	
1956	17.1	8.2	0.7	5.7	66.6	0.1	1.6	100.0	
1957	14.6	7.7	0.6	11.0	63.4	0.1	2.7	100.0	
1958	12.7	8.1	0.5	10.1	65.1	0.3	3.3	100.0	
1959	12.4	8.0	0.5	10.2	63.9	1.5	3.5	100.0	
1960	12.6	8.6	0.6	8.5	62.9	2.7	4.0	100.0	74.9
1961	12.4	8.3	0.6	8.3	63.3	3.2	3.9	100.0	75.2
1962	11.6	7.8	0.6	9.7	62.4	4.0	4.0	100.0	74.6
1963	11.0	7.3	0.7	10.0	62.3	4.9	3.8	100.0	74.6
1964	10.6	7.5	0.8	10.0	61.7	5.7	3.7	100.0	74.4
1965	11.0	7.1	0.7	10.2	60.7	6.6	3.7	100.0	73.6
1966	10.2	7.0	0.6	10.0	60.2	8.4	3.6	100.0	74.5
1967	9.8	6.5	0.7	9.0	60.4	10.3	3.4	100.0	75.4
1968	8.7	6.2	0.6	13.1	58.2	9.6	3.6	100.0	71.4
1969	7.3	5.5	0.5	11.6	63.0	9.2	3.0	100.0	74.2
1970	9.1	4.8	1.3	11.1	61.3	9.3	3.2	100.0	72.8
1971	8.1	5.3	1.7	10.2	60.2	9.5	5.0	100.0	71.4
1972	8.0	5.4	1.8	9.6	60.1	10.2	5.0	100.0	71.6
1973	7.6	5.0	3.0	8.9	59.6	10.5	5.4	100.0	71.4
1974	6.6	4.6	3.9	9.0	61.2	9.2	5.5	100.0	73.2
1975	6.1	5.6	4.0	9.5	59.7	9.1	6.0	100.0	71.4
1976	5.9	5.1	3.9	9.9	60.0	8.7	6.5	100.0	71.6
1977	5.6	6.3	4.3	9.7	59.2	8.4	6.3	100.0	70.4
1978	5.4	6.6	3.9	9.8	59.1	8.4	6.7	100.0	70.7
1979	5.8	9.0	3.8	9.9	57.1	8.4	6.1	100.0	68.5
1980	5.5	11.3	3.5	10.1	55.6	8.0	6.0	100.0	66.5
1981	5.5	12.6	3.4	9.8	54.5	8.4	5.9	100.0	65.4
1982	5.0	12.9	3.1	9.4	55.8	8.3	5.6	100.0	67.1
1983	4.9	12.9	3.1	9.2	56.3	8.1	5.5	100.0	67.4
1984	4.7	12.5	3.1	8.7	58.2	7.6	5.2	100.0	68.9
1985	4.4	15.5	2.9	8.4	56.4	7.3	5.0	100.0	70.0
1986	3.9	13.7	2.3	7.2	61.9	6.6	4.4	100.0	74.0
1987	3.8	13.6	1.9	6.7	63.2	6.4	4.4	100.0	75.1
1988	3.4	12.2	1.8	6.1	66.1	5.8	4.6	100.0	76.9
1989	3.3	12.1	1.7	6.0	66.3	5.9	4.8	100.0	77.0
1990	3.3	12.2	1.7	5.9	66.1	6.0	4.8	100.0	77.1
1991	3.1	12.4	1.7	5.8	65.9	6.2	4.8	100.0	77.2
1992	2.9	12.9	1.8	5.9	65.9	6.3	4.4	100.0	76.8
1993	3.1	13.0	1.8	5.9	65.7	6.3	4.3	100.0	76.7
1994	3.1	12.9	1.9	5.8	65.3	6.7	4.3	100.0	76.3
1995	3.1	12.8	2.1	5.7	64.7	7.3	4.3	100.0	76.3
1996	3.0	13.3	1.8	5.7	64.8	7.2	4.2	100.0	76.9
1997	3.0	13.4	1.7	5.7	64.8	7.2	4.2	100.0	77.1
1998	3.0	14.4	1.8	5.5	63.4	7.7	4.2	100.0	76.2

Table 5 – *World crude oil production by area, 1860-1998*[1]

Year	North America[2]	Latin America	Western Europe	Eastern Europe[3]	Middle East	Africa	Asia and Oceania	World total	OPEC[4]
					'000 bbl/d				
1860	1.4	-	..	..	-	-	-	1.4	
1865	7.1	-	..	0.3	-	-	-	7.4	
1870	15.1	-	..	0.8	-	-	-	15.9	
1875	33.4	-	..	2.6	-	-	..	36.0	
1880	73.0	-	..	9.2	-	-	0.1	82.2	
1885	60.6	-	..	40.1	-	-	0.1	100.7	
1890	127.7	-	..	81.8	-	-	0.5	210.0	
1895	146.9	-	..	132.4	-	-	4.8	284.1	
					mill. bbl/d				
1900	0.2	..	..	0.2	-	-	..	0.4	
1905	0.4	..	..	0.2	..	-	..	0.6	
1910	0.6	..	..	0.3	..	-	0.1	0.9	
1915	0.8	0.1	..	0.2	..	..	0.1	1.2	
1920	1.2	0.4	..	0.1	..	..	0.1	1.9	
1925	2.1	0.4	..	0.2	0.1	..	0.1	2.9	
1930	2.5	0.6	..	0.5	0.1	..	0.2	3.9	
1935	2.7	0.7	..	0.7	0.2	..	0.2	4.5	
1940	3.7	0.9	..	0.7	0.3	..	0.2	5.9	
1945	4.7	1.2	..	0.5	0.5	..	0.0	7.1	
1950	6.0	2.0	0.1	0.9	1.8	..	0.2	10.9	
1955	7.9	2.7	0.2	1.7	3.2	..	0.4	16.2	
1960	8.5	3.8	0.3	3.3	5.2	0.3	0.6	22.0	7.9
1961	8.8	3.9	0.3	3.6	5.6	0.5	0.7	23.5	8.5
1962	9.1	4.3	0.4	4.1	6.2	0.8	0.7	25.5	10.0
1963	9.4	4.4	0.4	4.5	6.8	1.2	0.7	27.4	10.9
1964	9.6	4.5	0.4	4.8	7.6	1.7	0.8	29.5	12.1
1965	10.0	4.7	0.4	5.3	8.4	2.3	0.9	31.8	13.2
1966	10.6	4.6	0.4	5.7	9.3	2.8	1.0	34.5	14.3
1967	11.3	5.0	0.4	6.2	10.0	3.2	1.0	37.0	15.3
1968	11.8	5.1	0.4	6.6	11.2	4.1	1.2	40.3	17.7
1969	12.1	5.1	0.4	7.0	12.4	5.2	1.4	43.6	20.3
1970	12.8	5.3	0.4	7.6	13.9	6.2	1.9	48.1	22.4
1971	12.7	5.2	0.4	8.1	16.3	5.8	2.4	50.8	25.5
1972	13.0	5.0	0.4	8.6	18.1	5.8	2.8	53.6	27.1
1973	13.1	5.3	0.4	9.1	21.3	6.0	3.3	58.5	30.9
1974	12.5	5.0	0.4	9.8	21.9	5.5	3.6	58.6	30.8
1975	11.7	4.5	0.5	10.4	19.7	5.1	3.7	55.7	27.5
1976	11.3	4.5	0.8	11.0	22.4	5.9	4.2	60.1	31.1
1977	11.5	4.7	1.3	11.5	22.5	6.3	4.7	62.6	31.7
1978	11.9	5.0	1.7	12.1	21.4	6.1	4.9	63.1	30.3
1979	11.9	5.5	2.3	12.4	21.9	6.8	5.0	65.8	31.5
1980	11.9	5.9	2.5	12.7	18.8	6.2	4.9	62.8	27.5
1981	11.8	6.3	2.3	12.8	18.6	5.0	4.9	61.6	23.4
1982	11.8	6.6	2.6	12.9	16.2	4.8	4.8	59.6	19.9
1983	11.9	6.5	3.5	12.9	12.1	4.7	5.2	56.8	18.5
1984	12.3	6.7	3.8	12.7	11.4	5.2	5.6	57.7	18.1
1985	12.4	6.6	4.1	12.3	10.6	5.5	5.9	57.4	17.2
1986	12.0	6.7	4.3	12.7	13.3	5.4	6.1	60.6	20.0
1987	11.9	6.8	4.4	12.9	13.2	5.4	6.2	60.7	19.5
1988	11.8	7.0	4.3	12.8	15.4	5.7	6.2	63.2	21.9
1989	11.1	7.1	4.2	12.6	16.6	6.2	6.5	64.2	23.5
1990	10.9	7.5	4.4	11.7	17.8	6.7	6.7	65.7	25.4
1991	11.1	7.9	4.6	10.6	17.3	6.9	6.9	65.3	25.4
1992	10.9	8.0	5.0	9.3	18.8	6.9	6.9	65.7	26.7
1993	10.8	8.2	5.3	8.3	19.6	6.9	6.9	65.9	27.1
1994	10.7	8.4	6.2	7.5	19.9	6.9	7.1	66.7	27.3
1995	10.7	8.7	6.5	7.4	20.1	7.0	7.3	67.5	27.6
1996	10.8	9.4	6.8	7.4	20.5	7.5	7.6	69.9	28.3
1997	10.8	9.9	6.8	7.5	21.7	7.8	7.7	72.2	29.9
1998	10.6	10.2	6.7	7.6	22.8	7.5	7.6	73.1	30.6

1. Includes condensates, natural gas liquids and bituminous shales. 2. United States and Canada. 3. Former Soviet Union and Eastern Europe. 4. Effective Member Countries.

Source: 1860-1955: API, *Basic Petroleum Data Book*, 1996; 1960-1997: BP, "Statistical Review of World Energy", various years; 1998: R.I.E. estimates.

Table 6 – *World crude oil production by area, 1860-1998 (%)*

Year	North America	Latin America	Western Europe	Eastern Europe	Middle East	Africa	Asia and Oceania	World total	OPEC
1860	98.4	-	..	1.6	-	-	-	100.0	
1865	96.0	-	..	4.0	-	-	-	100.0	
1870	95.0	-	..	5.0	-	-	-	100.0	
1875	92.6	-	..	7.3	-	-	..	100.0	
1880	88.7	-	..	11.2	-	-	0.1	100.0	
1885	60.1	-	..	39.8	-	-	0.1	100.0	
1890	60.8	-	..	38.9	-	-	0.2	100.0	
1895	51.7	..	..	46.6	-	-	1.7	100.0	
1900	43.3	0.2	..	53.7	-	-	2.8	100.0	
1905	62.9	0.3	..	30.6	-	-	6.2	100.0	
1910	64.0	1.5	..	28.6	-	-	5.8	100.0	
1915	65.1	8.5	..	20.1	0.8	..	5.4	100.0	
1920	64.3	23.8	..	5.6	1.8	0.2	4.2	100.0	
1925	71.5	14.6	..	7.1	3.2	0.1	3.4	100.0	
1930	63.7	16.2	..	12.4	3.3	0.1	4.1	100.0	
1935	60.3	15.2	..	15.1	5.2	0.1	3.9	100.0	
1940	63.3	14.6	..	12.7	4.8	0.3	3.8	100.0	
1945	66.4	17.3	..	7.5	7.5	0.4	0.6	100.0	
1950	54.8	18.0	0.6	7.9	16.1	0.4	2.2	100.0	
1955	49.0	16.9	1.1	10.3	20.1	0.2	2.3	100.0	
1960	38.7	17.1	1.4	14.8	23.8	1.3	2.9	100.0	35.9
1961	37.6	16.7	1.4	15.5	23.9	2.1	2.8	100.0	36.2
1962	35.6	16.9	1.4	15.9	24.2	3.2	2.8	100.0	39.2
1963	34.4	16.1	1.4	16.3	24.8	4.4	2.7	100.0	39.8
1964	32.6	15.4	1.4	16.2	25.8	5.8	2.7	100.0	41.0
1965	31.3	14.7	1.2	16.5	26.3	7.1	2.7	100.0	41.6
1966	30.8	13.5	1.1	16.6	27.0	8.3	2.8	100.0	41.5
1967	30.6	13.4	1.1	16.7	26.9	8.5	2.7	100.0	41.3
1968	29.2	12.7	1.0	16.3	27.7	10.1	2.9	100.0	43.9
1969	27.8	11.8	0.9	16.1	28.3	11.9	3.2	100.0	46.5
1970	26.5	11.0	0.8	15.7	29.0	12.9	4.0	100.0	46.5
1971	25.1	10.2	0.7	15.9	32.1	11.4	4.6	100.0	47.1
1972	24.3	9.3	0.7	16.0	33.7	10.9	5.1	100.0	47.2
1973	22.3	9.1	0.6	15.6	36.3	10.3	5.7	100.0	52.8
1974	21.3	8.5	0.6	16.6	37.4	9.4	6.1	100.0	52.5
1975	21.1	8.0	1.0	18.7	35.4	9.2	6.7	100.0	49.4
1976	18.9	7.5	1.4	18.3	37.2	9.8	7.0	100.0	51.8
1977	18.3	7.5	2.1	18.5	36.0	10.0	7.5	100.0	50.7
1978	18.8	7.9	2.8	19.2	34.0	9.7	7.7	100.0	48.1
1979	18.1	8.4	3.5	18.8	33.3	10.3	7.6	100.0	47.8
1980	19.0	9.4	3.9	20.2	29.9	9.8	7.8	100.0	43.7
1981	19.2	10.2	3.7	20.8	30.1	8.1	7.9	100.0	38.0
1982	19.8	11.0	4.3	21.7	27.2	8.0	8.0	100.0	33.4
1983	21.0	11.4	6.1	22.7	21.3	8.3	9.2	100.0	32.6
1984	21.3	11.6	6.6	22.0	19.7	9.0	9.8	100.0	31.4
1985	21.6	11.5	7.2	21.4	18.5	9.5	10.3	100.0	29.9
1986	19.9	11.1	7.0	20.9	22.0	9.0	10.1	100.0	33.0
1987	19.5	11.2	7.2	21.2	21.8	9.0	10.2	100.0	32.1
1988	18.6	11.0	6.8	20.2	24.4	9.0	9.9	100.0	34.6
1989	17.3	11.0	6.6	19.6	25.8	9.6	10.1	100.0	36.6
1990	16.6	11.4	6.7	17.9	27.1	10.2	10.2	100.0	38.7
1991	16.9	12.1	7.1	16.3	26.5	10.6	10.6	100.0	38.9
1992	16.6	12.1	7.6	14.1	28.6	10.5	10.5	100.0	40.6
1993	16.3	12.4	8.0	12.6	29.7	10.4	10.5	100.0	41.1
1994	16.0	12.6	9.3	11.3	29.8	10.3	10.7	100.0	41.0
1995	15.8	12.9	9.6	10.9	29.7	10.3	10.8	100.0	40.9
1996	15.5	13.4	9.7	10.6	29.3	10.7	10.9	100.0	40.5
1997	15.0	13.7	9.4	10.4	30.1	10.8	10.7	100.0	41.4
1998	14.5	14.0	9.2	10.4	31.2	10.3	10.4	100.0	41.9

Table 7 – *OPEC crude oil production by Member Country (mill. bbl/d)*

	Ecuador	Venezuela	UAE	Iran	Iraq	Kuwait	Qatar	Saudi Arabia	Algeria	Gabon	Libya	Nigeria	Indonesia	Total[1]	Total[2]
1960	..	2.8	..	1.1	1.0	1.7	0.2	1.3	0.2	..	..	..	0.4	7.9	8.7
1961	..	2.9	..	1.2	1.0	1.7	0.2	1.5	0.3	..	..	..	0.4	8.5	9.3
1962	..	3.2	..	1.3	1.0	2.0	0.2	1.6	0.4	..	0.2	0.1	0.5	10.0	10.5
1963	..	3.2	..	1.5	1.2	2.1	0.2	1.8	0.5	..	0.4	0.1	0.4	10.9	11.4
1964	..	3.4	0.2	1.7	1.3	2.3	0.2	1.9	0.6	..	0.9	0.1	0.5	12.1	13.0
1965	..	3.5	0.3	1.9	1.3	2.4	0.2	2.2	0.6	..	1.2	0.3	0.5	13.2	14.4
1966	..	3.4	0.4	2.1	1.4	2.5	0.3	2.6	0.7	..	1.5	0.4	0.5	14.3	15.8
1967	..	3.6	0.4	2.6	1.2	2.5	0.3	2.8	0.8	0.1	1.7	0.3	0.5	15.3	16.8
1968	..	3.6	0.5	2.8	1.5	2.6	0.3	3.0	0.9	0.1	2.6	0.1	0.6	17.7	18.7
1969	..	3.6	0.6	3.4	1.5	2.8	0.4	3.2	1.0	0.1	3.1	0.5	0.8	20.3	20.8
1970	..	3.8	0.8	3.8	1.6	3.0	0.4	3.9	1.0	0.1	3.3	1.1	0.9	22.4	23.5
1971	..	3.6	1.1	4.6	1.7	3.2	0.4	4.8	0.8	0.1	2.8	1.5	0.9	25.5	25.5
1972	0.1	3.3	1.2	5.1	1.5	3.3	0.5	6.1	1.1	0.1	2.2	1.8	1.1	27.1	27.1
1973	0.2	3.5	1.5	5.9	2.0	3.1	0.6	7.7	1.1	0.2	2.2	2.1	1.3	30.9	30.9
1974	0.2	3.1	1.7	6.1	2.0	2.6	0.5	8.6	1.0	0.2	1.5	2.3	1.4	30.8	30.7
1975	0.2	2.4	1.7	5.4	2.3	2.1	0.4	7.2	1.0	0.2	1.5	1.8	1.3	27.5	27.1
1976	0.2	2.4	1.9	5.9	2.4	2.2	0.5	8.8	1.1	0.2	1.9	2.1	1.5	31.1	30.6
1977	0.2	2.3	2.0	5.7	2.4	2.0	0.4	9.4	1.2	0.2	2.1	2.1	1.7	31.7	31.3
1978	0.2	2.2	1.8	5.3	2.6	2.2	0.5	8.6	1.2	0.2	2.0	1.9	1.6	30.3	29.8
1979	0.2	2.4	1.8	3.2	3.5	2.6	0.5	9.8	1.3	0.2	2.1	2.3	1.6	31.5	31.1
1980	0.2	2.2	1.7	1.5	2.6	1.7	0.5	10.3	1.1	0.2	1.8	2.1	1.6	27.5	27.1
1981	0.2	2.2	1.5	1.3	0.9	1.2	0.4	10.2	1.0	0.2	1.2	1.4	1.7	23.4	23.0
1982	0.2	2.0	1.3	2.4	1.0	0.9	0.3	6.9	1.0	0.2	1.1	1.3	1.4	19.9	19.6
1983	0.2	1.9	1.2	2.5	1.1	1.1	0.3	5.4	1.0	0.2	1.1	1.2	1.4	18.5	18.1
1984	0.3	1.9	1.3	2.1	1.2	1.2	0.4	4.6	1.1	0.2	1.0	1.4	1.5	18.1	17.6
1985	0.3	1.7	1.3	2.2	1.4	1.1	0.3	3.6	1.1	0.2	1.1	1.5	1.3	17.2	16.7
1986	0.3	1.9	1.6	2.1	1.9	1.4	0.4	5.2	1.2	0.2	1.1	1.5	1.4	20.0	19.5
1987	0.2	1.9	1.6	2.3	2.4	1.1	0.3	4.6	1.2	0.2	1.0	1.3	1.4	19.5	19.2
1988	0.3	2.0	1.6	2.3	2.8	1.5	0.4	5.7	1.3	0.2	1.1	1.4	1.4	21.9	21.4
1989	0.3	2.0	2.0	2.9	2.8	1.6	0.4	5.6	1.3	0.2	1.2	1.7	1.5	23.5	23.0
1990	0.3	2.2	2.3	3.3	2.2	1.3	0.4	7.1	1.3	0.3	1.4	1.8	1.5	25.4	24.9
1991	0.3	2.5	2.6	3.5	0.3	0.2	0.4	8.8	1.3	0.3	1.5	1.9	1.7	25.4	24.8
1992	0.3	2.5	2.5	3.5	0.5	1.1	0.5	9.1	1.3	0.3	1.5	1.9	1.6	26.7	26.1
1993	[3]	2.6	2.5	3.7	0.5	1.9	0.5	9.0	1.3	0.3	1.4	1.9	1.6	27.1	26.8
1994	[3]	2.7	2.5	3.7	0.5	2.1	0.5	8.9	1.3	0.3	1.4	1.9	1.6	27.3	27.0
1995	[3]	2.8	2.5	3.7	0.5	2.1	0.5	8.9	1.3	0.4	1.4	1.9	1.6	27.6	27.2
1996	[3]	3.1	2.6	3.7	0.6	2.1	0.6	9.0	1.4	[3]	1.4	2.2	1.6	28.3	28.3
1997	[3]	3.4	2.7	3.7	1.2	2.1	0.7	9.4	1.4	[3]	1.5	2.3	1.6	29.9	29.9
1998	[3]	3.3	2.7	3.7	2.2	2.0	0.7	9.4	1.4	[3]	1.4	2.1	1.5	30.6	30.6

1. Effective Member Countries. 2. Total 11 Member Countries as at end 1998. 3. Withdrawn from OPEC.
Source: BP, "Statistical Review of World Energy", various years; 1998: R.I.E. estimates.

Table 8 – *Surplus of oil producing capacity in the OPEC and non-OPEC Countries*[1], *1960-1998 (mill. bbl/d)*

	Production[2]			Producing capacity			Capacity surplus				
										as % of	
	OPEC[3]	non-OPEC	Total	OPEC	non-OPEC	Total	OPEC	non-OPEC	Total	capacity	production
1960	n.d.	n.d.	17.668	n.d.	n.d.	19.100	n.d.	n.d.	1.432	7.50	8.11
1961	9.364	9.319	18.683	10.150	10.050	20.200	0.786	0.731	1.517	7.51	8.12
1962	10.508	9.663	20.171	11.350	10.450	21.800	0.842	0.787	1.629	7.47	8.08
1963	11.514	10.036	21.550	12.450	10.850	23.300	0.936	0.814	1.750	7.51	8.12
1964	12.984	10.206	23.190	14.050	11.050	25.100	1.066	0.844	1.910	7.61	8.24
1965	14.339	10.552	24.891	15.500	11.400	26.900	1.161	0.848	2.009	7.47	8.07
1966	15.770	11.235	27.005	17.050	12.150	29.200	1.280	0.915	2.195	7.52	8.13
1967	16.850	12.155	29.005	18.200	13.150	31.350	1.350	0.995	2.345	7.48	8.08
1968	18.786	13.114	31.900	20.300	14.200	34.500	1.514	1.086	2.600	7.54	8.15
1969	20.907	13.569	34.476	22.600	14.650	37.250	1.693	1.081	2.774	7.45	8.05
1970	23.413	14.515	37.928	25.300	15.700	41.000	1.887	1.185	3.072	7.49	8.10
1971	25.326	14.575	39.901	27.400	15.750	43.150	2.074	1.175	3.249	7.53	8.14
1972	27.094	14.845	41.939	29.300	16.050	45.350	2.206	1.205	3.411	7.52	8.13
1973	30.989	14.942	45.931	33.500	16.150	49.650	2.511	1.208	3.719	7.49	8.10
1974	30.729	14.593	45.322	34.800	15.750	50.550	4.071	1.157	5.228	10.34	11.54
1975	27.155	14.418	41.573	36.100	15.600	51.700	8.945	1.182	10.127	19.59	24.36
1976	30.738	14.548	45.286	37.400	15.750	53.150	6.662	1.202	7.864	14.80	17.37
1977	31.278	15.595	46.873	38.700	16.850	55.550	7.422	1.255	8.677	15.62	18.51
1978	29.805	17.304	47.109	36.480	18.700	55.180	6.675	1.396	8.071	14.63	17.13
1979	30.928	17.708	48.636	34.260	19.150	53.410	3.332	1.442	4.774	8.94	9.82
1980	26.890	18.310	45.200	32.730	19.800	52.530	5.840	1.490	7.330	13.95	16.22
1981	22.624	18.853	41.477	31.200	20.400	51.600	8.576	1.547	10.123	19.62	24.41
1982	18.584	19.768	38.352	31.200	21.400	52.600	12.616	1.632	14.248	27.09	37.15
1983	17.317	20.718	38.035	31.200	22.400	53.600	13.883	1.682	15.565	29.04	40.92
1984	17.442	21.922	39.364	31.000	23.700	54.700	13.558	1.778	15.336	28.04	38.96
1985	16.039	22.727	38.766	30.000	24.400	54.400	13.961	1.673	15.634	28.74	40.33
1986	18.057	22.397	40.454	30.000	24.550	54.550	11.943	2.153	14.096	25.84	34.84
1987	17.465	22.905	40.370	28.000	24.700	52.700	10.535	1.795	12.330	23.40	30.54
1988	19.625	23.140	42.765	28.000	24.800	52.800	8.375	1.660	10.035	19.01	23.47
1989	21.155	22.745	43.900	28.000	25.000	53.000	6.845	2.255	9.100	17.17	20.73
1990	23.115	23.080	46.195	27.500	25.200	52.700	4.385	2.120	6.505	12.34	14.08
1991	22.995	23.620	46.615	25.000	25.300	50.300	2.005	1.680	3.685	7.33	7.91
1992	24.180	24.105	48.285	25.500	26.144	51.644	1.320	2.039	3.359	6.50	6.96
1993	24.860	24.380	49.240	26.500	26.443	52.943	1.640	2.063	3.703	6.99	7.52
1994	25.020	25.600	50.620	28.120	26.947	55.067	3.100	1.347	4.447	8.08	8.79
1995	25.170	26.285	51.455	28.800	27.380	56.180	3.630	1.095	4.725	8.41	9.18
1996	25.690	27.430	53.120	28.800	28.503	57.303	3.110	1.073	4.183	7.30	7.87
1997	27.100	28.000	55.100	29.117	29.159	58.275	2.017	1.159	3.175	5.45	5.76
1998	27.750	28.100	55.850	29.434	29.814	59.248	1.684	1.714	3.398	5.73	6.08

1. Excluding centrally planned economies. 2. Excluding natural gas liquids and condensates. 3. Effective Member Countries.
Source: 1960-1986: Pennwell, *Oil Industry Outlook 1988-1992*, 1987; 1987-1997: production: BP, "Statistical Review of World Energy", various years; Arthur Andersen, CERA, *World Oil Trends, 1995*; 1998: R.I.E. estimates; 1987-1991: producing capacity: "Petroleum Intelligence Weekly" e "Oil & Gas Journal"; 1992-1998: producing capacity: R.I.E. processing.

Table 9 – *World oil consumption (mill. bbl/d)*

Year	North America[1]	Latin America	Western Europe	Eastern Europe[2]	Middle East	Africa	Asia and Oceania	World total	OECD[3]
1860	..	-	-	..	-	-	-	..	
1870	..	-	..	..	-	-	-	..	
1880	..	-	..	..	-	-	..	0.1	
1890	..	-	..	..	-	-	..	0.2	
1900	0.2	-	..	0.2	-	-	..	0.4	
1910	0.6	-	0.1	0.2	-	-	0.1	0.9	
1920	1.3	n.d.	n.d.	n.d.	-	-	n.d.	1.9	
1930	2.6	n.d.	0.2	n.d.	..	..	n.d.	3.8	
1940	3.8	n.d.	0.5	n.d.	..	..	n.d.	5.9	
1950	6.9	0.7	1.0	1.4	0.1	0.2	0.4	10.7	
1955	8.9	1.2	2.3	1.6	0.2	0.3	0.7	15.2	
1960	10.7	1.6	3.9	2.7	0.6	0.4	1.7	21.4	
1961	10.8	1.7	4.5	3.1	0.6	0.4	1.9	23.1	
1962	11.3	1.9	5.1	3.4	0.7	0.4	2.1	24.9	
1963	11.7	1.9	5.9	3.7	0.7	0.5	2.5	26.9	
1964	12.1	2.0	6.7	3.9	0.7	0.6	2.9	29.0	
1965	12.4	2.0	7.8	4.5	0.7	0.6	3.2	31.1	24.0
1966	13.1	2.1	8.5	4.8	0.8	0.6	3.6	33.6	24.0
1967	13.6	2.2	9.2	5.3	0.8	0.7	4.3	36.1	25.7
1968	14.5	2.3	10.1	5.7	0.8	0.7	5.0	39.1	27.9
1969	15.3	2.5	11.3	6.2	0.9	0.8	5.8	42.8	30.5
1970	15.9	2.6	12.5	6.8	1.0	0.9	6.7	46.4	33.0
1971	16.4	2.8	13.1	7.4	1.1	0.9	7.5	49.1	34.6
1972	17.6	3.1	13.9	8.0	1.1	0.9	8.1	52.8	36.9
1973	18.6	3.4	14.9	8.6	1.2	1.0	9.3	57.0	39.7
1974	17.9	3.5	13.9	9.4	1.3	1.0	9.3	56.4	37.8
1975	17.6	3.5	13.2	9.8	1.3	1.1	9.2	55.7	36.6
1976	18.8	3.6	14.2	10.2	1.5	1.2	9.8	59.2	38.9
1977	19.7	3.8	13.9	10.7	1.6	1.2	10.2	61.2	39.8
1978	20.1	4.0	14.3	11.2	1.6	1.3	10.7	63.1	40.6
1979	19.8	4.4	14.7	11.0	1.9	1.4	11.2	64.5	40.7
1980	18.3	4.6	14.0	11.1	2.0	1.5	10.7	62.1	37.9
1981	17.3	4.8	13.1	11.1	2.1	1.5	10.3	60.2	35.8
1982	16.3	4.5	12.5	10.4	2.6	1.6	10.0	57.8	33.9
1983	16.2	4.4	12.2	10.2	2.7	1.6	10.1	57.4	33.5
1984	16.6	4.3	12.3	10.1	2.9	1.6	10.5	58.3	35.4
1985	16.7	4.4	12.7	9.6	3.0	1.7	10.5	58.4	35.1
1986	17.2	4.5	13.2	9.6	3.0	1.7	11.0	60.2	36.2
1987	17.6	4.7	13.3	9.7	3.1	1.8	11.3	61.5	36.8
1988	18.3	4.8	13.5	9.5	3.1	1.9	12.2	63.2	38.0
1989	18.4	5.0	13.6	9.5	3.2	1.9	12.9	64.5	38.5
1990	18.0	5.0	13.8	9.6	3.4	2.0	13.7	65.5	38.8
1991	17.6	5.1	14.0	9.0	3.5	2.0	14.3	65.5	38.8
1992	17.9	5.3	14.1	7.8	3.5	2.1	15.3	66.0	39.5
1993	18.2	5.4	14.1	6.4	3.6	2.1	15.9	65.6	39.7
1994	18.7	5.7	14.1	5.6	3.8	2.1	17.1	67.0	40.7
1995	18.7	5.7	14.3	5.3	3.9	2.2	17.9	67.9	40.8
1996	19.2	6.0	14.5	5.0	4.1	2.3	18.8	69.9	41.8
1997	19.5	6.3	14.7	5.1	4.2	2.4	19.5	71.7	42.5
1998	19.6	6.3	15.0	5.1	4.4	2.5	19.1	72.0	42.4

1. United States and Canada. 2. Former Soviet Union and Eastern Europe. Includes German Democratic Republic from 1965 to 1979. 3. Member Countries as at 1st January 1995.

Source: 1860-1955: our calculations; 1960-1964: API, *Basic Petroleum Data Book*, 1996; 1965-1997: BP, "Statistical Review of World Energy", various years; 1998: R.I.E. estimates.

Table 10 - *World oil consumption (%)*

Year	North America	Latin America	Western Europe	Eastern Europe	Middle East	Africa	Asia and Oceania	World total	OECD
1860	..	-	-	..	-	-	-	100.0	
1870	..	-	..	..	-	-	-	100.0	
1880	..	-	..	..	-	-	..	100.0	
1890	..	-	..	..	-	-	..	100.0	
1900	50.0	-	5.0	40.0	-	-	5.0	100.0	
1910	70.0	-	5.6	18.9	-	-	5.6	100.0	
1920	68.4	n.d.	n.d.	n.d.	-	-	n.d.	100.0	
1930	68.4	n.d.	5.3	n.d.	..	..	n.d.	100.0	
1940	64.4	n.d.	8.5	n.d.	..	..	n.d.	100.0	
1950	64.5	6.5	9.3	13.1	0.9	1.9	3.7	100.0	
1955	58.6	7.9	15.1	10.5	1.3	2.0	4.6	100.0	
1960	49.8	7.4	18.0	12.7	2.7	1.7	7.7	100.0	
1961	47.0	7.5	19.3	13.6	2.6	1.8	8.2	100.0	
1962	45.4	7.5	20.6	13.7	2.7	1.7	8.4	100.0	
1963	43.5	7.1	21.9	13.7	2.6	1.8	9.3	100.0	
1964	41.8	7.0	23.2	13.6	2.6	1.9	9.9	100.0	
1965	40.0	6.4	25.0	14.4	2.2	1.8	10.1	100.0	77.3
1966	38.9	6.3	25.4	14.4	2.3	1.9	10.8	100.0	71.6
1967	37.5	6.1	25.5	14.8	2.1	1.9	12.0	100.0	71.2
1968	37.0	5.9	25.8	14.6	2.0	1.9	12.7	100.0	71.5
1969	35.6	5.9	26.4	14.5	2.1	1.9	13.5	100.0	71.3
1970	34.2	5.7	26.9	14.7	2.1	1.8	14.5	100.0	71.0
1971	33.4	5.7	26.6	15.0	2.1	1.8	15.2	100.0	70.3
1972	33.4	5.8	26.4	15.1	2.1	1.7	15.3	100.0	70.0
1973	32.6	5.9	26.1	15.1	2.1	1.8	16.3	100.0	69.6
1974	31.8	6.2	24.7	16.6	2.3	1.8	16.5	100.0	67.1
1975	31.6	6.3	23.8	17.5	2.4	1.9	16.5	100.0	65.6
1976	31.7	6.0	24.0	17.3	2.5	2.0	16.5	100.0	65.7
1977	32.2	6.2	22.7	17.5	2.6	2.0	16.7	100.0	64.9
1978	31.8	6.3	22.7	17.7	2.6	2.0	16.9	100.0	64.3
1979	30.7	6.9	22.7	17.1	3.0	2.1	17.4	100.0	63.1
1980	29.5	7.4	22.5	17.8	3.3	2.3	17.2	100.0	61.1
1981	28.7	7.9	21.8	18.4	3.5	2.5	17.1	100.0	59.5
1982	28.2	7.8	21.6	18.0	4.4	2.8	17.2	100.0	58.6
1983	28.2	7.7	21.2	17.8	4.8	2.8	17.5	100.0	58.3
1984	28.4	7.4	21.1	17.3	5.0	2.7	18.0	100.0	60.7
1985	28.5	7.5	21.7	16.3	5.1	2.9	17.9	100.0	60.2
1986	28.6	7.5	21.9	15.9	5.0	2.8	18.2	100.0	60.2
1987	28.6	7.7	21.6	15.8	5.0	2.9	18.4	100.0	59.9
1988	28.9	7.6	21.4	15.1	4.9	2.9	19.2	100.0	60.1
1989	28.6	7.7	21.0	14.7	5.0	3.0	20.0	100.0	59.7
1990	27.5	7.6	21.1	14.7	5.2	3.1	20.9	100.0	59.2
1991	26.9	7.8	21.3	13.8	5.3	3.1	21.8	100.0	59.3
1992	27.1	8.0	21.4	11.9	5.4	3.1	23.2	100.0	59.8
1993	27.7	8.2	21.5	9.7	5.5	3.2	24.3	100.0	60.5
1994	27.9	8.5	21.0	8.4	5.7	3.1	25.4	100.0	60.7
1995	27.5	8.4	21.1	7.8	5.7	3.2	26.3	100.0	60.1
1996	27.5	8.6	20.7	7.2	5.9	3.3	26.9	100.0	59.8
1997	27.2	8.8	20.5	7.1	5.9	3.3	27.2	100.0	59.3
1998	27.2	8.8	20.8	7.1	6.1	3.5	26.5	100.0	58.9

Table 11 – *World refinery capacity, 1965-1998 (mill. bbl/d)*

Year	North America[1]	Latin America	Western Europe	Eastern Europe[2]	Middle East	Africa	Asia and Oceania	World total
1965	11.5	4.0	7.7	5.2	1.8	0.6	3.2	34.0
1966	11.7	4.1	9.6	5.6	1.8	0.6	3.7	37.1
1967	12.5	4.4	11.1	6.2	1.9	0.6	3.8	40.6
1968	12.9	4.7	12.0	7.1	2.1	0.7	4.5	43.9
1969	13.2	4.8	13.2	7.5	2.2	0.7	5.1	46.8
1970	14.2	5.4	14.2	8.4	2.3	0.7	5.9	51.1
1971	14.9	5.8	15.6	9.1	2.4	0.8	6.7	55.3
1972	15.4	6.4	17.0	9.9	2.4	1.0	7.8	59.9
1973	16.2	6.8	18.3	10.8	2.5	1.1	8.8	64.6
1974	17.0	7.4	19.0	11.7	2.7	1.2	9.2	68.2
1975	17.3	7.4	20.4	12.6	2.8	1.3	9.4	71.2
1976	18.6	7.8	20.8	13.4	3.0	1.4	9.6	74.5
1977	19.2	8.0	20.2	13.8	3.2	1.5	10.0	75.9
1978	19.7	8.2	20.4	14.6	3.4	1.7	10.1	78.0
1979	20.2	8.7	20.1	13.2	3.5	2.0	10.3	78.1
1980	20.4	9.0	20.1	13.4	3.6	2.1	10.8	79.3
1981	20.4	9.1	20.1	14.0	3.3	2.0	10.9	79.9
1982	19.1	8.3	18.1	14.2	3.6	2.4	12.8	78.5
1983	18.2	8.2	17.2	14.5	3.6	2.3	12.6	76.5
1984	17.7	7.8	15.7	14.4	3.5	2.5	12.7	74.4
1985	17.5	7.0	15.0	14.4	3.8	2.6	12.6	73.0
1986	17.6	7.0	14.7	14.8	3.9	2.6	12.5	73.0
1987	17.8	7.2	14.6	14.8	4.2	2.6	12.4	73.6
1988	17.5	7.3	14.4	14.9	4.3	2.8	12.3	73.5
1989	17.4	7.5	14.0	15.3	4.5	2.8	12.5	74.0
1990	17.6	7.6	14.0	15.3	5.0	2.9	13.2	75.5
1991	17.6	7.5	14.2	15.3	4.4	2.9	13.8	75.6
1992	17.1	7.6	14.3	12.9	4.8	2.8	14.6	74.0
1993	16.9	7.7	14.2	12.8	5.0	2.8	15.2	74.6
1994	17.1	7.7	14.4	12.9	5.3	2.9	15.5	75.6
1995	17.1	7.8	14.4	13.0	5.3	2.9	16.5	76.9
1996	17.2	7.7	14.6	12.2	5.6	2.9	17.6	77.9
1997	17.5	7.9	14.9	12.8	5.8	2.9	18.0	79.7
1998	17.9	7.8	15.2	12.1	5.9	3.0	20.0	81.9

1. United States and Canada. 2. Former Soviet Union and Eastern Europe.

Source: BP. "Statistical Review of World Energy", various years; 1998: R.I.E. estimates.

Table 12 – *Oil consumption. production and imports in OECD area, 1965-1998 (mill. bbl/d)*

	Consumption (A)				Production (B)				Net imports (A) – (B)				Dependence[2] (%)			
	North America[1]	Western Europe	Others	OECD	North America[1]	Western Europe	Others	OECD	North America[1]	Western Europe	Others	OECD	North America[1]	Western Europe	Others	OECD
1965	12.4	7.8	3.8	24.0	10.0	0.4	..	10.4	2.5	7.4	3.8	13.7	20	95	99	57
1966	13.1	8.5	2.5	24.0	10.6	0.4	..	11.0	2.5	8.1	2.4	13.0	19	95	99	54
1967	13.6	9.2	2.9	25.7	11.3	0.4	..	11.8	2.2	8.8	2.9	14.0	16	96	99	54
1968	14.5	10.1	3.4	27.9	11.8	0.4	0.1	12.3	2.7	9.7	3.3	15.7	18	96	98	56
1969	15.3	11.3	3.9	30.5	12.1	0.4	0.1	12.6	3.1	10.9	3.9	17.9	20	96	99	59
1970	15.9	12.5	4.6	33.0	12.8	0.4	0.2	13.4	3.1	12.1	4.4	19.6	20	97	96	60
1971	16.4	13.1	5.1	34.6	12.7	0.4	0.3	13.4	3.7	12.7	4.7	21.1	22	97	94	61
1972	17.6	13.9	5.4	36.9	13.0	0.4	0.4	13.8	4.6	13.6	5.0	23.2	26	97	93	63
1973	18.6	14.9	6.1	39.7	13.1	0.4	0.4	13.8	5.6	14.5	5.7	25.8	30	97	93	65
1974	17.9	13.9	6.0	37.8	12.5	0.4	0.4	13.3	5.5	13.5	5.6	24.6	30	97	93	65
1975	17.6	13.2	5.7	36.6	11.7	0.5	0.4	12.7	5.9	12.7	5.3	23.9	33	96	93	65
1976	18.8	14.2	5.9	38.9	11.3	0.8	0.4	12.6	7.4	13.3	5.5	26.3	40	94	93	68
1977	19.7	13.9	6.1	39.8	11.5	1.3	0.5	13.3	8.3	12.6	5.6	26.5	42	90	92	67
1978	20.1	14.3	6.2	40.6	11.9	1.7	0.5	14.0	8.2	12.6	5.7	26.6	41	88	93	65
1979	19.8	14.7	6.2	40.7	11.9	2.3	0.5	14.7	7.9	12.4	5.8	26.1	40	84	93	64
1980	18.3	14.0	5.7	37.9	11.9	2.5	0.4	14.8	6.4	11.5	5.2	23.1	35	82	93	61
1981	17.3	13.1	5.4	35.8	11.8	2.3	0.4	14.5	5.5	10.9	5.0	21.4	32	83	92	60
1982	16.3	12.5	5.1	33.9	11.8	2.6	0.4	14.8	4.5	9.9	4.7	19.1	28	79	92	56
1983	16.2	12.2	5.1	33.5	11.9	3.5	0.5	15.9	4.3	8.7	4.6	17.6	27	72	91	53
1984	16.6	12.3	6.5	35.4	12.3	3.8	0.5	16.6	4.3	8.5	6.0	18.8	26	69	92	53
1985	16.7	12.7	5.8	35.1	12.4	4.0	0.7	17.0	4.3	8.7	5.1	18.1	26	69	89	52
1986	17.2	13.2	5.8	36.2	12.0	4.1	0.6	16.7	5.2	9.1	5.3	19.5	30	69	90	54
1987	17.6	13.3	5.9	36.8	11.9	4.2	0.6	16.7	5.8	9.1	5.3	20.1	33	69	89	55
1988	18.3	13.5	6.2	38.0	11.8	4.2	0.6	16.5	6.5	9.3	5.6	21.5	36	69	90	57
1989	18.4	13.6	6.5	38.5	11.1	4.1	0.6	15.8	7.3	9.5	5.9	22.7	40	70	91	59
1990	18.0	13.8	7.0	38.8	10.9	4.2	0.7	15.8	7.1	9.6	6.3	23.0	40	69	91	59
1991	17.6	14.0	7.2	38.8	11.1	4.5	0.6	16.2	6.6	9.5	6.6	22.7	37	68	91	58
1992	17.9	14.1	7.5	39.5	10.9	4.8	0.6	16.4	6.9	9.3	6.9	23.1	39	66	92	58
1993	18.1	14.1	7.5	39.7	10.8	5.1	0.6	16.5	7.3	9.0	6.9	23.2	40	64	92	59
1994	18.7	14.1	7.9	40.7	10.6	6.1	0.6	17.3	8.0	8.1	7.3	23.4	43	57	92	57
1995	18.7	14.3	7.8	40.8	10.7	6.4	0.6	17.7	8.0	7.9	7.2	23.1	43	55	93	57
1996	19.2	14.5	8.1	41.8	10.8	6.8	0.6	18.2	8.4	7.7	7.5	23.6	44	53	92	56
1997	19.5	14.7	8.3	42.5	10.8	6.8	0.7	18.3	8.7	7.9	7.6	24.2	45	54	92	57
1998	19.6	15.0	7.8	42.4	10.6	6.7	0.7	18.0	9.0	8.3	7.1	24.4	46	55	91	58

1. United States and Canada. 2. Net imports as a share of total consumption.

Source: calculations made from previous tables.

Table 13 – *Oil consumption by product group in OECD area, 1965-1998*

	Gasoline	Middle distillates	Fuel oil	Others	Total	Gasoline	Middle distillates	Fuel oil	Others	Total
	Million tonnes					%				
1965	321	280	307	188	1,096	29	26	28	17	100
1966	345	302	329	201	1,177	29	26	28	17	100
1967	369	326	357	206	1,259	29	26	28	16	100
1968	395	362	386	227	1,369	29	26	28	17	100
1969	416	396	430	249	1,491	28	27	29	17	100
1970	438	432	485	259	1,614	27	27	30	16	100
1971	464	456	504	266	1,690	27	27	30	16	100
1972	497	493	535	285	1,810	27	27	30	16	100
1973	524	530	583	302	1,939	27	27	30	16	100
1974	508	500	550	288	1,846	28	27	30	16	100
1975	509	495	502	269	1,776	29	28	28	15	100
1976	538	536	535	283	1,891	28	28	28	15	100
1977	547	558	533	292	1,929	28	29	28	15	100
1978	572	581	523	299	1,974	29	29	26	15	100
1979	561	583	518	312	1,974	28	30	26	16	100
1980	530	539	460	290	1,818	29	30	25	16	100
1981	513	515	398	277	1,703	30	30	23	16	100
1982	507	490	342	268	1,606	32	30	21	17	100
1983	514	493	302	275	1,583	32	31	19	17	100
1984	521	519	291	287	1,618	32	32	18	18	100
1985	523	525	256	285	1,589	33	33	16	18	100
1986	539	547	265	293	1,644	33	33	16	18	100
1987	561	563	250	306	1,681	33	34	15	18	100
1988	577	587	258	320	1,742	33	34	15	18	100
1989	585	591	261	319	1,756	33	34	15	18	100
1990	587	598	260	324	1,768	33	34	15	18	100
1991	583	606	252	326	1,766	33	34	14	18	100
1992	591	615	251	344	1,801	33	34	14	19	100
1993	595	628	238	344	1,805	33	35	13	19	100
1994	602	645	242	356	1,845	33	35	13	19	100
1995	613	658	223	357	1,851	33	36	12	19	100
1996	628	688	218	368	1,901	33	36	11	19	100
1997	641	698	212	376	1,927	33	36	11	20	100
1998	649	696	208	371	1,925	34	36	11	19	100

Source: BP, "Statistical Review of World Energy", various years; 1998: R.I.E. estimates

Table 14 – *Spot oil product prices in Italy*[1] *(doll./tonn.)*

Year	Premium gasoline	Regular gasoline	Virgin Naphtha	Jet kerosine	Gas oil	Fuel oil 1% sulphur	Fuel oil 3,5% sulphur
1969	27.78	22.99	n.d.	n.d.	22.40	n.d.	9.68
1970	22.10	17.76	n.d.	n.d.	18.36	n.d.	9.63
1971	24.29	19.31	n.d.	26.02	23.36	17.31	16.86
1972	30.23	22.40	n.d.	29.46	27.43	18.11	16.93
1973	35.60	27.32	23.10	28.29	25.03	16.40	12.82
1974	87.63	79.31	58.89	79.69	74.15	33.84	27.02
1975	133.86	124.26	114.99	100.35	90.77	72.90	70.19
1976	145.54	130.51	125.46	115.36	105.22	70.10	67.52
1977	136.02	124.11	121.06	124.80	117.71	84.29	76.69
1978	157.62	136.14	140.62	141.50	128.20	81.64	73.72
1979	335.33	323.65	275.23	340.66	307.33	144.51	130.98
1980	353.73	343.11	314.82	348.43	308.46	189.85	173.76
1981	356.85	341.36	318.36	332.06	298.48	201.38	183.28
1982	322.43	306.18	286.23	313.82	288.61	179.51	162.25
1983	287.15	277.12	268.05	277.69	244.58	171.40	162.60
1984	260.61	253.56	239.46	264.44	235.87	180.60	174.63
1985	262.87	254.13	235.04	260.91	233.80	153.41	146.17
1986	151.96	136.62	118.74	152.68	130.30	76.88	66.22
1987	174.65	160.96	151.03	166.51	149.77	106.02	92.68
1988	158.84	143.99	130.47	146.59	129.16	79.72	60.31
1989	193.19	176.50	152.20	177.99	156.88	103.57	79.87
1990	254.61	[2]	213.67	243.61	207.59	120.88	89.56
1991	232.64	[2]	197.61	209.34	194.44	94.40	70.20
1992	209.86	[2]	175.08	183.46	171.20	94.93	73.97
1993	186.28	[2]	151.28	171.20	160.02	88.69	59.33
1994	169.07	[2]	138.58	152.00	139.70	90.22	75.63
1995	175.37	[2]	144.42	157.25	144.79	100.63	88.17
1996	205.11	[2]	174.74	200.07	183.79	114.93	95.95
1997	202.69	[2]	177.07	174.59	164.38	100.46	84.85
1998	149.13	[2]	119.47	118.13	111.35	71.14	57.68

1. Annual average. cargoes fob Mediterranean. 2. This quotation ceases from 1990.

Source: "Platt's Oilgram Price Service"

Table 15 – *OPEC oil revenues, 1960-1998 (mill. doll.)*

	Algeria	Ecuador	Gabon	Indonesia	Iran	Iraq	Kuwait	Libya	Nigeria	Qatar	Saudi Arabia
1960	106	-	9	221	723	445	855	-	12	103	682
1961	201	-	9	260	722	499	853	11	32	106	740
1962	105	0	9	216	706	500	961	132	48	113	809
1963	465	0	10	269	799	579	1,029	327	57	118	869
1964	394	0	12	267	1,109	634	1,146	606	90	132	922
1965	493	1	14	272	1,138	660	1,207	785	191	141	1,073
1966	508	1	15	203	1,149	698	1,305	990	258	180	1,298
1967	532	1	35	240	1,748	610	1,313	1,164	199	202	1,384
1968	583	1	42	298	1,688	755	1,392	1,860	109	211	2,020
1969	606	0	58	383	1,857	769	1,476	2,162	422	222	2,107
1970	681	0	62	446	2,358	788	1,582	2,379	718	231	2,418
1971	614	1	89	565	3,494	1,083	2,412	2,693	1,375	304	3,803
1972	1,030	60	100	913	3,638	1,027	2,834	2,931	1,803	382	5,477
1973	1,522	250	212	1,609	5,617	1,842	3,574	3,984	3,049	612	8,956
1974	4,267	613	773	5,211	20,904	6,534	10,566	8,257	8,997	1,979	35,476
1975	4,295	516	918	5,311	19,634	8,227	8,593	6,767	7,744	1,754	29,473
1976	4,791	565	924	6,004	22,923	9,201	9,090	9,553	9,444	2,137	38,157
1977	5,562	494	900	7,298	23,599	9,560	8,918	11,394	11,561	2,055	43,308
1978	5,856	604	921	7,439	21,684	10,913	9,557	9,885	9,452	2,305	40,332
1979	8,746	1,161	1,283	8,871	19,186	21,382	17,294	15,941	15,624	3,693	62,855
1980	12,971	1,551	1,745	15,595	11,693	26,096	18,935	21,906	24,931	5,372	108,175
1981	13,066	1,709	1,786	18,164	11,491	10,039	14,229	14,930	17,291	5,496	118,998
1982	11,149	1,304	1,802	15,493	20,168	9,933	9,066	13,692	11,883	4,214	78,119
1983	9,655	1,578	1,652	13,558	21,150	7,816	10,069	12,341	9,941	2,993	44,830
1984	9,778	1,779	1,684	12,477	16,726	8,863	10,996	10,970	11,534	4,386	36,285
1985	9,668	1,926	1,629	9,083	13,710	10,097	9,451	12,132	12,568	3,068	25,937
1986	5,161	983	723	5,501	6,255	6,905	6,378	8,202	4,770	1,720	18,061
1987	6,555	724	896	6,157	10,755	9,416	7,523	6,293	7,024	1,829	20,427
1988	5,725	976	779	5,042	9,673	9,312	6,840	6,070	6,267	1,709	20,205
1989	6,815	1,147	1,200	6,062	12,031	11,876	10,432	6,573	7,470	1,955	24,095
1990	9,588	1,258	1,967	7,404	17,906	9,594	6,385	10,715	13,265	3,273	40,130
1991	8,464	1,029	1,740	6,714	15,767	351	874	10,211	11,792	2,828	43,701
1992	7,885	937	1,712	6,619	16,802	482	6,224	9,326	11,642	2,870	44,754
1993	6,902	[3]	1,506	5,693	14,251	425	9,708	7,689	11,510	2,811	38,621
1994	6,335	[3]	1,668	6,005	14,801	421	10,482	7,170	11,040	2,623	38,586
1995	6,938	[3]	1,771	6,443	14,944	461	12,054	7,763	11,512	2,987	42,502
1996	8,830	[3]	[3]	7,879	17,960	680	14,099	9,543	14,888	3,801	54,270
1997	8,350	[3]	[3]	7,240	17,662	4,280	13,470	8,905	15,840	4,655	52,140
1998	5,690	[3]	[3]	4,600	11,100	5,110	8,970	5,610	9,850	3,180	33,400

1. Effective Member Countries. 2. Total 11 Member Countries as at end 1998. 3. Withdrawn from OPEC.

Source: OPEC, "Annual Statistical Bulletin", various years; 1998: R.I.E. estimates.

	UAE	Venezuela	Total OPEC		Exports				Revenue[2]		
			1	2	Products	Crude oil mill. bbl/d	Total	Total mill.bbl	Average per bbl	Per capita	Population[2] mil.
1960	-	1,983	4,688	5,130	1.2	6.7	8.0	2,908	1.8	27.4	187
1961	-	2,033	4,953	5,457	1.2	7.4	8.6	3,144	1.7	28.4	192
1962	9	2,156	5,593	5,755	1.3	8.3	9.7	3,523	1.6	29.2	197
1963	31	2,155	6,145	6,698	1.5	9.2	10.7	3,909	1.7	33.1	202
1964	128	2,305	7,121	7,733	1.5	10.6	12.1	4,421	1.7	37.2	208
1965	190	2,280	7,556	8,430	1.6	11.7	13.4	4,884	1.7	39.5	213
1966	241	2,190	8,013	9,020	1.6	13.5	15.1	5,513	1.6	41.1	220
1967	262	2,307	8,968	9,961	1.7	14.2	15.8	5,781	1.7	44.2	226
1968	335	2,330	10,889	11,581	1.7	16.0	17.7	6,466	1.8	50.0	232
1969	407	2,279	12,268	12,690	1.7	17.9	19.6	7,160	1.8	53.3	238
1970	523	2,371	13,777	14,495	2.0	20.1	22.1	8,066	1.8	59.2	245
1971	842	2,882	20,067	20,067	1.8	21.9	23.7	8,664	2.3	79.8	251
1972	1,111	2,857	24,003	24,003	1.8	23.9	25.7	9.374	2.6	92.9	258
1973	1,740	4,328	36,833	36,833	2.0	27.2	29.2	10.655	3.5	138.7	266
1974	6,327	10,548	119,679	119,066	1.9	26.9	28.8	10.509	11,3	435,5	273
1975	6,806	8,324	108,362	106,928	1.6	23.7	25.3	9,222	11.6	380.4	281
1976	8,383	8,763	129,935	128,446	1.8	27.1	28.9	10,565	12.2	446.1	288
1977	9,258	9,110	143,017	141,623	1.7	27.3	29.1	10,607	13.4	478.6	296
1978	8,661	8,740	136,349	134,824	1.9	25.8	27.7	10,095	13.4	443.7	304
1979	12,862	13,633	202,531	200,087	2.1	26.5	28.6	10,433	19.2	641.7	312
1980	19,390	17,562	285,922	282,626	2.0	22.6	24.6	8,972	31.5	879.3	321
1981	18,761	18,609	264,569	261,074	1.9	18.3	20.3	7,398	35.3	790.0	330
1982	15,956	15,633	208,412	205,306	2.2	14.0	16.3	5,933	34.6	604.5	340
1983	13,016	13,857	162,456	159,226	2.0	12.2	14.2	5,189	30.7	455.3	350
1984	12,037	14,824	152,339	148,876	2.3	11.7	14.0	5,108	29.1	413.6	360
1985	10,896	12,956	133,121	129,566	2.4	10.6	12.9	4,716	27.5	350.2	370
1986	6,865	7,178	78,702	76,996	2.6	12.6	15.2	5,553	13.9	202.4	380
1987	7,900	9,054	94,553	92,933	3.0	11.8	14.8	5,401	17.2	238.1	390
1988	7,627	8,158	87,407	86,628	3.4	13.0	16.4	5,986	14.5	216.3	400
1989	10,215	10,001	109,872	107,525	3.6	14.9	18.4	6,734	16.0	262.0	410
1990	14,846	13,953	150,284	147,059	3.6	16.1	19.6	7,155	20.6	351.8	418
1991	14,356	12,302	130,129	127,360	3.1	17.0	20.1	7,331	17.4	298.6	427
1992	14,251	11,208	134,712	132,063	3.4	17.4	20.8	7,595	17.4	302.5	437
1993	12,118	10,565	121,799	120,293	3.6	17.9	21.5	7,844	15.3	269.4	446
1994	11,683	11,307	122,121	120,453	3.8	18.0	21.8	7,953	15.1	263.7	457
1995	12,822	13,739	133,936	132,165	4.2	18.1	22.2	8,119	16.3	284.1	465
1996	14,980	18,520	165,450	165,450	4.3	18.3	22.6	8,232	20.1	348.3	475
1997	15,269	18,190	166,001	166,001	4.5	19.2	23.7	8,644	19.2	342.3	485
1998	11,130	11,510	110,150	110,150	4.6	19.6	24.2	8,847	12.4	222.5	495

REFERENCES

Adelman, M.A.

1964a "The World Oil Outlook", in Clawson M. (ed. by) *Natural Resources and International Development*, Baltimore: Johns Hopkins University Press.

1964b *Efficiency of Resources Use in Crude Petroleum*, in "Southern Economic Journal", October.

1964c *Oil Prices in the Long Run (1963-75)*, in "Journal of Business of the University of Chicago", April, pp. 143-161.

1972a *The World Petroleum Market*, Baltimore: Johns Hopkins University Press.

1972b *Is the Oil Shortage Real? Oil Companies As OPEC-Tax Collectors*, in "Foreign Policy", Winter, pp. 69-108.

1976 *The World Oil Cartel: Scarcity, Economics, and Politics*, in "Quarterly Review of Economics and Business", April, no. 2.

1983 "The Multinationals in the World Oil Market: the 1970s and 1980s", in Kindleberger C.P., Andretsch D.B. (ed. by) *The Multinational Corporation in the 1980s*, Cambridge (Mass.): MIT Press.

1986 *The Competitive Floor to World Oil Prices*, in "The Energy Journal", no. 4, pp. 4-31.

1990 *The 1990 Oil Shock is Like the Others*, in "The Energy Journal", vol. 11, no. 4, pp. 1-13.

1991 *Oil Fallacies*, in "Foreign Policy", Spring, pp. 2-16.

Adelman, M.A. et al.

1991 *User Cost in Oil Production*, in "Resources and Energy", vol. 13, pp. 217-240.

Adelman, M.A., Friis, S.

1974 *Monopoli in via di trasformazione e rifornimenti europei di petrolio: si sposta nel panorama petrolifero mondiale la bilancia del potere economico*, in "Economia Internazionale delle Fonti di Energia", January-April, pp. 59-94.

Ailleret, P.

1963 *Énergétique, les besoins d'énergie*, Paris: Eyrolles ed.

Al Chalabi, F.

1980a *OPEC and the International Oil Industry: a Changing Structure*, Oxford: Oxford University Press.

1980b "Oil: Prices, Costs, Taxes", in Mabro R. (ed. by) *World Energy Issues and Policies*, Oxford: Oxford University Press, pp. 143-156.

Alchian, A.

1959 "Costs and Outputs", in Abramovitz M. (ed. by) *The Allocations of Economic Resources: Essays in Honor of B.F. Haley*, Stanford: Stanford University Press.

American Petroleum Institute (API)
1976 *Witnesses for Oil: the Case Against Dismemberment,* Washington D.C.
1995 *Are we Running out of Oil?*, Discussion Paper, Washington D.C.
(various years) *Basic Petroleum Data Book*, Washington D.C.

Andreatta, F.
1998 *Interdipendenza e geopolitica del petrolio,* in "Energia", no. 1, pp. 32-39.

Angelier, J.P. et al.
1983 *Rente et structures des industries de l'énergie*, Grenoble: Presse Universitaire.

Appert, O.
1996 "Le faits porteurs d'avenir", in IFP *Panorama 1996, Actes du colloque*, pp. 7-43.
1997 *Sfide e opportunità tecniche per l'Europa,* in "Energia", no. 1, pp. 38-45.

Arthur Andersen & Co., Cambridge Energy Research Ass.
1984 *The Future of Oil Prices: the Perils of Prophecy*, Cambridge.
1985 *World Oil Trends*, Cambridge.

Babusiaux, D., Boy de La Tour, X.
1999 *Industria petrolifera: impatto della tecnologia sui costi*, in "Energia", no. 1, pp. 16-22.

Bain, J.S.
1941 *The Profit Rate as a Measure of Monopoly Power*, in "Quarterly Journal of Economics", vol. 55.

Banks, F.E.
1980 *The Political Economy of Oil*, Toronto: Lexington Books.

Barnett, H.J.
1979 "Scarcity and Growth Revisited", in Smith V.K. (ed. by) *Scarcity and Growth Reconsidered*, Baltimore: Johns Hopkins University Press for RFF.

Barnett, H.J., Morse, C.
1963 *Scarcity and Growth: the Economics of Natural Resources Availability*, Baltimore: Johns Hopkins University Press for RFF.

Baumol, W.J.
1968 *Teoria economica e analisi operativa*, Milan: Franco Angeli.

Becattini, G. (edited by)
1981 *Marshall*, Bologna: Il Mulino.

Biondi, L.
1983 *Il sistema energetico: fonti, vettori, utenze*, Seminario Università di Modena (unpublished).

Blair, J.M.
1977 *The Control of Oil*, London: McMillan.

Bradley, P.
1987 "Cost and Output Analysis in Mineral and Petroleum Production", in Gordon R.L. et al. (ed. by) *Energy: Markets and Regulation*, Cambridge (Mass.): MIT Press, pp. 281-303.

British Petroleum (BP)
(various years) "Statistical Review of World Energy".

Brown, L.R. et al.
1987 *State of the World 1987*, WorldWatch Institute, New York: Norton & Co.
1993 *Vital Signs 1993. The Trends that are shaping our Future,* WorldWatch Institute, New York: Norton & Co.

Brown, L.R., Jacobson, J.
1987 "Assessing the Future of Urbanization", in Brown L.R. et al. *State of the World 1987*, WorldWatch Institute, New York: Norton & Co., pp. 38-56.

Bruni, L.
1964 *International Economies of Scale with a given Technique*, in "Journal of Industrial Economics".

Burns, J.M.
1983 "Future Markets and Market Efficiency", in Streit M.E. (ed. by) *Futures Markets*, Oxford: Basil Blackwell.

Burrows, J.C., Domencich, T.A.
1970 *An Analysis of the United States Oil Import Quota*, Lexington (Mass.): D.C. Heath and Co.

Butter, S.
1990a *Price Hikes, Fears of Oil Supply Trail Iraq Attack on Kuwait* in "Oil & Gas Journal", 13 August.
1990b *The Seven Pillars of Oil Market Wisdom,* in "Financial Times", 10 October.

Central Intelligence Agency (CIA)
1979 *The World Oil Market in the Years Ahead*, Washington D.C.

Cabet, R.
1966 *L'économie des transports par conduite*, Paris: Technip.

Campbell, J.M.
1959 *Oil Property Evaluation*, New York: Prentice Hall Inc., Englewood Cliffs.
1991 *Golden Century of Oil, 1950-2050*, Dordrecht: Kluwer.

Campbell, C.J., Laherrère, J.H.
1995 *The World Oil Supply, 1930-2050,* Petroconsultants Report, October.
1998 *La fine del petrolio a buon mercato*, in "Le Scienze", no. 357, May, pp. 78-84.

Cattan, H.
1967a *Evolution of Oil Concessions in the Middle East and North Africa*, New York: Dobbs Ferry.

1967b *The Law of Oil Concessions in the Middle East and North Africa*, New York: Oceania Publications.

Chandler, A.D.
1994 *Dimensione e diversificazione,* Bologna: Il Mulino.

Chase Manhattan Bank (CMB)
1982 *Financial Analysis of a Group of Petroleum Companies*, in "Petroleum Situation", special issue, vol. 6, n. 1, New York.
(various years) *Capital Investments of the World Petroleum Industry*, New York.

Chevalier, J.M.
1975 *La Nuova Strategia del Petrolio*, Milan: Il Formichiere.

Cleveland, C.J., Costanza, R., Hall, C.A.S, Kaufmann, R.
1984 *Energy and the U.S. Economy: A Biophysical Perspective*, in "Science", vol. 225, August, pp. 890-897.

Clô, A.
1973 *Politica energetica americana e prezzi del petrolio*, Ipalmo.
1978 *Le prospettive energetiche dell'Europa a cinque anni dalla crisi*, in "Rivista di Economia e Politica Industriale", no. 2, pp. 137-158.
1979 *La situazione petrolifera mondiale e i suoi riflessi sulla politica dei prezzi in Italia*, in "Energia e Materie Prime", no. 7-8, pp. 29-36.
1980a *Un equilibrio instabile*, in "Energia", no. 1, pp. 32-45.
1980b *La situazione energetica italiana: problemi vecchi e nuovi*, in "Energia", no. 1, pp. 48-58.
1982 *Mutamenti nel mercato del petrolio e nuove condizioni di equilibrio*, in "Energia", no. 1, pp. 16-36.
1984 *Mutamenti di breve e di lungo periodo nel sistema energetico internazionale e nazionale*, in "Rivista Internazionale di Scienze Economiche e Commerciali", no. 8, pp. 734-757.
1986 *Storia e cronaca del prezzo del petrolio*, in "Energia", no. 2, pp. 20-26.
1987 *Il quadro energetico nel 1986/1987*, in "Energia", no. 2, pp. 64-79.
1988 "Instabilità e interdipendenza dei sistemi energetici", in Guerci C.M., Zanetti G. (ed. by) *Sviluppo economico e vincolo energetico*, Bologna: Il Mulino, pp. 17-38.
1989a *Il mito delle risorse energetiche rinnovabili*, in "Le Scienze", no. 245, pp. 17-25.
1989b *Motivi di riflessione dalla congiuntura energetica del 1988*, in "Energia", no. 1, pp. 5-18.
1990a *La congiuntura energetica in Italia nel 1989*, in "Energia", no. 1, pp. 58-69.
1990b *Evitare gli errori del 1979 sulle scorte di petrolio*, in "Il Giornale", 31 August.
1991a *Petrolio: l'illusione svanita*, in "Relazioni Internazionali", March, pp. 108-116.
1991b "La crisi del Golfo Persico tra cronaca e prospettive", in Bruni F., Onida F. (ed. by) *L'economia italiana nel mondo che cambia*, ISPI, Bologna: Il Mulino, pp.37-78.
1993 "Energia, fonti di", in *Enciclopedia delle Scienze Sociali*, Roma: Enciclopedia Treccani, pp. 570-585.

Clô, A., Peredo Roman, H.
1972 *Le rendite differenziali nel mercato petrolifero internazionale*, in "Economia Internazionale delle Fonti di Energia", no. 2, pp. 77-125.

Clô, A., Prodi, R.
1975 *The Oil Crisis in Perspective: Europe*, in "Daedalus", Fall.

Colitti, M.
1979 *Energia e sviluppo in Italia: la vicenda di Enrico Mattei*, Bari: De Donato.
1988 *The Structure of Oil: the Inevitable Partnership*, Roma (manuscript).

Colitti, M., Simeoni, C.
1996 *Perspectives of Oil and Gas: the Road to Interdependence*, London: Kluwer Academic Publishers.

Cross, J.S.
1953 *Vertical Integration in the Oil Industry*, in "Harvard Business Review", no. 31, pp. 69-77.

Darby, M.R.
1982 *The Price of Oil and World Inflation and Recession*, in "American Economic Review", no. 4, pp. 438-451.

Darmstadter, J., Landsberg, H.H., Morton, H.C., Coda, M.J.
1986 *Vivere con l'incertezza, il problema dell'energia nel mondo contemporaneo*, Bologna: Il Mulino.

De Chazeau, M.G., Kahn, A.E.
1959 *Integration and Competition in the Petroleum Industry*, New Haven: Yale University Press.

De Latour, P.
1990 *Du pétrole à bon marché*, in "Le Monde", 2 October.
1991 *Petrolio e povertà: proposte di aiuto internazionale*, in "Energia", no. 1, pp. 12-17.

Department of Energy (DOE)
1987 *Energy Security*, Washington D.C.: U.S. Government Printing Office.
1991 *National Energy Strategy - Powerful ideas for America*, Washington D.C.: U.S. Government Printing Office.

Deschert, C.R.
1963 *Eni: a profile of a State Corporation*, Leiden Brill.

Desprairies, P.C., Boy de la Tour, X., Lacour, J.J.
1984 *La mobilisation progressive de ressources pétrolières, facteur de hausse modérée des prix*, in "Revue de l'Energie", no. 367, pp. 627-642.

Develle, M.M.
1987 *Si le prix du pétrole était rationnel?*, in "Bullettin de l'Institut Pétrolier", 25 May.

Dirlam, J.B.
1966 "The Petroleum Industry", in Adams W. (ed. by) *The Structure of American Industry*, New York: MacMillan.

Drollas, L.P.
1986 *Oil Supplies and the World Oil Market in the Long Run*, 8th International Association of Energy Economics Conference, Tokyo, 7 June.

Durand, D.
1962 *La politique pétrolière internationale*, Paris: Presses Universitaires de France.

Eckstein, O.
1978 *The Great Recession,* Amsterdam: North Holland.

El Sayed, M.
1967 *Organisation des pays exportateurs du pétrole*, Paris.

Energia
1990 *Cronaca della crisi del Golfo*, no. 4, pp. 80-85.
1991 *Cronaca della crisi del Golfo,* no. 1, pp. 66-72.

Energy Modeling Forum (EMF)
1982 *World Oil*, Stanford (Ca.): EMF.

ENI
1983 *Strategie delle Compagnie Petrolifere - Elementi di Analisi*, Roma.

Essley, P.H.
1965 *What is Reservoir Engineering?* in "Journal of Petroleum Technology", January, pp. 20-21.

Eswaran, M.
1996 *Cartel Unity over the Business Cycle*, in "Canadian Journal of Economics", vol. 30, no. 3, pp. 644-672.

Ezzati, A.
1976 *Future OPEC Price and Production Strategies as affected by its Capacity to absorb Oil Revenues*, in "European Economic Review", no. 8, pp. 107-138.

Fassina, E.
1981 *Un'analisi del comportamento delle imprese petrolifere multinazionali nell'ultimo trentennio*, in "Rivista Internazionale di Scienze Sociali", vol. 89, April-June.

Federal Trade Commission (FTC)
1952 *Staff Report: The International Petroleum Cartel*, Washington D.C.: U.S. Government Printing Office.

Fellner, W.
1949 *Competition Among the Few*, New York: Knopf.

Ferderer, J.P.
1996 *Oil Price Volatility and the Macroeconomy*, in "Journal of Macroeconomics", vol. 18, no. 1, pp. 1-26.

Ferrier, R.W.
1982 *The History of the British Petroleum Company*, in "The Developing Years, 1901-1932", vol. 1, Cambridge: Cambridge University Press.

Fesharaki, F., Isaak, T.D.
1983 *OPEC, The Gulf, and the World Petroleum Market*, Westview Press.

Fischer, D., Gately, D., Kyle, J.F.
1975 *The Prospects for OPEC - A Critical Survey of Models of the World Oil Market*, in "Journal of Development Economics", vol. 2, December, pp. 363-386.

Fisher, S.
1982 *Relative Price Variability and Inflation in the United States and Germany*, in "European Economic Review", vol. 18, pp. 171-196.

Frank, H.J.
1966 *Crude Oil Price in the Middle East - A Study in Oligopolistic Behavior*, New York: Praeger.

Frankel, P.H.
1946 *Essentials of Petroleum - A Key to Oil Economics*, London: Frank Cass.
1962 *The Significance of Marginal Capacity in the Oil Industry*, in "The Journal of Industrial Economics", no. 2, pp. 93-99.
1966 *Mattei: Oil and Power Politics*, London: Faber & Faber.
1982 *La influenza del neo-liberalismo sui mercati energetici e sulla strategia delle compagnie petrolifere e dei governi*, in "Energia", no. 2, pp. 6-13.

Frankel, P.H., Newton, W.L.
1961 *The Location of Refineries*, in "Institute of Petroleum Review", July, pp. 197-201.
1968 *Economics of Petroleum Refining - Present State and Future Prospects*, in "Journal of the Institute of Petroleum", February, pp. 23-35.

Funke, M.
1997 *The Nature of Shocks in Europe and in Germany*, in "Economica", vol. 64, no. 255, pp. 461-469.

Gately, D.
1979 *The Prospects for OPEC Five Years After 1973-'74*, in "European Economic Review", October, pp. 369-379.
1984 *A Ten-Years Retrospective: OPEC and the World Oil Market*, in "Journal of Economic Literature", September, pp. 1,100-1,114.

1986a *The Prospects for Oil Prices Revisited*, in "Annual Review of Energy".
1986b *Lessons from the 1986 Oil Price Collapse*, in "Brookings Papers on Economic Activity", no. 2, pp. 237-284.

Georgescu-Roegen, N.
1982 *Energia e Miti Economici*, Turin: Boringhieri.

Gever, J., Kaufmann, R., Skole, D., Vorosmarty, C.
1986 *Beyond Oil, the Threat to Food and Fuel in the Coming Decades*, Cambridge (Mass.): Ballinger Publishing Co.

Ginzburg, A.
1991 *Le cause economiche della guerra*, in "Politica ed Economia", no. 4, pp. 3-11.

Girod, J.
1977 *La domande d'énergie: methodes et techniques de modelisation*, Paris: CNRS.

Goldoni, G.
1987 *Tecnica ed economia della raffinazione*, Dissertation, Università degli Studi di Modena.

Gordon, R.
1977 *World Inflation and Monetary Accomodation in Eight Countries*, in "Brookings Papers on Economic Activity", no. 2, pp. 409-477.

Graham, R.
1990 *Beginning of The Real Phoney War*, in "Financial Times", 23 October.
1991 *Saudis Tighten up as Costs Wipe out Windfall Profits*, in "Financial Times", 25 January.

Griffin, J.M.
1979 *Energy Conservation in the OECD: 1980 to 2000*, Cambridge (Mass.): Ballinger Publishing Co.

Griffin, J.M., Steele, H.B.
1980 *Energy Economics and Policy*, Orlando (Flor.): Academic Press Inc.

Griffin, J.M., Teece, D.J. (ed. by)
1982 *OPEC Behavior and the Oil Price*, London: George Allen & Unwin.

Griffin, J.M., Xiong, W.
1997 *The Incentive to Cheat: An Empirical Analysis of OPEC*, in "Journal of Law and Economics", vol. 40, no. 2, pp. 289-316.

Hafele, W. (Program Leader)
1981 *Energy in a Finite World*, Cambridge (Mass.): Ballinger Publishing Co.

Harrigan, K.
1983 *Strategies for Vertical Integration*, Toronto: Lexington Books.

Hartshorn, J.E.
1966 *Oil Companies and Governments*, London: Faber & Faber.

1980 "The Special Characteristics of OPEC and Importing Countries National Oil Companies", in Mabro R. (ed. by) *World Energy Issues and Policies*, Oxford: Oxford University Press, pp. 157-165.

1985 "Government Sellers in a Re-Structured Crude Oil Market", in Hawdon D. (ed. by) *The Changing Structure of the World Oil Industry*, Croom Helm, pp. 59-69.

Hawdon, D. (ed. by)

1985 *The Changing Structure of the World Oil Industry*, Croom Helm.

Helm, D.R.

1991 *The Assessment: Energy Policy*, in "Oxford Review of Economic Policy", vol. 7, no. 2, pp. 1-16.

1993 *Energy Policy and Market Doctrine*, in "The Political Quarterly".

Hieronymus, T.

1982 *Economics of Future Trading*, New York: Commodity Research Bureau.

Hooker, M.A.

1996 *What Happened to the Oil - Price Macroeconomy Relationship*, in "Journal of Monetary Economics", vol. 38, no. 2, pp. 195-213.

Hotelling, H.

1931 *The Economics of Exhaustible Resources*, in "Journal of Political Economy", vol. 39, pp. 137-175.

Hubbert, M.K.

1956 *Drilling and Production Practice*, in "Nuclear Energy and the Fossil Fuels", American Petroleum Institute, Spring Meeting, St. Antonio, pp. 7- 25.

1979 *Le risorse energetiche della terra*, in "Energia Oggi e Domani, Letture da Le Scienze", pp. 29-39.

Hume, I. (Chairman)

1986 *Long-term Investments in the Energy Sector: Needs, Contraints and Proposals*, Working Group 7, 13th Congress of World Energy Conference, Cannes.

Institut Français du Pétrole (IFP)

1996 *Panorama 1996 - Actes du colloque*, Paris: IFP.

1998 *Panorama 1998 - Prospettive sugli idrocarburi,* Italian edition ed. by RIE, Bologna: Editrice Compositori.

International Petroleum Encyclopedia

(various years) Tulsa: Pennwell Publishing.

Jenkins, G.

1986 *Oil Economists' Handbook*, London: Elsevier Applied Science Publ.

1989 *Oil Economists' Handbook*, London: Elsevier Applied Science Publ.

Johany, A.

1980 *The Myth of the OPEC Cartel*, New York: Wiley & Sons.

Kahn, A.E.
1964 *The Depletion Allowance in the Contest of Cartelization*, in "American Economic Review", June.

Kissinger, K.
1982 "Foreword", in Ebinger et al. *Energy and National Security in the 1980s*, Cambridge (Mass.): Ballinger Publishing Co.

Koutsoyannis, A.
1981 *Microeconomia*, Milan: Etas Libri.

Landsberg, H.H.
1987 *Possibili shocks petroliferi: quali strategie?*, in "Energia", no. 2, pp. 88-94.

Lawson, N.
1989 "Energy Policy", in Helm D.R. et al. (ed. by) *The Market for Energy*, Oxford: Clarendon Press, pp. 23-29.

Lee, K., Shawn, N., Ratti, R.A.
1995 *Oil Shocks and the Macroeconomics: the Role of Price Variability*, in "The Energy Journal", vol. 16, no. 4, pp. 39-56.

Leeman, W.T.
1962 *The Price of Middle East Crude Oil*, New York: Cornell University Press.

Lerner, A.P.
1934 *The Concept of Monopoly and the Measurement of Monopoly Power*, in "Review of Economic Studies", April.

Levy, W.J.
1974 *World Oil Cooperation or International Caos*, in "Foreign Affairs", July, pp. 691-713.

Longrigg, S.H.
1968 *Oil in the Middle East*, Oxford: Oxford University Press.

Lovejoy, W.F., Homan, P.T., Galvin, C.O.
1963 *Cost Analysis in the Petroleum Industry*, in "Journal of the Graduate Research Center", no. 1-2, pp. 12-30.

Lowinger, T.L., Ram, R.
1984 *Product Value as a Determinant of OPEC's Official Crude Oil Prices: Additional Evidence*, in "The Review of Economics and Statistics", no. 4, pp. 691-695.

Mabro, R.
1990 *Political Dimensions of The Gulf Crisis*, Gulf and World Oil issues series, paper 1, Oxford: Oxford Institute for Energy Studies.

Mac Avoy, P.
1982 *Crude Oil Price*, Cambridge (Mass.): Ballinger Publishing Co.

Machlup, F., Taber, M.
1960 *Bilateral Monopoly, Successive Monopoly, and Vertical Integration*, in "Economica", vol. 27, May.

Mainguy, Y.
1967 *L'économie de l'énergie*, Paris: Dunod.

Malmgren, H.B.
1961 *Information, Expectations and the Theory of the Firm*, in "The Quarterly Journal of Economics", August, pp. 399-412.

Manne, A., Schrattenholzer, L.
1983 *International Energy Workshop: A Summary of the 1983 Poll Responses*, Luxemburg: IIASA.

March, J., Simon, H.A.
1958 *Organizations*, New York: John Wiley & Sons.

Marr, P.
1985 *The Modern History of Irak*, Boulder (Color.): Westview Press.

Marshall, A.
1961 *Principles of Economics*, London: McMillan.

Martin, J.M.
1992 *Économie e politique de l'énergie*, Paris: Armand Colin.
1999 *Viewpoint: concerning the 'end of cheap oil'*, in "Energy Policy", no. 27, pp. 69-72.

Masera, R.S.
1987 *Aspetti macroeconomici dell'interazione fra sviluppo ed energia*, Temi di discussione, Banca d'Italia, no. 87.

Masseron, J.
1982 *L'économie des hydrocarbures*, Paris: Technip.
1992 *Petroleum Economics*, Paris: Technip.

Masters, C.D. et al.
1983 *Distribution and Quantitative Assessment of World Crude Oil Reserves and Resources*, U.S. Department of the Interior, Geological Survey.
1987 *World Resources of Crude Oil, Natural Gas, Natural Bitumen, and Shale Oil*, U.S. Geological Survey, 12th World Petroleum Congress.

Mc Kelvey, V.
1973 "Mineral Estimates and Public Policy", in Brobst D., Pratt W. (ed. by) *United States Mineral Resources*, U.S. Geological Survey, Professional Paper, pp. 9-19.

Mc Gee, J.
1958 *Predatory Price Cutting: The Standard Oil (N.J.) Case*, in "Journal of Law and Economics", October, pp. 137-169.

Mc Lean, J.G., Haigh, R.W.
1954 *The Growth of Integrated Oil Companies*, Boston: Harvard University Press.

Mead, W.J.
1986 *The OPEC Cartel Thesis Reexamined: Price Constraints from Oil Substitutes*, in "Journal of Energy and Development", vol. 11, no. 2, pp. 213-242.

Melvin, G., De Chezeau, M.G., Kahn, A.E.
1959 *Integration and Competition in the Petroleum Industry*, Yale: University Press.

Menger, C.
1981 *Principi di Economia Politica*, Turin: UTET.

Mikdashi, Z.
1966 *A Financial Analysis of Middle Eastern Oil Concessions: 1901-1965*, New York: F.A. Praeger.
1972 *A Financial Analysis of Middle Eastern Oil Concessions*, London: George Allen & Unwin.
1975 *The OPEC Process*, in "Daedalus", Fall.

Montagnon, P.
1990 *Sharp Blow to The Most Vulnerable*, in "Financial Times", 7 September.

Morse, E.L.
1984 *I costi e i benefici della riduzione del prezzo del petrolio*, in "Energia", no. 1, pp. 4-12.

Mory, J.F.
1993 *Oil Prices and Economic Activity: Is the Relation Symmetric?*, in "The Energy Journal", vol. 14, no. 4, pp. 151-162.

Murat, D.
1969 *L'intervention de l'État dans le secteur pétrolier en France*, Paris: Technip.

National Petroleum Council (NPC)
1972 *U.S. Energy Outlook: An Initial Appraisal 1971-1985*, December, Washington D.C.
1986 *U.S. Oil and Gas Outlook*, Interim Report, Washington D.C.
1987 *Factors Affecting U.S. Oil and Gas Outlook*, Washington D.C.

Nevis, A.
1953 *Study in Power: John D. Rockefeller*, New York: Scribner's.

Nordhaus, W.D.
1980 *Oil and Economic Performance in Industrial Countries*, in "Brookings Papers on Economic Activity", no. 2, pp. 341-399.

Odell, P.R.
1986 *Oil and World Power*, London: Penguin Books (1st ed. 1970).
1994 *World Oil Resources, Reserves, Production*, in "The Energy Journal", special issue on *The Changing World Petroleum Market*, pp. 89-114.
1997 *The Global Oil Industry: the Location of Production - Middle East Domination or Regionalization?*, in "Regional Studies", vol. 31, no. 3, pp. 311-322.

OECD
1974 *Perspectives énergétiques jusqu'en 1985*, Paris.

OECD-IAE
1982 *World Energy Outlook*, Paris.
1987 *Renewable Sources of Energy*, Paris.
1996 *Global Offshore Oil Prospects to 2000*, Paris.
1998 *World Energy Outlook*, Paris.

Oil & Gas Journal (OGJ)
1990 *Prices Hikes, Fears of Oil Supply Trail Irak Attack on Kuwait,* in "Oil & Gas Journal", 13 August.

OPEC
1967 *Collective influence in the recent trend towards the stabilization of international crude and product price*, 4th Arab Petroleum Congress.

Parra, A.A.
1986 *OPEC: the New Challenge*, in "Middle East Economic Survey", September.

Pauwels, J.P.
1997 *Géopolitique de l'approvisionnement énergétique de l'UE au XXI siècle*, Bruxelles: Bruylant.

Pearce, J. (ed. by)
1983 *The Third Oil Shock: the Effects of Lower Oil Prices*, London: The Royal Institute of International Affairs.

Penrose, E.T.
1959 *Profit Sharing between Producing Countries and Oil Companies in the Middle East*, in "The Economic Journal", no. 274, pp. 238-254.
1960 *Middle East Oil: The International Distribution of Profits and Income Taxes*, in "Economica", August, pp. 203-213.
1964 "Monopoly and Competition in the International Petroleum Industry", in *The Yearbook of World Affairs*, London Institute of World Affairs, pp. 151-177.
1965 *Vertical Integration with Joint Control of Raw Material Production: Crude Oil in the Middle East*, in "Journal of Development Studies", vol. 1, no. 3, April.
1968 *The Large International Firm in Developing Countries - The International Petroleum Industry*, London: George Allen & Unwin.
1969a *OPEC and the Changing Structures of the International Petroleum Industry*, in "Middle East Economic Survey", 7 March.
1969b *The Role of OPEC in Changing Circumstances*, in "Middle East Economic Survey", 3 October.
1975 *The Development of Crisis*, in "Daedalus", Fall, pp. 39-57.
1985 "Downstream Implications of Structural Change", in Hawdon D. (ed. by) *The Changing Structure of the World Oil Industry*, Croom Helm.

Petroleum Economist
1984 *Changing Management Strategies*, June.

Petroleum Intelligence Weekly (PIW)
1983 *Corporate Strategies up to Unsettling Times*, 5 September.

1990 *How Vulnerable are Oil Facilities in The Gulf?*, 27 August.

Pinchot, G.
1910 *The Fight for Conservation*, New York.

Pindyck, R.S.
1978 *Gains to Producers from fhe Cartelization of Exhaustible Resources*, in "Review of Economics and Statistics", May, pp. 238-251.
1982 "OPEC Oil Pricing, and the Implications for Consumers and Producers", in Griffin J., Teece D. (ed. by) *OPEC Behaviour and World Oil Prices*, London: George Allen & Unwin, pp. 175-185.

Pinzani, C.
1991 *Lezioni di una guerra*, in "Politica Internazionale", no. 2-3, pp. 6-15.

PIRINC
1971 *Oil Consumers and the Recent Tax Increases in Foreign Producing Countries,* New York.

Polanyi, G.
1966 *The Taxation of Profits from Middle East Oil Production: some Implications for Oil Prices and Taxation Policy*, in "The Economic Journal", December, pp. 768-785.

Pratten, C.F.
1965 *The Economics of Large-Scale Production in British Industry: An Introductory Study*, Cambridge: Cambridge University Press.
1971 *Economies of Scale in Manufacturing Industries*, Cambridge: Cambridge University Press.

Quadrio Curzio, A.
1988 "Le 'scarsità relative' nella storia della Teoria Economica e rilevanza attuale", in Guerci C.M., Zanetti G. (ed. by) *Sviluppo Economico e Vincolo Energetico*, Bologna: Il Mulino, pp. 39-53.

Ramachandra, S.G. (Chairman)
1986 *Energy Problems in Developing Countries*, Round Table 4, 13th Congress of World Energy Conference, Cannes.

Ridker, R.
1980 *Resource and Environmental Conseguences of Population and Economic Growth*, Washington: Resources For the Future.

Rifaï, T.
1973 *La crise pétrolière internationale 1970-'71: un essai d'interpretation*, in "Pétrole & Gaz Arabes", pp. 21-37.
1974 *Le prix du pétrole,* Paris: Technip.

Rodman, P.W.
1991 *Middle East Diplomacy After The Gulf War*, in "Foreign Affairs", Spring, pp. 1-18.

Rouhani, R.
1970 *Concession Agreements: Survey and Future Trends*, Proceedings of OPEC Seminar, Wien.

Roushdi, M.H.
1972 *The Proper Concept of Direct Relations*, 8th Arab Petroleum Congress, Algiers.

Safer, A.E.
1979 *International Oil Policy*, Toronto: Lexington Books.

Salant, S.W.
1982 *Imperfect Competition in the World Oil Market: a Computerized Nash-Cournot Model*, Toronto: Lexington Books.

Sampson, A.
1976 *Le Sette Sorelle*, Milan: Mondadori.

Sanderson, F.H.
1986 *Etanolo, l'infelice esperienza americana*, in "Energia", no. 4, pp. 14-14.

Sarkis, N.
1972 *L'accord de New York et ses incidences pétrolières, économiques, et politiques sur les pays arabes*, Conférence au Centre Arabe d'Etudes Pétrolières, Kuwait, 18 November.

Scherer, F.M.
1985 *Economia Industriale*, Milan: Unicopli.

Schlesinger, A. jr.
1991 *Zio Sam va alla guerra per niente*, in "Il Sole 24 Ore", 9 January.

Schneider, H.K., Schulz, W.
1986 *Investments Requirements of the World Energy Industries 1980-2000*, Working Group 7, 13th Congress of World Energy Conference, Cannes.

Schuler, G.H.M.
1974 *The International Oil "Debacle" since 1971*, in Special Supplement "Petroleum Intelligence Weekly", 22 April.

Schurr, S., Homan, P.
1971 *Middle Eastern Oil and the Western World,* New York: Elsevier Press.

Scuola Superiore Enrico Mattei (SSEM)
1998 *Glossario dell'Industria Petrolifera,* Biblioteca ENI, Milan.

Seymour, Y.
1985 "OPEC and Structural Change", in Hawdon D. (ed. by) *The Changing Structure of the World Oil Industry*, Croom Helm.

Shaffer, E.H.
1968 *The Oil Import Program of the United States - An evolution,* New York: Praeger.

Shenoy B.V. et al.
1999 *Caspian Oil Export Choices Clouded by Geopolitics, Project Economics*, in "Oil & Gas Journal", 19 April.

Simon, J.L.
1981 *The Ultimate Resources*, Princeton: Princeton University Press.

Smith, A.
1970 *The Wealth of Nations*, London: Penguin.

Solow, R.M.
1974 *The Economics of Resources or the Resources of Economics*, in "American Economic Review", vol. 64, May, pp. 1-14.

Stauffer, T.
1994 *Trends in Oil Production Costs in the Middle East, Elsewhere,* in "Oil & Gas Journal", 21 March, pp. 105-107.
1998 *La geoeconomia delle esportazioni asiatiche: l'"Iranian connection"*, in "Energia", no. 1, pp. 40-52.

Stevens, P.
1983 *Il Futuro dei Prezzi Mondiali del Petrolio: la Fine di un'Era?*, in "Energia", no. 3, pp. 62-75.
1985 "A Survey of Structural Changes in the International Oil Industry", in Hawdon D. (ed. by) *The Changing Structure of the World Oil Industry*, Croom Helm, pp. 18-51.

Stournaras, Y.
1985 *Are Oil Price Movements Perverse? A Critical Explanation of Oil Price Levels 1950-1985*, Oxford: Oxford Institute for Energy Studies.

Sutowo, I.
1962 *The Role of Oil in the National Life*, Djakarta: Hong Bon Printing Co.

Sylos Labini, P.
1964 *Oligopolio e Progresso Tecnico*, Turin: Einaudi.
1981 *Prezzi rigidi, prezzi flessibili e inflazione*, in "Moneta e Credito", December.

Tanzer, M.
1969 *The Political Economy of International Oil and the Underdeveloped Countries*, Boston: Beacon Press.

Terzian, P.
1985 *OPEC, The Inside Story*, London: Zed Books.

The Economist
1991 *Oil and War-Fairly Flammable*, 12 January.

The Wall Street Journal
1987 *Cartel's Comeback?*, 21 August.

Thur, O.
1960 *Le funzioni economiche dei prezzi*, in "Economia Internazionale delle Fonti di Energia", January-February.

Toman, M.A.
1990 *What Do We Know About Energy Security?*, in "Resources", Fall.

Tourk, K.
1977 *The OPEC Cartel: A Revival of the Dominant Firm Theory*, in "Journal of Energy and Development", vol. 2, no. 2, pp. 321-328.

Tugendhat, C.
1968 *Oil the Biggest Business*, Boston: Beacon Press.

Turner, L.
1978 *Oil Companies in the International System*, London: George Allen & Unwin.

Verdross, A.
1964 "Quasi International Agreements and International Economic Transactions", in *The Yearbook of World Affairs*, London Institute of World Affairs, pp. 230-249.

Verleger, P.K. jr.
1982 *The Determinants of OPEC Crude Oil Prices*, in "Review of Economics and Statistics", no. 2, pp. 177-183.
1987 "The Evolution of Oil as a Commodity", in Gordon R.L. et al. (ed. by) *Energy: Markets and Regulation*, Cambridge (Mass.): MIT Press, pp. 161-186.
1990a *Understanding the 1990 Oil Crisis*, in "The Energy Journal", vol. 11, no. 1, pp. 15-33.
1990b *U.S. Oil Crisis and Management Falters*, in "Petroleum Intelligence Weekly", 13 August.
1990c *The Oil Vortex: A Way Out*?, in "Economic Insight", Sep.-Oct., pp. 3-7.
1991 "Structural Change in the 1980s: Effects and Permanence", in Kohl W.L. (ed. by) *After the Oil Price Collapse*, Baltimore: Johns Hopkins University Press, pp. 85-107.

Viani, M.
1992 *Strategie di integrazione verticale nel mercato petrolifero internazionale*, Dissertation, Università degli Studi di Modena.

Vuono, C.E.
1991 *Desert Storm and Conventional Forces*, in "Foreign Affairs", Spring, pp. 49-68.

Williamson, H.F., Daum, A.
1959 *The American Petroleum Industry - The Age of Illumination*, Eraston (Illin.).

Williamson, O.E.
1971 *The Vertical Integration of Production: Market Failure Consideration*, in "American Economic Review", no. 61, May.

Wilson, J.W.
1975 *Market Structure and Interfirm Integration in the Petroleum Industry*, in "Journal of Economic Issues", June.

World Bank
1986 *World Development Report*, Oxford: Oxford University Press.

World Energy Conference (WEC)
1978 *World Energy Resources 1985-2020*, New York.

World Resources Institute
1986 *World Resources*, New York: Basic Books.

Yamani, A.Z.
1986 *Oil Markets: Past, Present and Future*, in "Middle East Economic Survey", 15 September.
1990 *Se gli stati entrano in riserva*, in "La Repubblica", 15 October, "Affari e Finanza".

Yergin, D.
1993 *Il premio*, Milan: Sperling & Kupfer.

Zannetos Zenon, S.
1987 "Oil Tanker Markets: Continuity amidst Change", in Gordon R.L. et al. (ed. by) *Energy: Markets and Regulation*, Cambridge (Mass.): MIT Press, pp. 235-257.